Emerging Trends in Industrial Engineering and Management

Emerging trends in industrial engineering and management (IEM) refer to the new and transformative developments, practices, and technologies that are currently gaining prominence in the field of industrial engineering and management. Trends in industrial engineering and management can encompass a wide range of topics such as utilization of Industry 4.0 strategies like Industrial Internet of Things, artificial intelligence, and theoretical, numerical, and computational approaches to model the methods and process of IEM.

This book:

- Provides a comprehensive discussion of industrial engineering and management.
- Includes principles of continuous improvement, encouraging readers to adopt a mindset of ongoing optimization and innovation in industrial engineering and management.
- Presents multiobjective optimization, stochastic optimization, and metaheuristic optimization algorithms for solving complex optimization problems in industrial engineering.
- Aligns with the needs of various industries, addressing specific challenges faced by manufacturing, healthcare, logistics, service, and other sectors.
- Highlights the importance of using digital technological tools like the Internet of Things, Industrial Internet of Things, big data, and artificial intelligence in practices of industrial management to enhance competitiveness, decision-making, and operations efficiency.

It is primarily written for senior undergraduates, graduate students, and academic researchers in the fields of industrial engineering, production engineering, mechanical engineering, operation management, industrial management, quality engineering, and engineering management.

Industrial and Manufacturing Systems and Technologies: Sustainable and Intelligent Perspectives

Series Editors: Ajay Kumar *and* Parveen Kumar

This book series provide a platform of single and multidisciplinary advance research on industrial and manufacturing engineering to introduce a requirement for a synergy from management and engineering. It provides the readers a paradigms and systems of intelligence and sustainability in industrial and manufacturing engineering. It will contribute towards dissemination of 5.0 industrial revolution techniques towards academia, industry and various practical communities involved in mechanical, industrial, computer science, automobile, electrical, electronics, aerospace engineering, supply chain, logistics, banking, finance management, circular economy, medical, pharmaceuticals and healthcare industries. It will focus on important topics such as additive manufacturing, virtual reality in industrial engineering, industrial internet of things, intelligent engineering asset management system, and modeling of intelligent industrial engineering systems.

Smart Electric and Hybrid Vehicles
Fundamentals, Strategies and Applications
Edited by Ajay Kumar, D. K. Rajak, Parveen Kumar and Sarita Rathee

Smart Electric and Hybrid Vehicles
Design, Modeling, and Assessment by Industry 4.0 Approaches
Edited by Ajay Kumar, D. K. Rajak, Parveen Kumar and Sarita Rathee

Emerging Trends in Industrial Engineering and Management
Ajay Kumar, Parveen Kumar, Rakesh Kumar Phanden, Mario Schmidt, Ayon Chakraborty

For more information about this series, please visit: www.routledge.com/Industrial-and-Manufacturing-Systems-and-Technologies-Sustainable-and/book-series/CRCIAMSAT

Emerging Trends in Industrial Engineering and Management

Edited by
Ajay Kumar, Parveen Kumar,
Rakesh Kumar Phanden, Mario Schmidt,
and Ayon Chakraborty

CRC Press
Taylor & Francis Group
Boca Raton London New York

CRC Press is an imprint of the
Taylor & Francis Group, an **informa** business

Designed cover image: Shutterstock

First edition published 2026
by CRC Press
2385 NW Executive Center Drive, Suite 320, Boca Raton FL 33431

and by CRC Press
4 Park Square, Milton Park, Abingdon, Oxon, OX14 4RN

CRC Press is an imprint of Taylor & Francis Group, LLC

ISBN: 978-1-032-84404-6 (hbk)
ISBN: 978-1-032-84406-0 (pbk)
ISBN: 978-1-003-51258-5 (ebk)

DOI: 10.1201/9781003512585

Typeset in Sabon
by Deanta Global Publishing Services, Chennai, India

Contents

Preface

Emerging trends in industrial engineering and management (IEM) refer to the new and transformative developments, practices, and technologies that are currently gaining prominence in the field of industrial engineering and management. Trends in industrial engineering and management can encompass a wide range of topics such as utilization of Industry 4.0 strategies like Industrial Internet of Things, artificial intelligence, theoretical, numerical, and computational approaches to model the methods and process of IEM, sustainable practices in IEM, automation in IEM, human-centric approach to IEM, digital transformation, agile and lean management perspectives, employee empowerment, data-driven decision-making strategies, and so on. IEM is an ever-evolving field that plays a crucial role in shaping the success of modern organizations. Over the years, it has continuously adapted to the changing business landscape, incorporating cutting-edge technologies, innovative methodologies, and novel management approaches. This book delves into the fascinating realm of emerging trends in industrial engineering and management, aiming to provide readers with a comprehensive understanding of the latest developments that are reshaping the industry. As the world undergoes rapid technological advancements, global markets become increasingly interconnected, and consumer preferences evolve, industrial engineers and managers face a myriad of challenges and opportunities. This book intends to serve as a guiding light, illuminating the path toward harnessing the power of these emerging trends to drive efficiency, productivity, and sustainable growth. The academicians and students will get a comprehensive update on the state of the art in this area and ample ideas for further research and innovation in industrial engineering and management. This book has the following key features:

- This book clearly identifies and defines the emerging trends in IEM. It helps readers recognize the challenges posed by these trends, such as implementing new technologies, managing data overload, or adapting to changing market demands.

- The approach of this book provides in-depth analysis and explanations. Readers can gain a thorough understanding of the underlying principles, processes, and potential implications.
- This book considers various perspectives within IEM, such as technological, managerial, and human-centered aspects.

The book consists of 17 chapters that describe perspectives of emerging trends in industrial engineering and management.

- Chapter 1, "Literature Review on Balanced and Unbalanced Industrial Assignment Problems," presents a review of specific types of assignment problems discussed in the literature to understand more about them. At present time, the problem of assignment arises a different role in the efficiency, allocation, and personnel selection of the country as it enables trade between people.
- Chapter 2, "Systems for Detecting Intruders in Internet of Things Environments Using the Cloud: A Survey," explores various methodologies, technologies, and approaches for detecting and mitigating intrusions within Internet of Things (IoT) ecosystems by leveraging cloud computing resources.
- Chapter 3, "Utilizing Advanced Machine Learning to Improve Network Security in Industries against Changing Cyberthreats," underscores the pivotal role of advanced machine learning in fortifying network security against the dynamic landscape of evolving cyberthreats, offering a promising path toward more resilient and adaptive cybersecurity measures.
- Chapter 4, "Big Data Analytics and FinTech in BFSI Industries: Emerging Trends," attempts to study the evolution, tools, and applications of big data and FT in the BFSI sectors, along with the financial engineering applications in this process and the emerging trends in the process.
- Chapter 5, "Internet of Things Products: Management of Youth's Purchase Intention," considers four independent variables for analyzing the impact of intelligence, interaction, social interaction, and privacy on emotional and functional experience and then on purchase intention.
- Chapter 6, "Artificial Intelligence and Its Futuristic Approach in Industries," attempts to address and exemplify the scope and use of AI across multiple domains.
- Chapter 7, "Analysis on a Resource-Based Population Model with Caputo Fractional Derivative," focuses on the impact of foreign species on the population of native species and distribution of resources in that particular area and how it creates an ecological imbalance.

- Chapter 8, "Selection of Teaching Pedagogy for Engineering Classroom Learning from Instructor's Viewpoints Using Fuzzy TOPSIS approach: A Case Study of Selected Higher Education Institutions in Northern India," addresses the crucial challenge of selecting the most appropriate teaching pedagogy for classroom learning of higher education in engineering.
- Chapter 9, "A Bibliometric Analysis and Management of Big Data and Machine Learning Research: Current Trends and Future Directions," explains how by combining big data with machine learning enables to evaluate data and identify explanatory models that are fundamental to the symmetry principles but would be unable to do so using conventional methods.
- Chapter 10, "Water Scarcity and Its Management Strategies in India" discusses the issues and ensures a more water-secure future for all, immediate measures are made to manage water resources sustainably, encourage water conservation, and implement effective water management strategies.
- Chapter 11, "Group Search Optimization and Neural Network–Based Security and Management of Digital Data," projects a security model for securing the images by embedding watermark at random locations as a watermark verifies the authenticity of an image.
- Chapter 12, "Building a Smart Contract-Based Escrow Platform for Management of Crowdfunding and Public/Private Tenders on Blockchain," proposed a comprehensive framework that empowers fundraisers with complete financial control.
- Chapter 13, "Analysis of Water Quality for Sustainable Resource Management Using Machine Learning Techniques," discusses different approaches such as Random Forest, Logistic Regression, Decision Tree Classifier, GaussianNB, LinearSVC, and XGBClassifier computations to estimate the immaculateness of the drinking water.
- Chapter 14, "Management of Crop Recommendation System Using Predictive Analytics," compared decision tree, discriminant analysis, naive Bayes, SVM, KNN, and ensemble learning models. The naive Bayes classifier is highly accurate, with a precision rate of 99.58%.
- Chapter 15, "Analysis of Creditworthiness and Risk Assessment Management for Enterprises Using Machine Learning," understands various machine learning models used in creditworthiness over time.
- Chapter 16, "Intersection of Artificial Intelligence in Agriculture Sector Management: A Critical Analysis," emphasizes an existing successful research effort and a good potential for application, albeit emphasizing the difficulty of translating machines and algorithms tested in laboratory conditions to real contexts.

- Chapter 17, "A Study on Management of Employee Perception toward the Role of Artificial Intelligence Tools in Promoting Employee Engagement: An Empirical Study of the Banking Sector in Kumaun Region," comprises a study on the perspectives of bank employees in Uttarakhand's Kumaun region concerning the impact of AI on engagement, identifying effective and ineffective aspects, and providing future research routes for better integrating AI into HR practices.

This book is intended for both the academia and the industry. The postgraduate students, Ph.D. students, and researchers in universities and institutions, who are involved in the areas of industrial engineering and management, will find this compilation useful.

The editors acknowledge the professional support received from CRC Press and express their gratitude for this opportunity.

Reader's observations, suggestions, and queries are welcome.

Ajay Kumar
Parveen Kumar
Rakesh Kumar Phanden
Mario Schmidt
Ayon Chakraborty

Acknowledgments

The editors are grateful to CRC Press for publishing this book in the buzz area of emerging trends in industrial engineering and management. The editors express their gratitude to Gauravjeet Singh Reen (Executive Editor), CRC Press, for giving consent to publish our work. He and his methodical staff helped the editors to compile and finalize the manuscript. The editors also extend their gratitude to Mehnaz Hussain, CRC Press, for her support during her tenure.

The editors wish to thank all the chapter authors who contributed their valuable research and experience to compile this volume. The chapter authors, and corresponding authors in particular, deserve special acknowledgment for bearing with the editors, who persistently kept bothering them for deadlines, and with their remarks.

Dr. Ajay also wishes to express his gratitude to his parents, Shri Jagdish and Smt. Kamla, and his loving brother Shri Parveen for their true and endless support. They have made him walk tall before the world regardless of sacrificing their happiness and living in a small village. He cannot close these prefatory remarks without expressing his deep sense of gratitude and reverence to his life partner Sarita Rathee for her understanding, care, support, and encouragement by keeping his moral high all the time. No magnitude of words can ever quantify the love and gratitude he feels in thanking his daughters, Sejal Rathee and Mahi Rathee, and son Kushal Rathee who are the world's best children.

Finally, the editors dedicate this work to the divine creator and express their indebtedness to the "ALMIGHTY" for gifting them power to yield their ideas and concepts into substantial manifestation. The editors believe that this book will enlighten readers about each feature of emerging trends in industrial engineering and management.

Ajay Kumar
Parveen Kumar
Rakesh Kumar Phanden
Mario Schmidt
Ayon Chakraborty

About the editors

Ajay Kumar is a Professor (Research Track), Department of Mechanical Engineering, School of Core Engineering, Faculty of Science, Technology & Architecture, Manipal University, Jaipur, Rajasthan, India. He earned his Ph.D. in the field of advanced manufacturing and automation from Guru Jambheshwar University of Science & Technology, Hisar, India, after B.Tech. (Hons.) in mechanical engineering and M.Tech. (Distinction) in manufacturing and automation. His areas of research include Biomedical Engineering, Incremental Sheet Forming, Artificial Intelligence, Sustainable Materials, Robotics and Automation, Additive Manufacturing, Mechatronics, Smart Manufacturing, Industry 4.0, Industrial Engineering Systems, Waste Management, and Optimization Techniques. He has over 150 publications in international journals of repute, including Scopus, Web of Science, and SCI indexed database and refereed international conferences. He has organized various national and international events, including an international conference on Mechatronics and Artificial Intelligence (ICMAI-2021) as conference chair. He is currently organizing an international conference on Artificial Intelligence, Advanced Materials, and Mechatronics Systems (AIAMMS-2023) as conference chair. He has more than 20 national and international patents to his credit. He has supervised more than eight M.Tech. and Ph.D. scholars and numerous undergraduate projects/thesis. He has 15 years of experience in teaching and research. He is guest editor of many reputed journals. He has contributed to many international conferences/symposiums as a session chair, expert speaker, and member of the editorial board. He has won several proficiency awards during the course of his career, including merit awards, best teacher awards, and so on. He has also co-authored and co-edited more than 20 books. He has also authored many in-house course notes, lab manuals, monographs, and invited chapters in books. He has organized a series of faculty development programs, international conferences, workshops, and seminars for researchers, Ph.D., UG, and PG-level students. He is associated with many research, academic, and professional societies in various capacities.

Parveen Kumar is a Skill Instructor (Tool and Die) in the Skill Department of Automotive Studies, Shri Vishwakarma Skill University, Palwal, Haryana, India. Currently, he is pursuing his Ph.D. from the National Institute of Technology, Kurukshetra, Haryana, India. He completed his B.Tech. (Hons.) from Kurukshetra University, Kurukshetra, India, and M.Tech. (Distinction) in Manufacturing and Automation from Maharshi Dayanand University, Rohtak, India. His areas of research include Intelligent Manufacturing Systems, Materials, Mechatronics Systems, Advanced Materials, Industry 4.0, Die-Less Forming, Design of Automotive Systems, Additive Manufacturing, CAD/CAM, Artificial Intelligence, Machine Learning, and Internet of Things in Manufacturing, Multiobjective Optimization Techniques. He has over 50 publications in international journals of repute, including Scopus, Web of Science, and SCI indexed database and refereed international conferences. He has eight national and international patents to his credit. He has supervised more than two M.Tech. scholars and numerous undergraduate projects/theses. He has 13 years of experience in teaching and research. He has co-authored/co-edited several books with reputed publication house such as CRC Press, Elsevier, Springer, Wiley, DeGruyter, IGI Global and has organized a series of faculty development programs, workshops, and seminars for researchers and UG-level students. He is also the series editor of the series *Industrial and Manufacturing Systems and Technologies: Sustainable and Intelligent Perspectives* (CRC Press).

Dr. Rakesh Kumar Phanden is a seasoned researcher, academic leader, and accomplished administrator with expertise in project management, research outcomes and quality assurance in the higher and vocational education sector. He earned his PhD from the National Institute of Technology, Kurukshetra, India, and has an extensive publication record that includes edited books, special issues, and over 120 research papers. His significant editorial contributions include roles as EiC, guest editor, associate editor, reviewer, and editorial board member for prestigious international journals. Dr. Phanden's teaching and research interests encompass Lean Six Sigma, Circular Economy, Risk Management, Operations Management, Sustainability issues, and the Application of various Industry 4.0 Tools and technologies. His commitment to mentoring stands out as a defining aspect of his career, having successfully guided numerous PhD, UG, and PG students, thereby shaping the future generation of scholars and industry professionals.

Mario Schmidt is Managing Director of Lingel Windows and Doors Technologies Pvt Ltd, both in India and Germany. He completed his undergraduate alongside a certification for "Master Craftsman for Windows and Furniture Design" in 1994 and an MBA in Business Economist from Handwerkskammer Chemnitz, Germany. He has an honorary doctorate for his contribution to skill development. He has also participated in the AIMA course in global management from the Berkley University of California. He was the founding member of the window association uWDMA India in 2008 and technical head from 2008 until 2016 and president of the association since 2016. He has various specialties such as general management, corporate development, product development, business operations, strategy, and customer relations. He is also working closely with Indian government bodies BIS, BEE, MSDE, as well as the German government bodies of GIZ and DEG. He delivered many keynotes, talks, and expert lectures at international events.

Ayon Chakraborty is Associate Professor in Engineering Project Management at the Institute of Innovation, Science and Sustainability at Federation University, Australia. Prior to joining the university, he was Associate Professor in Operations Management and Decision Sciences area at the Indian Institute of Management Tiruchirappalli, India. Dr. Chakraborty was also Post-Doctoral Research Fellow, Services Science Discipline, Information Systems School at the Queensland University of Technology, Brisbane, Australia. He earned his Ph.D. in Industrial and Systems Engineering from the National University of Singapore in the area of Service Quality. He has more than 11 years of experience in teaching and consulting business students and executives at various institutes such as Queensland University of Technology and James Cook University Australia (Singapore Campus). Dr. Chakraborty's main research interests are in the area of Quality Management such as on Lean, Six Sigma, and other quality tools and techniques. In recent years, he also worked on practices related to sustainable development and circular economy, specifically in micro, small, and medium enterprises (MSMEs). He is also involved in international collaborative research projects such as Erasmus Plus and Global Challenges Research Funding (GCRF) with Aston University in the UK. Both these projects focus on the sustainability and well-being of MSMEs and are sponsored and funded by the UK Government.

Contributors

Harsh Agarwal
Department of Computer Science and Engin
Vellore Institute of Technology (VITeering)
Bhopal University
Astha, Madhya Pradesh, India

Ankita Arora
School of Computer Science and Engineering
Lovely Professional University
Phagwara, Punjab, India

Manish Kumar Bansal
Department of Mathematics
Jaypee Institute of Information Technology
Noida, Uttar Pradesh, India

Barsarani
Department of Sciences
Manav Rachna University
Faridabad, Haryana, India

Beena Bundela
Department of Mathematics
JECRC University
Jaipur, Rajasthan, India

Venkamaraju Chakravaram
Jindal Global Business School
OP Jindal Global University
Sonipat, Haryana, India

Roshan Chandel
Department of Computer Science and Engineering
JECRC University
Jaipur, Rajasthan, India

Harpreet Kaur Channi
Department of Electrical Engineering
Chandigarh University
Chandigarh, Punjab, India

Avnish Chauhan
Department of Computer Science and Engineering
Manav Rachna International Institute of Research and Studies
Faridabad, Haryana, India

R.P. Chauhan
Department of Mathematics
Amrita Vishwa Vidyapeetham
Amaravati, Andhra Pradesh, India

Gagandeep Singh Cheema
Mittal School of Business
Lovely Professional University
Phagwara, Punjab, India

Ria Choudhari
Department of Computer Science and Engineering
SRM University Delhi-NCR
Sonepat, Haryana, India

Deepti Dave
Department of Computer Science and Engineering
Vellore Institute of Technology (VIT)
Bhopal University
Astha, Madhya Pradesh, India

Shalini Gautam
Department of Mechanical Engineering
Guru Gobind Singh Indraprastha University
Delhi, India

Gitanjali
Department of CSE
Chandigarh University
Chandigarh, Punjab, India

Astha Gupta
Amity International Business School
Amity University
Noida, UP, India

Avni Gupta
Department of Computer Science and Engineering
Manav Rachna International Institute of Research and Studies
Faridabad, Haryana, India

Ruchi Gupta
Department of Sciences
Manav Rachna University
Faridabad, Haryana, India

Inzimam Ul Hassan
Department of Computer Science and Engineering
Chandigarh University
Mohali, Punjab, India

Sugandha Joshi
Department of Management
Graphic Era Hill University
Dehradun, Uttarakhand, India

Megha Kamble
School of Computer Science & Technology
LNCT University
Bhopal, Madhya Pradesh, India

Amritpal Kaur
Department of Computer Science and Engineering
Amity School of Engineering and Technology
Gurgaon, Haryana, India

Inderdeep Kaur
Department of Computer Science and Engineering
Chandigarh University
Mohali, Punjab, India

Seema Kharb
Department of Computer Science and Engineering
SRM University Delhi-NCR
Sonepat, Haryana, India

Swati Domaji Kharabe
Department of Mathematics
JECRC University
Jaipur, Rajasthan, India

Smitha Kothari
School of Management
Graphic Era Hill University
Dehradun, Uttarakhand, India

Ajay Kumar
Department of CSE
JECRC University
Jaipur, Rajasthan, India

Kamlesh Lakhwani
Department of Computer Science and Engineering
JECRC University
Jaipur, Rajasthan, India

Ashwini Jagannath Mahamuni
School of Education
Lovely Professional University
Phagwara, Punjab, India

Priyanka Malik
Galgotias University
Greater Noida, Uttar Pradesh, India

Bhavika Mathur
Department of CSE
JECRC University
Jaipur, Rajasthan, India

Prithviraj Mutalik
Department of Computer Science and Engineering
Vellore Institute of Technology (VIT)
Bhopal University
Astha, Madhya Pradesh, India

Naman Nagi
Department of Computer Science and Engineering
Manav Rachna International Institute of Research and Studies
Faridabad, Haryana, India

Ayush Patel
Department of CSE
SRM University Delhi-NCR
Sonepat, Haryana, India

Subhash C. Patel
Department of Computer Science and Engineering
Vellore Institute of Technology (VIT)
Bhopal University
Astha, Madhya Pradesh, India

Rakesh Kumar Phanden
Amity School of Engineering and Technology
Amity University
Noida, Uttar Pradesh, India

Vaibhav Puri
Department of Computer Science and Engineering
Manav Rachna International Institute of Research and Studies
Faridabad, Haryana, India

Pawan Singh Rajput
School of Computer Science & Technology
LNCT University
Bhopal, Madhya Pradesh, India

Sunitha Ratnakaram
Jindal Global Business School
OP Jindal Global University
Sonipat, Haryana, India

Pranjal Rawat
School of Management
Graphic Era Hill University
Dehradun, Uttarakhand, India

Deepali Rani Sahoo
Symbiosis Law School
Noida, Symbiosis International (Deemed University)
Pune, Maharashtra, India

Ramandeep Sandhu
School of Computer Science and Engineering
Lovely Professional University
Phagwara, Punjab, India

Amanjeet Singh Sethi
Faculty of Commerce & Business Management
Amrapali University
Haldwani, Uttarakhand, India

Gajanand Sharma
Department of CSE
JECRC University
Jaipur, Rajasthan, India

Parveen Sharma
School of Mechanical Engineering
Lovely Professional University
Phagwara, Punjab, India

Sonia Sharma
School of Education
Lovely Professional University
Phagwara, Punjab, India

Suruchi Sharma
School of Management
Graphic Era Hill University
Dehradun, Uttarakhand, India

Aditya Singh
Department of Computer Science and Engineering
Manav Rachna International Institute of Research and Studies
Faridabad, Haryana, India

Meeta Singh
Department of Computer Science and Engineering
Manav Rachna International Institute of Research and Studies
Faridabad, Haryana, India

Tushar Kumar Singh
Department of Computer Science and Engineering
Manav Rachna International Institute of Research and Studies
Faridabad, Haryana, India

Shweta Sinha
Department of Computer Science and Engineering
Amity School of Engineering and Technology
Gurgaon, Haryana, India

Shekhar Kumar Swarnkar
Department of Computer Science and Engineering
JECRC University
Jaipur, Rajasthan, India

Rajesh Tiwari
School of Management
Graphic Era Hill University
Dehradun, Uttarakhand, India

Rajesh Kumar Tyagi
Department of Computer Science and Engineering
Amity School of Engineering and Technology
Gurgaon, Haryana, India

Prashant Vats
Department of Computer Science and Engineering
Amity School of Engineering and Technology
Gurgaon, Haryana, India

Shilpa Wadhwa
Department of Management
Graphic Era Hill University
Dehradun, Uttar Pradesh, India

Part I

Advanced trends in industrial engineering

Chapter 1

Literature review on balanced and unbalanced industrial assignment problems

Ruchi Gupta and Barsarani

1.1 INTRODUCTION

The assignment problem is a well-known optimization problem in operations research and combinatorial optimization. It requires allocating a number of tasks or jobs to a number of people or agents in an effective way, and the aim is to minimize the sum of all costs or maximize total profit. In the assignment problem, each task should be assigned to exactly one resource, and each and every person will be assigned a maximum of one task. This is an assignment problem and its function is to optimize a certain objective, such as minimizing the entire cost of the assignment, maximizing the total profit, and accomplishing all tasks within the stipulated time.

The assignment problem can be defined as a matrix, where the tasks are rows and the resources/people are columns. Each cell in the matrix contains the cost, profit, or time associated with assigning a particular resource. The objective is to get a match between the task and the resource that minimizes or maximizes the total value of the objective function. There are different methods and algorithms to resolve the assignment problems, including the Hungarian method developed by Hungarian mathematicians, the enumeration method, the transportation method, and linear programming approaches. These algorithms aim to find the optimal assignment combinations and evaluate the objective function for each assignment. The assignment problem has various applications in real-world scheduling, production planning, resource allocation, and matching problems in economics and computer science. It provides a systematic approach to allocating resources efficiently and optimizing performance in various domains.

DOI: 10.1201/9781003512585-2

1.2 FLOWCHART OF AN ASSIGNMENT PROBLEM (HUNGARIAN METHOD)

1.2.1 Descriptions of the flowchart of assignment problem

Step 1: Row reduced matrix chooses the smallest value for each row and subtract it from other elements of the row.
Step 2: Column reduced matrix chooses the value by subtracting the smallest value of the column from the other values of that column.
Step 3: If there is only one zero in the row and cross (X) other zeros in that column, then row-wise assign a zero.
Step 4: If there is only one zero in that column and cross other zeros in that row, then assign column-wise a zero.
Repeat Steps 3 and 4 until all zeros are either assigned or crossed.
Step 5: Examine the uncovered elements. Select the minimum.

a) Put minimal number of lines for all 0's.
b) Select uncover elements.
c) Subtract uncover element and sum it with the element of the intersection of lines.

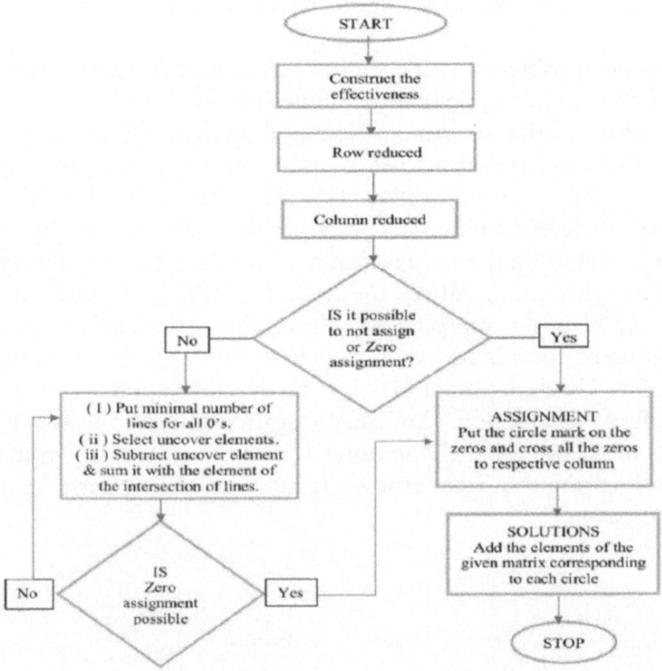

Figure 1.1 Flow chart of an assignment problem (Hungarian method).

Step 6: Repeat Steps 5, if zero assignment is not possible, else put the circle mark on the zeros and cross all the zeros to the respective column.
Step 7: Add the elements of the given matrix corresponding to each circle [1–4].

1.3 MATHEMATICAL MODEL FOR ASSIGNMENT PROBLEM

There are k employees and k jobs with the price of each employee for each job, we need to assign at least one and only one job to each employee so on to optimize the full cost.

Let α_{ij} be the assigning if ith to jth job, x_{ij} be the assignment of ith employee to jth job and $x_{ij} = 1$, if ith employee is assigned to jth job = 0, otherwise.

Thus, an assignment will be represented by $k \times k$ matrix which covers all the $k!$ possible ways of constructing assignments (Table 1.1).

The assignment problem will be expressed symbolically as follows:

$$\text{Minimize } z = \sum_{i=1}^{k}\sum_{j=1}^{k} cijxij \tag{1.1}$$

Subject to constraints

$$\sum_{j=1}^{k} xij = 1, \qquad i = 1,\ 2,\ 3\ldots\ k \tag{1.2}$$

(Exactly one job is assigned to only one worker)

Table 1.1 Model for assignment problem

	Jobs					
Worker	1	2	3	...	...	K
1	α_{11}	α_{12}	α_{13}	...	...	α_{1k}
2	α_{21}	α_{22}	α_{23}	...	...	α_{2k}
.	.	.	.	.	.	.
.	.	.	.	.	.	.
.	.	.	.	.	.	.
k	α_{k1}	α_{k2}	α_{k3}	...	...	α_{kk}

$$\sum_{i=1}^{k} xij = 1, \qquad j = 1,\ 2,\ 3\ldots\ k \tag{1.3}$$

(Exactly one employee is assigned to h job where takes a worth 0 or 1.)

1.4 SUMMARY OF OPTIMIZATION APPROACHES

Prof. Ramashankar Prajapati et al. [5]

This method is to solve very common assignment problems that arise by workers while doing their jobs. The aim of the study is to reduce time and find optimal solutions in fewer steps.

We are discussing two methods in the study: the MOA method and the Hungarian method. The results of an optimal solution are the same as compared to the Hungarian method and the MOA method.

Kadhim et al. [6]

This proposed a way to resolve assignment problems by finding an optimized technique to resolve the assignment problems with various kinds of minimization.

This approach could be used to resolve two assignment problems.

(i) The number of people to be assigned is equal to the number of machines and is called a balanced assignment problem.
(ii) The assigned number of people is not equal to the number of machines and is called an unbalanced assignment problem.

Many examples have been solved using this approach, and almost all the optimized solutions are the same as the Hungarian method optimized solutions. It realized that this new technique gives an optimal solution in simple and few steps compared to the H.M. (Hungarian method) because it contains simple and minimum steps, so effort and duration are significantly saved.

Helena Gaspars-Wieloch [7]

This proposed a mathematical model of an assignment that is applied to solve problems for new projects (innovation projects) or unstable projects. Nondeterministic models and algorithms are required by human resource project management and optimization.

The method was created to resolve assignment problems in case of uncertain situations. It depends on selected features of the binary programming, Hurwitz, standard deviation analysis, and Bayes decision rules. This enables and considers diverse decision-makers and allows them to solve problems without the sophisticated mathematical skills of managers.

Faudzi et al. [8]

This presented a case study on the domain of education and the problem of assignment classified into two parts:

(i) A problem that provides for a particular event time is called a timetabling problem (i.e., further classified examination, course, etc.).
(ii) A problem that is allocated to an individual of specific cases or projects depending on the interest is called an allocation problem (i.e., student–project allocation, space allocation, etc.).

To solve the problem, we get an optimized solution based on the complexity of the problem. The meta-heuristic and heuristic techniques were used to solve the problems of the top trend of assignment because this is the respective approach procedure, but this is not an optimized solution. This will be an infinity puzzle problem assignment for the upcoming future.

Sudrajat Supian et al. [9]

Here we discussed the workers' traveling duration from the source point (central post office Bandung) to the destination point for delivering the package. It was enhanced by using the Hungarian method. The analysis sensitivity of the assignment issues can be illustrated by changes in the coefficient range value that allowed functional objective, so that does not change the optimal.

Reena et al. [10]

They developed a new solution method to solve the optimization problem, where the purpose is to reduce the total time of the problem. In other solution methods, we find the possibility of optimization and then compare it to determine the best solution. In this case, we solve only balanced assignment problems, but some methods are very time-consuming.

The proposed algorithm is in the following steps:

(i) Construct the cost matrix based on the problem.
(ii) (1) Find the first row's smallest element and subtract it from the first row's each element.
    (2) Find the first column's smallest element only and subtract it from the first columns of each element.
(iii) Now get the smallest value of each row and then subtract each element of the row except the first row.
(iv) Find the assignment based on the number of zeros in rows and columns.
(v) Test of optimality.
(vi) Revising the optimality cost of the matrix.

Humayra Dil Afroz et al. [11]

They compute by existing two methods and discuss the new method. We have compared the optimal solution between the existing two methods and the new method. This method proposed a systematic procedure, assignment problem, and was easy to apply. We get the same optimal solutions as the HA method and MOA method. The main goal of the study is to solve the assignment problem easily by consuming the minimum time.

Akpan **et al.** [12]

They discussed about problem of Bids (in Millions of Naira) of tender to different projects and developed to solve an assignment problem, by using "TORA 2.0" statistical software which gave the same results of 881 (in the Millions of Naira). In the study, two methods are used for solving the problem of assignment (the Hungarian method and the Alternate assignment method) and doing result comparisons. After comparison, the results indicate the same optimal solution for both methods, but optimal solutions were found by the Alternate method in a few steps and this saved our time.

Priyadharshini L. S. et al. [13]

They created a method to find optimized solutions to assignment problems directly. It requires logical calculations and very simple arithmeticals. This is the most attractive feature of this method. This technique provides an optimized solution in minimum iteration, for the assignment problem solution. The main proposal of this technique takes minimal time and is very helpful and understandable easily by logistics and supply chain problem decision-makers. In this study, the case of analysis sensitivity to change objective function coefficient from constraints assignment and single-valued function is used. It also allows function objectively, and because of that the optimal assignment won't change the initial problem.

Rai et al. [14]

They proposed a method for a classic problem assignment, which is an LPP special case. This was made on one basis. The paper is an observation of assignment problem variation. This problem shows the matching optimality of the elements of more than one set, where problem dimension refers to the number of element sets that should be matched. The aim is to find that variation and it is easy for researchers to develop some variations of assignment problems for certain applications.

1.5 CONCLUSION

This chapter is an illustration of mathematical models, methods, and algorithms for solving different types of assignment problems to get optimal solutions. For solving assignment problems, we need constructive systematic methods. By systematic method, we mean a method having less number of steps, less time-consuming maximizes the profit, gives the best optimal solution, easy to solve the problem, and is advantageous for the users such as the Hungarian method. For future work, we are looking forward to improving some more algorithms and techniques to solve such problems and to find solution of optimal solutions.

1.6 FUTURE SCOPE OF STUDY

For future work, we are looking forward to improving some more algorithms and techniques to solve such problems and to find solution of optimal solutions. The assignment problem study has a favorable future, and it will use the advancement of technology. Here are some potential areas of future scope for studying assignment problems:

The assignment problems solution has very good future scope in different fields as follows:

- Research: We can use this method in resource allocation, optimize logistics, and time scheduling.
- Network design: The HM (Hungarian method) can use to resolve routing and network flow problems in network communication and transportation.
- Machine learning: The solutions/techniques used to integrate the algorithms to associate data, computer vision tracking, and natural language processing.
- Healthcare: Various methods are used in patient–doctor assignment optimization and hospital resource management optimization.
- Algorithmic advances: The research and analysis into more methodical algorithms to solve assignment problems can make fast and more flexible solutions. It explores and optimizes advanced techniques, various methods, and algorithms, e.g., genetic algorithms, simulated annealing, and ant colony optimization.

REFERENCES

1. Kamble, P. D., Giri, J., Makki, E., Sunheriya, N., Sahare, S. B., Chadge, R., Mahatme, C., Giri, P. and Panchal, H., An application of hybrid Taguchi-ANN to predict tool wear for turning EN24 material. *AIP Advances*, 14(1), 015206 (2024). https://doi.org/10.1063/5.0186432
2. Narasimhamu, K. L., Natarajan, M., Thejasree, P., Makki, E., Giri, J., Sunheriya, N., Chadge, R., Mahatme, C., Giri, P. and Sathish, T., Development of Hybrid Optimization Model Using Grey-ANFIS-Jaya Algorithm for CNC Drilling of Aluminium Alloy. *Journal of Engineering*, 2024, 1–12 (2024). https://doi.org/10.1155/2024/1476770

3. Natarajan, M., Pasupuleti, T., Abdullah, M. M. S., Mohammad, F., Giri, J., Chadge, R., Sunheriya, N., Mahatme, C., Giri, P. and Soleiman, A. A., Assessment of Machining of Hastelloy Using WEDM by a Multi-Objective Approach, *Sustainability*, 15(13), 10105 (2023). https://doi.org/10.3390/su151310105
4. Natarajan, M., Pasupuleti, T., Giri, J., Sunheriya, N., Katta, L. N., Chadge, R., Mahatme, C., Giri, P., Mallik, S. and Ray, K., Machinability of Titanium Grade 5 Alloy for Wire Electrical Discharge Machining Using a Hybrid Learning Algorithm. *Information*, 14(8), 439 (2023). https://doi.org/10.3390/info14080439
5. Prajapati, R., Haque, A., Jain, J. and Singh, S., A study on solving assignment problem, *VIVA-Tech International Journal for Reasearch and Innovation*, 10, 2581–7280 (2021).
6. Kadhim, H. J., Shiker, M. A. and Al-Dallal, H. A., A New Technique for Finding the optimal Solution to Assignment Problems with Maximization Objective Function, *Journal of Physics: Conference Series*, 1963, 012104 (2021).
7. Gaspars-Wieloch, H., The Assignment Problem in Human Resource Project Management under Uncertainty, *Risks*, 9, 25 (2021).
8. Faudzi, S., Abdul-Rahman, S. and Abd Rahman, R., An Assignment Problem and its application in Education Domain: A Review and Potential Path, *Advances Operational Research*, 19, 8958393 (2018).
9. Supian, S., Wahyuni, S. and Nahar, J., Optimization of personnel Assignment Problem Based on Travelling Time by using Hungarian methods: Case study on the Central Post Office Bandung, *IOP Conference Series: Materials Science and Engineering*, 300(1), 012005 (2018). https://doi.org/10.1088/1757-899X/300/1/012
10. Reena, G. P., Bhavin, S. P. and Bhathawala, P. H., Optimal Solution of an Assignment Problem as a Special Case of Transportation Problem, *Journal of Emerging Technologies and Innovative Research (JETIR)*, 5(1) (January 2018).
11. Afroz, H. D. and Hossen, M. A., New Proposed Method for Solving Assignment Problem and Comparative Study with the Existing Methods, *IOSR Journal of Mathematics (IOSR-JM)*, 13, 84–88 (2017).
12. Akpan, N. P. and Abraham, U. P., A Critique of the Hungarian Method of Solving Assignment Problem to the Alternate Method of Assignment Problem, *International Journal of Sciences: Basic and applied Research (IJSBAR)*, 29 (2016).
13. Priyadharshini, L. S. and Devi Abirami, P., A New Method for Finding an Optimal Solution of Assignment Problem, *International Journal for Research in Applied Science & Engineering Technology (IJRASET)*, 10 (2022).
14. Rai, N. and Khan, A. A Brief Review on Classic Assignment Problem and its Applications, *IOSR Journal of Engineering*, 9 (2019).

Chapter 2

Systems for detecting intruders in Internet of Things environments using the cloud

A survey

Amritpal Kaur, Rajesh Kumar Tyagi, Shweta Sinha, and Prashant Vats

2.1 INTRODUCTION

Future cities can be built with inherently sustainable infrastructure, thanks to the IoT which holds a central role in the next-generation infrastructure [1, 2]. Major sensors and devices in the IoT often collect and analyze environmental and temporal statistics to solve various challenges [3, 4]. With advancements in IoT devices, medical care has become more intelligent, and communication has improved [5, 6]. As a result, IoT is extensively utilized in various sectors, including home automation, healthcare, finance, tourism, smart cities, education, entertainment, and transportation [7, 8]. It is also employed in energy distribution. It is estimated that by 2022, 1 trillion objects will be connected to the Internet [9, 10]. Cloud services based on IoT data are used to analyze sensor data streams in real time [11, 12].

In the area of IT (information and technology), the subject of cloud computing has been a rapidly growing topic [13]. Internet-based computing that uses virtual shared servers, or data centers, to provide systems, regulations, software, infrastructure, and a variety of facilities is known as cloud computing [14, 15, 95–97]. Nevertheless, security and privacy are crucial considerations in its achievement due to the open and largely distributed design, which is open to attacks [16, 17, 98–100]. For instance, cryptographic approaches cannot be used to prevent attacks like the Sybil attack, DoS, black hole, and others [18, 19]. IDS is now the most frequently used part of compliance and the procedures to secure the computer systems for defending cloud-based environments from various risks and attacks [20, 21]. To keep pace with the development of crimes related to technology, it employs a variety of reaction strategies to pinpoint weaknesses, detect criminal activity, and put preventative measures in place [22, 23].

In this chapter, we provide an overview of research pertinent to IDS for the issues of IoT security. We focus on potential solutions and support for cybersecurity in IoT challenges. The development of IDS to serve the IoT environment raises serious safety concerns.

DOI: 10.1201/9781003512585-3

2.2 RESEARCH BACKGROUND

Cloud computing is a novel IT technique because of its open and dispersed nature. The majority of industries and IT organizations are implementing the paradigm of cloud by shifting their machines to the networks on cloud, owing to its attractive qualities, like a huge accessibility of the resources in the variety of services, the cost-effectiveness, dynamicity, including other features [24, 25]. In an environment of cloud computing, on-demand network right to use is provided to a common collection of adaptable processor resources, such as applications, storage, servers, services, and networks. Cloud security is a crucial issue to take into consideration because these weaknesses in the environment with cloud computing are an attraction for attackers [26].

IDSs which are designed for the IoT environments are essential for preventing security risks related to IoT that take benefit of a few of these glitches in the security [27]. Traditional IDSs do not need to constitute a suitable alternative for IoT environments because of the inadequate CPU and the storage abilities of the instruments of IoT as well as the precise protocols which are used [28, 29]. To protect the stored data, organizations must employ ID, firewalls, and various prevention methods like authentication, encryption, and other advanced security measures for both software and hardware. Attackers search for security holes to take advantage of. Enterprises utilize IDPS to guarantee security along with privacy in the cloud environment [30, 31].

The list of network ID (NID)-related articles that have undergone systematic reviews in recent years is provided below. The goal of this task is to identify flaws and holes, as well as the accessible cracks that can be taken care of. The research articles are located, and their rationale as well as the conclusions are outlined.

The first reviews on IDS were produced by prior research projects and are compiled in Table 2.1. None of these articles have been investigated and the IDS is analyzed with a vision to become "learning based," "rule based," and "pattern based," despite the fact that they are all organized and include useful information. In order to carry out such a categorization in IDS, this work offers a thorough review. In addition, our study consists of research papers from the literature since July 2021, which none of the literature evaluations have done as yet.

The following queries will be addressed in this study:

1. What is the distribution of experiments by year? This query will be addressed in Section 2.4.
2. Which type of data collection is most common among researchers? Section 2.5 has the answer.
3. What issues and challenges do you encounter when developing IDS? These topics are covered in Section 2.7.

Table 2.1 Latest research outcomes

Author	*Publisher*	*Research outcomes*	*Limitations*
Goncalves et al. [18]	IEEE	A thorough SLR (systematic literature review) based on using IDS that are intelligent for VANET (vehicular ad hoc network) is being presented. The majority of them had not adequately revealed the procedures used to construct their datasets and attacks, it was discovered during the evaluation of the study. None of them published the data they collected online for peer assessment. Some utilize popular datasets that are accessible to the general public, similar to the Kyoto dataset as well as the NSL-KDD. Nevertheless, since they may not be gathered by VANETs, they might alter the outcomes	Lacks a section specifically for future work
Ramaki et al. [32]	Wiley	A detailed analysis of ID methods on the basis of concealed Markov models, together with a debate on the main problems and open problems in the field. The results showed that ID techniques based on hidden Markov models have key six advantages: 1. Discovering recent and unanticipated invasions 2. Reliable ID 3. Projecting the possible steps that will be taken by the intruders 4. Using a variety of sources of data as inputs. 5. Using real-time applications which interact with data streams instantaneously 6. An assessment of visual information depiction using various machine learning techniques	Has no recommendations for further work
Perker and Oney [33]	IEEE	The following conclusions were reached after reviewing publications on the ANN techniques in NID for pinpointing the major trends: 1. The amount of research on this subject is expected to increase greatly in the following years given the current increase in rates of publication and the good outcomes of ANN techniques for NID 2. Deep NN along with its versions have been the majorly well-liked and frequently applied artificial neural networks in relation to the NID system (NIDS) industry	Lacks A section specifically for future work

(Continued)

Table 2.1 (Continued)

Author	*Publisher*	*Research outcomes*	*Limitations*
		3. The results demonstrated which dataset of KDD'99 is utilized the most commonly in studies on NID utilizing ANNs	
Daniya and Ganeshan [34]	IOP	Examining different anomaly-based IDSs used for cybersecurity, categorizing these on the basis of multiple standards, and rating them depending on the datasets utilized, the acquired precision, the metrics for assessment, and the tools for their deployment The intrusion detection system was most likely developed through a number of trials using MATLAB and the KDD CUP99 dataset	There is no separate area for potential work, and the process is not clearly described
Pandya and Mishra [35]	IEEE	The following conclusions can be drawn from an examination of a multi-layered investigation of numerous vulnerabilities in security, which are prevalent in IoT levels, such as support, network, perception, and application layers, especially focusing on DDoS (Distributed Denial of Service) threats: 1. Discussing several DDoS types of attacks, their impacts, and how the IoT devices are affected by them, along with the improvement techniques 2. Examining distinct ID and defense models to stop DDoS attacks 3. Considering IDS classification, different IDS dataset-based models, different detection of anomalies strategies, and numerous DL and ML methods for preprocessing of data and detection of malware 4. Examining issues from the study, possible fixes, and intended future results	There are no suggested study categories
Salo et al. [36]	IEEE	Performing an SLR to look that DataMining (DM) techniques fall under IDS. The inspiration came from pertinent experimental research that was presented at conferences and published in journals at the time. It presents the following conclusions: 1. In IDSs, the most popular DM techniques are SVM (support vector machines), DT, BN, and ANN 2. Keeping in mind the various attack types, like User to Root (U2R), DoS, Probe and Remote to Local (R2L), that IDSs are intended to counter	Lacks A Section Specifically for future work

(*Continued*)

Table 2.1 (Continued)

Author	*Publisher*	*Research outcomes*	*Limitations*
		3. There has been a dearth of real-time IDS studies 4. Learning that IDS uses 25 private and public datasets 5. Highlighting the benefits and drawbacks of the adopted DM techniques	
Singh and Sangher [37]	IEEE	A method is suggested that employs web robots and a P2P layout system to discover threats and log the data in a manner which will be helpful for technical analysis and investigation according to the findings of an extensive amount of research that analyses the newest IDS. DL-based solutions reduce the limitations of huge data processing in cyberattacks and forensics while also delivering insights with strong analytical capabilities	There is not a clear depiction of the method
Salo et al. [38]	ARXIV	The following results were obtained from an SLR of DM methods used in IDS-based solutions from 2013 to 2018: 1. Among the most often used DM methods for IDS for big data include BN, DT, ANN, k-means, as well as SVM 2. The frequently encountered DSF was Hadoop, which has a frequency of 12 trials 3. Listing the metrics for performance and efficiency evaluation 4. The majority of the research works have categorized attacks as being either malevolent or normal	Has not posed any open questions for additional study
Pateriya and Kaur [39]	ACADEMIA	Reviewing the many mobile IDS already in use, their benefits, and their drawbacks the finding is that the capabilities of detection of IDS in use currently are limited. They need tools that can control malware or data-related threats and recognize patterns over a wider area.	No clear statement of the method

(Continued)

Table 2.1 (Continued)

Author	*Publisher*	*Research outcomes*	*Limitations*
Idrissi et al. [40]	IEEE	The study uses the Cisco IoT reference model framework to categorize certain well-known security concerns in order to focus on the present state of IoT security threats and weaknesses. Recently, the area of security in IoT has been a central research topic as it is of prime importance. There is an important use of deep learning (DL) in a number of scholarly fields, including fields of cybersecurity as well as IDS. The use of DL in cybersecurity has produced some encouraging outcomes, paving the path for enhanced effective security in environments using IoT. However, at this time, the majority of the researches in this field is just in theory or may be just a way of benchmarking of dataset using DL methodologies. It takes time to create a true IDS for the Internet of Things, thus researchers should include their prototypes into existing systems of IDS and then subsequently test these by using actual network data	No separate segment for possible future research
Puente and Aludhilu [41]	UCI	A comparison of the benefits and drawbacks of various approaches using multiple ID, including the statistical anomaly, matching pattern, deep learning, and machine learning Machine learning is the greatest alternative to develop IDS services because it can run as an automated process with little interaction from humans The usage of neural networks with recurrent patterns are the most effective technique for creating this kind of technology	Lacks a section, for work
Ferdiana [42]	IEEE	From January 2016 to May 2020, research developments in the methods, databases, and methods employed for IDS subjects are studied and recognized. Currently, there are seven approaches to study in the IDS sector: categorization, association, estimate, statistic, clustering, prediction, and dataset analysis, according to a survey of a few key studies	Has not posed any open questions for additional study

In many different problem categories, anomaly detection systems are researched or explored. They haven't, however, been thoroughly studied in a cloud context. The detection of anomaly methods in computing that are cloud-based is currently being thought about and growing since it causes issues. The safety of the machines and their owners will depend on continuous evaluation of the execution of IDS at every tier of service and comparisons among systems. The goal of this chapter is to pull the focus of renowned researchers to possible methods for developing IDSs by consolidating the most recent, varied research initiatives that focus on protection of the recently prevalent resources and cloud services.

Moreover, it outlines the model IDS requirements in cloud computing as well as explores the implications in real life. Additionally, the main goal this research is to highlight the strengths and weaknesses of the modern state-of-the-art Intrusion Recognition, developed for IoT and aid the researchers in their future work by outlining crucial study directions. By combining, comparing, and unifying various research efforts, these objectives support IoT security experts.

2.2.1 Research methodology and statistics of data

This chapter aims to offer a thorough evaluation of the content of the available literature for identifying gaps in research and limitations in the knowledge. The primary advantage of this study is that it provides a methodical research process for consolidating previous research results and examining and evaluating prior research.

This study employs an organized method of analysis to offer an exhaustive, comprehensive, and impartial review that may enable to (a) gather and analyze all current studies in an area; (b) investigate all researches that are easily accessible and include the data about the topic being studied; and (c) avoid study of a project being researched, either unintentionally or intentionally, for directing the findings of the research in the direction of a specific goal. We introduce many reliable, methodical research techniques that have been widely employed by scholars across numerous disciplines [43, 44].

Using keywords like "Intrusion Detection System in the Cloud," "Intrusion Detection Methods," "Evaluation Criteria of Intrusion Detection Systems," "Intrusion Detection AND IoT," "Intrusion Detection in cloud," "Intrusion Detection AND learning based," "Intrusion Detection AND pattern based," or "Intrusion Detection AND rule based," we methodically evaluate the ongoing studies relevant to this work.

The English-language resources are examined during the study process. The majority of the research sources, nevertheless, are dependable publications. The abstracts and titles of the selected papers were evaluated to see if they contained relevant information. Finally, our conclusions based on the previously conducted research in this area have been compiled in Table 2.1. The fact that using the search term "IoT AND Intrusion Detection" returned results for around 25,000 documents indicates that academics are keenly intrigued for the security of IoT devices. The dates of publication of all of these research are shown in Figure 2.1. The graph demonstrates how the number of researches on IoT (intrusion detection systems) has almost increased by 1,000× over the last ten years.

The studies have appeared in numerous journals. The journals have independently expressed the number of these articles published in the top journals in the past ten years: 7,500 articles on the topic are expected to be published in various magazines in 2020, according to the graph in Figure 2.1. The bottom of the chart is curved because it is still early in 2021, and a reversal will undoubtedly occur by the end of the year.

Approximately 21.46% of all articles that were published in 2020 have appeared in IEEE journals.In accordance with the overall quantity of research articles in 2020 (shown in the above figure), which includes 7,500 articles. This chapter includes more recent studies in order to concentrate on the most recent research. If the goal is to review 25 articles, the articles in reputed periodicals are crucial. Articles with a high number of citations will be considered for examination if a certain subject has not been previously studied.

Finally, 27 articles are shortlisted to be examined. They are split into majorly three categories:

- Mechanisms on the basis of learning
- Mechanisms on the basis of patterns
- Mechanisms based on rules

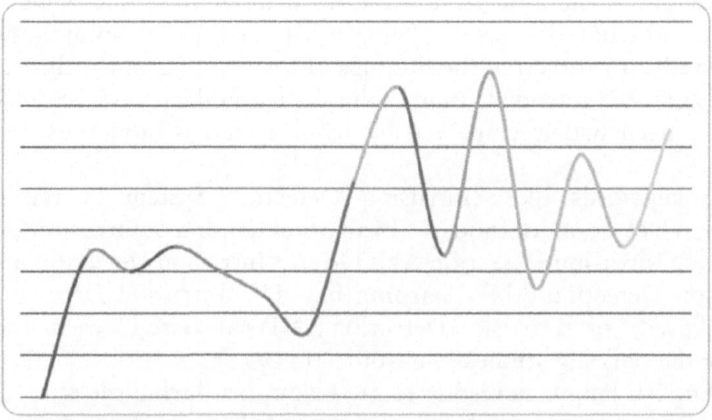

Figure 2.1 Year-wise research growth on IDS architecture.

2.3 INTRUSION DETECTION SYSTEMS

The ultimate purpose of a planning process is to cover extensive data and have a wider range of applications to make security control and IDS efficient. The three distinct approaches to process data are biological principles approach, processing-based approach, and artificial intelligence approach, which have mostly been taken into consideration while classifying IDS techniques. However, this classification makes it difficult to recognize all of the characteristics of detection methods [45]. This section looks at all 23 chosen articles in an effort to comprehend the tendency of IDS on the cloud. Three categories, namely, learning-based, pattern-based, and rule-based mechanisms, have been used to group these publications. Additionally, these methods' variations, benefits, and drawbacks have been covered in detail.

2.3.1 Learning-based mechanisms

Artificial learning techniques are used in the IDS approach to blend learning models and learning competencies [46]. Recently, learning techniques have been used extensively in anomaly identification because they can automatically form feedback on a person's typical behavior, on whether or not they have been watched over. The research by Ma et al. proposed an original framework for multi-party deep learning (DL) on the cloud which preserves privacy, where an extensive quantity of data for training is distributed across several parties. Actually, Ma et al. offered a new, interactive deep learning model that protects privacy and depends on aggregate signatures, Diffie Hellman exchange of keys, and El Gamal encryption. These methods allowed different users to use the same neural network (NN) model as a learning tool, using the entire dataset together [47].

The machine learning approach is practical for automatically creating the ID framework from data used for training, which will spare human effort from detecting the signs of imminent intrusions or to determine the sensor node's typical behavior [48]. On the labeled data sections, the supervised anomaly ID trains to generate the typical network profiles. However, the unsupervised IDS aids in identifying attacks even when there is no prior knowledge of them or natural examples. When there is an anomaly or it is discovered that the ML alters its behavior, it continues with the new order/rule and confirmed quantities, or the natural behavior of the data is retained [49]. The approach described above offers a mechanism for improving performance by gaining knowledge from the past outcomes [50].

However, the cloud server does not give access to the local dataset or learning model. An extensive review reveals that their solutions pass the privacy and verifiability security requirements. Their approach also has a suitable computing component competence and is possible to be applied in various DL applications that require privacy protection. In an effort for increasing the precision of the anomaly recognition system, Al Jamal et al. [51] also provided an anomaly detection system, which is network-based and at the level of cloud hypervisor, that used a combination of techniques to combine the K-mean clustering technique with the algorithm for SVM, based on classification. The proposed method and subsequent results were evaluated with the use of information from the research on UNSW-NB15 and evaluated against the prior investigations.

With the model of K-means clustering, which they propose, the result is a little more accurate than others. Nevertheless, Aljamal et al.'s [51] SVM model's precision proved lower than that of supervised techniques. A novel feature selection wrapper for NID that consolidates the whale-swarm optimizer and Pearson's correlation methodology was proposed by Ravindranath et al. [52] to be used in cloud-based computing environments. An attribute selection approach based on wrappers was developed to give the model the necessary features. The findings demonstrated that feature selection wrappers increased the accuracy of model forecasting and the level of dataset relevance. In addition, when it is compared to previous annealing by binary Whale simulation, the selection wrapper by Whale Pearson feature delivered better results in 100 iterations with substantially less CPU time.

The recommended swarm intelligence method employed a mutation to address the function for correlation bias, together with a random initialization approach. In contrast to the majority of recent investigations, Salman et al. [53] focused on finding and classification of anomalies instead of just revealing them. Using a freely accessible dataset, Salman et al. developed and evaluated models of machine learning for identifying and categorizing a variety of assaults. As ML techniques which were supervised, Salman et al. used random forest and linear regression. Salman et al. showed that attack similarities could cause the categorization to be less precise even when recognition is perfect. The results showed that certain assaults could not be identified, with an accuracy of detection above 99% and an accuracy of classification 93.6%. Additionally, Salman et al. asserted that similar approaches of ML could be utilized for categorization of multi-cloud setups.

A unique DL-based IDS was proposed by Liu, Otoum, and Nayak [54] as a means of spotting major anomalies in ever-expanding IoT-based networks. To achieve the highest level of detection accuracy, their suggested module combines the polynomial network, which is stacked-deep and the Spider Monkey Optimization method (SMO). A network which is stacked-deep and polynomial classifies the information as standard or abnormal using the best features from the datasets selected by SMO. U2R attacks, DoS attacks, probing attacks, and R2L attacks are a few examples of anomalies that DL-IDS has discovered. The results show that accuracy, recall, F-score, and accuracy all are improved by the suggested DL-IDS. Additionally, research was conducted on IDS which is anomaly-based and perfect for defending IoT against DoS attacks [55].

The random forests, as well as the gradient-boosted machines, along with AdaBoost, regression and classification trees, highly randomized trees, as well as multilayer perceptrons have been used for evaluating the functioning of seven ML classification approaches. Utilizing more common metrics and techniques for validation, the classifier's performance is evaluated. Additionally, a method for selecting the ideal classifier for a specific application was provided by Verma and Ranga. The extreme gradient boosting classifier and regression trees together are apt for the development of IoT-specific IDS, built on anomaly, because they give the better trade-off among critical metrics as well as the response time, according to the results of the statistical analyses and performance tests. The main objective of the chapter was to influence experts on IoT security to create IDSs on the basis of ensemble learning and to propose suitable techniques to evaluate statistically the performance of the classifiers.

For IoT networks, a unique IDS which classifies traffic flow according to DL principles was proposed by Ge et al. [56]. Using an IoT dataset that was recently released, Ge et al. [56] developed general features using the field database at the packet level. Ge et al. developed a NN-based model, which acted as network to forward feed for multiclass and binary sorting, as well as DDoS, reconnaissance, DoS, and theft of information operations against IoT machines. The proposed layout's examination with the use of the dataset that was processed revealed that it had acquired a satisfactory accuracy in the process of classification. Wang et al. [57] have presented a DBN (deep belief network) and augmented GA (genetic algorithm)-based models for detecting intrusion.

The GA was continually iterated to create the perfect amount of layers as well as the neurons that were concealed in every layer when exposed to multiple forms of assault, which results in an ID model on the basis of the DBN that has a high rate of detection and a structure that is compact. Finally, the dataset of NSL-KDD was used to evaluate and for the testing of the model and methodology. Due to the empirical findings, a better detection of intrusion method, in combination with DBN, may efficiently boost the frequency at which intrusion attacks are detected, while also decreasing the design complexity of the neural network.

The viability and benefits of divesting the ongoing job of detection of intrusion on the basis of DL were demonstrated by Loukas et al. [58] using a small four-wheeled robotic means of transportation on land, as a study case. It used time-series data that was input into a neural network structure as real-time input from each of the physical and online processes. A deep perceptron which is multilayer and a neural network design which is recurrent, both were used by Loukas et al. The former takes advantage of the hidden layer of memory that is long–short and which was especially important for comprehending the chronological situation of various attacks. Loukas et al.'s use of malware, assaults of denial of service, and denial of command injection are examples of cyberattacks that are pertinent to autonomous vehicles.

Using detection delay as the criterion, Loukas et al.'s [58] arithmetic model was developed for determining if the calculation of offloading had any benefit, given the variables in relation to the operations that the network has and the needs to process for the DL model. The latency of detection is increasingly reduced with the help of offloading, the reliability of the network is increased, and more processing power is required. Nine articles were reviewed in this section. Each method's benefits and drawbacks were looked at. A precise framework, strategy, or algorithm was taken from the papers and applied in some of these researches.

2.3.2 Pattern-based mechanisms

When the intrusion patterns are modeled, matched, and discovered in accordance with the package head, content, or both, pattern-based IDSs are frequently utilized. In host-oriented IDSs, the intrusion patterns could also be identified by joining the words which indicate the calls of the system in a trail of the system. This collection, for instance, employs more systems that are fuzzy [59]. The steady creation of novel types and varied forms of incursions cause signatures to increase, which results in a more expensive pattern in terms of computational costs. In a networked environment, the payload data in the current is tracked independently from the traffic flow. A collection of descriptions of previously identified harmful patterns of traffic have been compared to the present traffic pattern.

An attempt is made to reach up to the level of some of the patterns of traffic, with the supplied malicious descriptions of traffic [60]. A visual sensor NID technique was presented by Huang et al. [61]. In order to understand traffic patterns and spot breaches, they offered a traffic framework that generates useful features which are subsequently given as an input to a neural network that is hierarchical and has a map that is self-organizing. They also offered a method of active learning to overcome the challenge of dealing with huge and uneven datasets for training their self-organizing map learning strategy. Based on empirical data, the recommended method performs well in real time and has a high detection accuracy. Aparicio-Navarro et al. recommended to take into account an all-encompassing information pertaining to the monitored network, like the prior knowledge of the network administrator and the pattern-of-life, while assessing the occurrence or absence of an attack in the network traffic [62]. This effectively deals with the issue of class imbalance and quickens training Table 2.2, shows review of pattern based articles.

Aparicio-Navarro et al. added two novel approaches to the approach originally shown in Ref. [63], which merged a FCM (fuzzy cognitive map) including an IDS to include related data in the method of detection. The information, which included both normal traffic and signs of port scanning attacks, was gathered over the course of nine days from an actual LAN at a research facility. The results demonstrate that adding the FCM into the IDS

Table 2.2 Pattern-based articles review

Article	*Pros/Cons*	*Algorithm Used*
Huang et al. [61]	High performance in real-time and detecting accuracy The issue of class imbalance can be somewhat addressed by this learning strategy	ASS, N-gram
Navarro et al. [62]	Lowering the overall frequency of false alerts Limited ability to detect multistage attacks. Ineffective in actual use	BPA
Liu et al. [64]	Increase in recognition accuracy. Fewer alarms that were false Uncertain role in the physical network that is distributed	ANID
Wang et al. [65]	Results show how successful the methodology is, and the runtime is substantially shorter than that of previous algorithms	C-Average
Raju and Rao [66]	High speed plus shorter search times plus running trials on actual data sets	MapReduce
Calix and Iqbal [67]	The hardware only assessed up to 4,096 samples for training	CogniBlox modules
Rajagopalan and Manickram [68]	Optimized organization and improved performance	Hybrid GSO TS, GSO- ANN

increases its effectiveness. Additionally, the results show that the most efficient way to use pattern of life in the detection process is to change the basic BPA (probability assignments) before fusion of data. Their objective was to determine the most effective strategy to use an FCM in an IDS combo.

An ANID (adaptive NID) technique was presented by Liu et al. [64]. It combines global optimal GMM-based pattern recognition with greedy algorithm-based fuzzy rough set feature selection. By evaluating the gain rate of information for each characteristic that is potentially sensitive, selection of fuzzy rough set feature, on the basis of the fuzzy rough set theory, selects the optimum feature subset. Using an incremental learning technique, GA-GOGM is able to generate the clusters in an automatic way, mitigating the harmful effects of the cluster centers initially. They are done so that the best pattern features can be automatically extracted from instances of normal and malicious network connections for NID based on pattern matching.

Along with the latest online findings of NID, an approach for updating the libraries of learned normal and attack pattern was established. This strategy involves mining frequent patterns and rejecting out-of-date patterns. The suggested NID model may be able to adapt to dynamic shifts in the network design while achieving high accuracies in NID for both recognized and unidentified threats with a low false alarm and missed rates of reporting. The worth and singularity of the ANID technique that is suggested were depicted using a self-built network simulation environment, which is NIDS bench-based and detailed analysis of comparison and validation, with the results on the NSL-KDD benchmark dataset. Physical simulations can be widely used to monitor cybersecurity in the real world since they can achieve higher detection accuracies on network components which are actual and have a low number of false alarms.

In Ref. [65], the researcher looked into increased sequential pattern mining-based intrusion detection and performance simulation. The IDS was planned by Wang et al. with the use of DM techniques; the results of the simulation show the method's effectiveness. The simulation showed that prefix span executed far faster than the other way when the minimum support was very small, and the disparity between them was evident. The method for IDS mining is primarily autonomous of precise data and the systems; therefore, the DM-based IDS to the source of data is incredibly minor. Rao et al. [66] also tried the matching of the patterns by parallelly implementing a matching method using MapReduce on a unit for processing graphics.

Rao et al. parallelized a pattern-matching method using the MapReduce framework. Due to the utilization of the GPU, the suggested solution initially showed a reduction in searching time. The information within the file was searched for signatures using the popular two-way and two-window algorithms for pattern matching. A huge quantity of data was tested, and a major discrepancy in GPU and CPU timings was found. The GPU found use in taking the process of MapReduce to parallelism's next level. The study in Ref. [67] set out to examine the precision and the speed that a network which is based on payload on IDS has, using a pattern-recognition engine and a method for extracting unigram features.

According to the results of Ref. [67], using processors for pattern recognition for NIDS improved presentation in terms of speed. The pattern identification engine demonstrated that regardless of the size of dataset used for training, detection might be accomplished efficiently. Additionally, the outcomes demonstrated that payload-based IDS processing efficiency employing parallel-based processors for pattern recognition is superior to that of the CPU processors which are sequential based. The current IDS system uses a single rule set with several attached patterns that have been logged in various databases. To prevent any further unauthorized or illegal operations, the entire network traffic is swiftly matched against this.

As a result, in Ref. [68], the multilayer ANN that is structure-optimized in the case of the supplied cloud relied on the IDS. The framework was optimized while accelerating convergence and overcoming long-standing issues, including the entrapment of local optima and associated premature convergence using the GSO-TS, an approach which is hybrid, for glow-swarm optimization search. According to the results, the ANN with a single hidden layer had an average rate of detection that was about 0.39% and 0.73% lower than those with two and three hidden layers, respectively. In comparison to the GSO-Tabu-ANNs with two and three hidden layers, the average detection rate for the ANN which is GSO-Tabu with one hidden layer is at approximately 0.65% lower level.

2.3.3 Rule-based mechanisms

These types of irregularity identification techniques have traditionally been suggested for identifying the use of nonconformities in the audit trails. Instead of using statistical formulas to identify consumption patterns, this approach applies sets of guidelines to display and save the patterns of consumption in audit data [69]. One of the main methods for preventing ID misuses is rule-oriented ID systems. These methods encrypt intrusive circumstances in accordance with a set of rules coordinated with respect to network traffic data or audit data.

Kumar et al. [70] presented a special IDS which is misuse-based and assessed how effective it was on both offline data that is conventional and the data that is real time and online in order to recognize five groups in a given network: Exploit, DoS, Probe, generic, and normal. Using the standard dataset UNSW-NB15 and a real-time information set (RTNITP18), the proposed model surpassed other current methods vis-à-vis accuracy, mean F-measure attack accuracy, detection rate, average accuracy, as well as the rate of false alarm. The suggested model of IDS keeps an eye on the network and searches for various threats like a dog walker. Additionally, with regard to the topic of misuse detection mechanisms, Papamartzivanos et al. [71] presented dendron, a novel method for the development of a decision tree classifier using GA, aiming at producing detection rules. They tried to boost the productivity of security administrators and make their duties easier by implementing regulations that were easy to grasp linguistically.

Additionally, their method takes care of a number of issues brought on by the nature of traffic on the network and combats ML algorithms' inclination to primarily ignore trivial types of attacks. When it was tested on three ID sets of data – NSL-KDD, KDDCup'99, and UNSW-NB15 – the suggested technique outpaced the others in terms of average values like MFM, AttAcc, and AvgAcc.

Yazidi et al. [72] proposed an inventive design for a parallel NIDS on the basis of function virtualization and network rule distribution. Open vSwitch and Snort-running Docker-based containers were used to isolate the IDS features, allowing for flexibility and adaptability in the face of shifting traffic on the network. Haugerud et al. [72] developed two dynamic strategies that utilize a node-level parallelism method to dynamically update and distribute the signature rules among NIDS nodes in order to achieve intelligent rule ordering (Table 2.3 shows Rule Based Articles).

Table 2.3 Rule-based articles

Article	*Pros/Cons*	*Dataset Used*
Kumar et al. [70]	Increased precision plus increased rate of attack detection. Unable to categorize new, unknown attacks; unable to sense any zero-day attacks; unable to detect the rate on real-time set of data	RTNITP 18
Papamartzivanos et al. [71]	Accurate attack detection, capable of noticeably detecting uncommon intrusive situations. Able to surpass other contemporary and antiquated approaches on a variety of classification metrics. Some methods are unable to address the issue of identifying infrequent invasive events	UNSW-NB15
Haugerud et al. [72]	To enable the system to scale, more virtual elements that analyze network traffic should be added. Using containers, technology-defined networks, and virtual hardware in different infrastructures is not yet competent	Real-life
Lou et al. [73]	The recommended method performs better than the competition with respect to the F-measure, recall, and accuracy and also outperforms current algorithms in terms of speed. Real-time ID has a lag time	KDD-99
Ganapathy and Riyaz [74]	The proposed approach outperforms other ways in terms of making decisions. Yet, not each and every attack has the best attack detection precision	NSLKDD
Shaikh and Kshirsagar [75]	The decision table outperforms different algorithms that are rule- based in terms of performance. Only DoS attacks are defined for this technique	DoS GoldenEye
Mishra et al. [76]	As far as the NSL-KDD and ISCXIDS2012 datasets are concerned, the PART classifier was found to be the most accurate. The suggested architecture has not yet been created	NSL-KDD

By creating an operational prototype that utilized a number of recent networking technologies, their theories were put to the test in actual-world circumstances. The IDS network function virtualization exhibited by the prototype offered an elastic system that employed Snort being run on Open Docker and vSwitch containers. According to [73] security managers must use the security log rule basis to carry out vulnerability and intrusion detection. A method for developing a cyberID rule base is presented that is based on an adaptive-miner algorithm. Effective security principles were obtained via the analysis of multisource logs which are heterogeneous [73].

Their method utilized a cluster environment of Apache Spark and offered a distributed MapReduce calculation approach based on association rule mining to accelerate algorithm computation. Tests simulating assaults were carried out for evaluation of the proposed method. In terms of computation speed, the technique surpasses the conventional a priori approach as well as two additional rule mining algorithms having a parallel association. Real-time intrusion detection was also carried out. The experiments showed that the proposed strategy surpassed alternative approaches in terms of memory, precision, and F-measure. The most important characteristics that were utilized to classify the data as attacks and conventional traffic were chosen using a unique fuzzy rule and information gain ratio-based feature selection technique presented by Riyaz and Ganapathy [74]. This method contrasted with unsupervised and supervised learning models.

Additionally, well-known classifiers like SVM and least-square support vector machines were used for effective classification. On the NSLKDD dataset, multiple experiments were run to evaluate the suggested methodology. The results of their analysis revealed that the proposed method outperforms current feature selection algorithms in terms of classification. Additionally, Kshirsagar and Shaikh [75] proposed an intrusion detection strategy that included preprocessing of data, selection of features, and rule-based classifiers. With the aid of information gathering and a ranker, the feature choice was made. The method was created and put to test in CICIDS2018 utilizing classifiers, which are based on the basic dataset of the Golden Eye tool. The analysis of classifiers that are rule-based was finished, and the effectiveness of each was evaluated.

In order to evaluate which rule-based classifier might be most suited to serve as the IDS's base learner, [76]investigated 12 rule-based classifiers in a high-class imbalance situation. To create a class imbalance situation, three sets of data of invasions of class imbalance were examined. The classifier's efficiency was assessed using the receiver operating classification accuracy, characteristic value, and false alarm rate. The projective adaptive resonance theory classifier had the best results on the ISCXIDS2012 and NSL and KDD datasets. A decision table with naive Bayes hybrid outperforms other classifiers in the CICIDS2017 dataset. The CICIDS2017 dataset's ID architecture was developed utilizing a hybrid of naive Bayes and decision tables using the research findings [76].

Seven articles from recent studies were chosen and reviewed for this part. These studies' test datasets for their methodology are retrieved. Additionally, some of these methods' benefits and drawbacks are listed.

2.4 RESULTS AND DISCUSSION

In this chapter, overall 23 studies were examined and distributed into three groups. Nine publications were assessed for the "based on education" group, and their approaches were taken note of. In both the "model-based" and "method-based" groups, the identical procedure was followed. The results of these studies were as follows. In addition to the speed and accuracy with which an IDS can identify and locate threats, reality that false alarms occur frequently, each of them has a function in how much success it can have.

In order to construct an IDS with a high rate of detection and accuracy, our work aims to merge GA and extreme training machines. To ensure that an exceptionally effective classifier is chosen for the second test, we suggest splitting the training mode into virtual training and virtual testing. IDS is created in a number of ways. The IDS ought to theoretically be able to identify many intrusion types, including U2R, DoS, R2L, and Probing, Web Attack, Heartbleed, and PortScan. The benefits and drawbacks of various ID schemes differ. ML is a fully automated technique with minimal human involvement.

2.5 CHALLENGES

Although the field of research in cybersecurity has developed various distributed architectures and ID algorithms that are able to safeguard enormous networks, cloud-specific problems still remain. These problems include some of the following:

1. Despite the different recommended techniques, concepts, and implemented systems, there are no criterion which can be universally accepted or measures to evaluate an IDPS. Even though receiver operating characteristic (ROC) is frequently used to assess accuracy, it is not a reliable tool of evaluation because the outcomes are more often insufficient and deceptive [77, 78].
2. It is challenging to develop a self-adaptation ability that will enhance and drastically decrease operator involvement [79].
3. The design and implementation of NIDS involves a number of steps that increase computational complexity, including model deployment and training, data preparation and reduction of feature, and particularly NIDS [80, 81] based on DL and ML and its designing and neural P systems with a low complexity of computing. The needs are a bigger issue and a potential research topic [19, 82].
4. It is a challenging topic that has grown increasingly crucial as the number and variety of security threats to such systems have drastically expanded.. In light of the unique properties of these networks, such as battery life, bandwidth and processing overheads, and network dynamics, ID for IoT is a problem that should take the balance into account, as it must be struck between efficiency excess and identification precision [83].
5. These include lack of pertinent security data, resource limitations, experimental rigor, attack complexity, and experimental rigor.

6. DL modules' accuracy is constrained while dealing with confusing data. In real-time circumstances, we have a ton of lesser known facts [54].
7. Among the largest obstacles to the extensive deployment of IDPS is the enormously large number of false warnings generated [78].
8. IDSs need to be more reliable, capable of identifying various intrusions, and less prone to alarms that are false [84].
9. Most techniques of intrusion detection need more time to train, and this has a detrimental effect on the model's accuracy to the point where, in order to reduce training time, the performance of the entire system must be sacrificed. Because DL models have so many layers, it turns into a more serious setback [35].
10. To distinguish attacks in an offline state, the majority of IDSs were taught and expanded utilizing datasets that are virtualized or a benchmark. IDSs have not traditionally been designed with online (real-time) detection in mind. Academics face a computational hurdle when it comes to real-time network monitoring because of the annual exponential growth in network data [36].
11. Any IDS in this arena must overcome three major obstacles: appropriately classifying these assaults; recognizing unique attack patterns with considerable precision; and improving the human-readability of detection rules [71].

2.6 FUTURE OF RESEARCH

There is a major scope for future study in this area due to IoT security solutions still being in their infancy.

1. Deciding on the ideal placement technique and detecting system should be the aim of future study. Future studies ought to concentrate on managing developing IoT technologies and varying threat detection techniques [9, 85].
2. Future work should concentrate more on identifying unidentified and zero-day hacks in IoT networks as well as creating IDSs which can continuously update the list of possible risks [86].
3. Machine learning and deep learning techniques, as well as large data techniques, must be mastered by NIDS if they are to update their training model in real time in a streaming detection context [87, 88].
4. There hasn't yet been any technical or theoretical research on how to integrate blockchain, despite some recent publications advocating doing so to enhance the presentation of errors on the basis of IoT or wireless sensor networks [89, 90]. Therefore, more researches are needed to tackle this problem. Researchers can design a new protocol

with dynamic consensus on the basis of a collaborative technique to detect intrusion, using the paradigm for blockchain [91, 92].
5. The security threats associated with edge computing and IoT-enabling protocols and technologies should be included in the IDS design [80, 93].
6. Showcasing various ML and DL techniques as well as theoretical and technical components of performance evaluation [].
7. Developing a system that can identify intelligent attacks using intelligent systems.
8. Making datasets big enough to support efficient ML algorithm training should be one of the key objectives of this research.
9. Additionally, a thorough explanation of the attacks, standard messages, and how the dataset was created should be offered. It must also be made available to the general public for peer-review [94].

2.7 CONCLUSION

As the IoT turns out to be more widespread, additional efforts that deal with data will be made. Cloud security is therefore its top priority. Many security solutions are useless because they require a lot of resources. As a result, second-line protections are valued in IoT networks. This chapter offers a research strategy for the development of effective IoT intrusion detection. In this chapter, the most recent work is meticulously assessed to prevent intrusion in a cloud context. Three categories were used to group the papers that were subject of this investigation. The analysis's results demonstrate the following:

1. We think that the study on IoT IDS is still in its early phases. The works that were looked at do not include a wide variety of IoT technologies and cannot recognize a larger variety of threats.
2. As IDPS uses a variety of ID and prevention techniques, integrating several technologies of IDPS into a single solution of protection, which greatly lowers the cost of management; as a result, while developing a strategy for security, it must be extensive to stay ahead of the next security risk.
3. A number of most recent articles suggest that integrating IDS and blockchain may enhance the efficiency of detection of anomaly in an IoT setting or sensor network which is wireless. A summary of the recently researched literature in the area of detection of anomaly was presented in this chapter.
4. The chapter claims that many researchers employ a detection method which is anomaly-based, to track network traffic in order to find both known and undiscovered dangers.

We believe that the study has identified promising areas for further investigation into effective IoT intrusion detection. Additionally, articles written in languages other than English were excluded from this study.

REFERENCES

1. Arshad, Junaid, Muhammad Ajmal Azad, Roohi Amad, Khaled Salah, Mamoun Alazab, and Razi Iqbal. "A review of performance, energy and privacy of intrusion detection systems for IoT." *Electronics* 9, no. 4 (2020): 629.
2. Lv, Zhihan, Dongliang Chen, Ranran Lou, and Houbing Song. "Industrial security solution for virtual reality." *IEEE Internet of Things Journal* 8, no. 8 (2020): 6273–6281.
3. Gubbi, Jayavardhana, Rajkumar Buyya, Slaven Marusic, and Marimuthu Palaniswami. "Internet of Things (IoT): A vision, architectural elements, and future directions." *Future Generation Computer Systems* 29, no. 7 (2013): 1645–1660.
4. Singh, Dhananjay, Gaurav Tripathi, and Antonio J. Jara. "A survey of Internet-of-Things: Future vision, architecture, challenges and services." In *2014 IEEE World Forum on Internet of Things (WF-IoT)*, pp. 287–292. IEEE, 2014.
5. Lv, Zhihan, Ranran Lou, Jinhua Li, Amit Kumar Singh, and Houbing Song. "Big data analytics for 6G-enabled massive internet of things." *IEEE Internet of Things Journal* 8, no. 7 (2021): 5350–5359.
6. Lv, Zhihan, Liang Qiao, and Houbing Song. "Analysis of the security of internet of multimedia things." *ACM Transactions on Multimedia Computing, Communications, and Applications (TOMM)* 16, no. 3s (2020): 1–16.
7. Ahadi, Amir, Noradin Ghadimi, and Davar Mirabbasi. "An analytical methodology for assessment of smart monitoring impact on future electric power distribution system reliability." *Complexity* 21, no. 1 (2015): 99–113.
8. Chaabouni, Nadia, Mohamed Mosbah, Akka Zemmari, Cyrille Sauvignac, and Parvez Faruki. "Network intrusion detection for IoT security based on learning techniques." *IEEE Communications Surveys & Tutorials* 21, no. 3 (2019): 2671–2701.
9. Santos, Leonel, Carlos Rabadao, and Ramiro Gonçalves. "Intrusion detection systems in Internet of Things: A literature review." In *2018 13th Iberian Conference on Information Systems and Technologies (CISTI)*, pp. 1–7. IEEE, 2018.
10. Lv, Zhihan, Liang Qiao, and Ilsun You. "6G-enabled network in box for internet of connected vehicles." *IEEE Transactions on Intelligent Transportation Systems* 22, no. 8 (2020): 5275–5282.
11. Gendreau, Audrey A., and Michael Moorman. "Survey of intrusion detection systems towards an end to end secure internet of things." In *2016 IEEE 4th International Conference on Future Internet of Things and Cloud (FiCloud)*, pp. 84–90. IEEE, 2016.
12. Zanbouri, Kouros, and Nima Jafari Navimipour. "A cloud service composition method using a trust-based clustering algorithm and honeybee mating optimization algorithm." *International Journal of Communication Systems* 33, no. 5 (2020): e4259.

13. Singh, Shefali, Krati Saxena, and Zubair Khan. "Intrusion detection based on artificial intelligence techniques." *International Journal of Computer Science Trends and Technology* 2, no. 4 (2014): 31–35.
14. Kene, Snehal G., and Deepti P. Theng. "A review on intrusion detection techniques for cloud computing and security challenges." In *2015 2nd International Conference on Electronics and Communication Systems (ICECS)*, pp. 227–232. IEEE, 2015.
15. Souri, Alireza, Amir Masoud Rahmani, Nima Jafari Navimipour, and Reza Rezaei. "A hybrid formal verification approach for QoS-aware multi-cloud service composition." *Cluster Computing* 23 (2020): 2453–2470.
16. Heidari, Arash, Mohammad Ali Jabraeil Jamali, Nima Jafari Navimipour, and Shahin Akbarpour. "Internet of things offloading: ongoing issues, opportunities, and future challenges." *International Journal of Communication Systems* 33, no. 14 (2020): e4474.
17. Chen, Xiaotian, Tao Wang, Ruixuan Ying, and Zhibo Cao. "A fault diagnosis method considering meteorological factors for transmission networks based on P systems." *Entropy* 23, no. 8 (2021): 1008.
18. Gonçalves, Fábio, Bruno Ribeiro, Oscar Gama, Alexandre Santos, António Costa, Bruno Dias, Joaquim Macedo, and Maria Joao Nicolau. "A systematic review on intelligent intrusion detection systems for VANETs." In *2019 11th International Congress on Ultra Modern Telecommunications and Control Systems and Workshops (ICUMT)*, pp. 1–10. IEEE, 2019.
19. Wang, Tao, Xiaoguang Wei, Jun Wang, Tao Huang, Hong Peng, Xiaoxiao Song, Luis Valencia Cabrera, and Mario J. Pérez-Jiménez. "A weighted corrective fuzzy reasoning spiking neural P system for fault diagnosis in power systems with variable topologies." *Engineering Applications of Artificial Intelligence* 92 (2020): 103680.
20. Zouhair, Chiba, Noreddine Abghour, Khalid Moussaid, Amina El Omri, and Mohamed Rida. "A review of intrusion detection systems in cloud computing." *Security and Privacy in Smart Sensor Networks* (2018): 253–283.
21. Zhao, Can, Xinzhi Liu, Shouming Zhong, Kaibo Shi, Daixi Liao, and Qishui Zhong. "Secure consensus of multi-agent systems with redundant signal and communication interference via distributed dynamic event-triggered control." *ISA Transactions* 112 (2021): 89–98.
22. Azeez, Nureni Ayofe, Taiwo Mayowa Bada, Sanjay Misra, Adewole Adewumi, Charles Van der Vyver, and Ravin Ahuja. "Intrusion detection and prevention systems: an updated review." *Data Management, Analytics and Innovation: Proceedings of ICDMAI 2019*, 1 (2020): 685–696.
23. Wang, Tao, Wei Liu, Junbo Zhao, Xiaokang Guo, and Vladimir Terzija. "A rough set-based bio-inspired fault diagnosis method for electrical substations." *International Journal of Electrical Power & Energy Systems* 119 (2020): 105961.
24. Rani, Manisha. "A review of intrusion detection system in cloud computing." In *Proceedings of International Conference on Sustainable Computing in Science, Technology and Management (SUSCOM)*, Amity University Rajasthan, Jaipur-India, 2019.
25. Krishnaraj, N., Mohamed Elhoseny, E. Laxmi Lydia, K. Shankar, and Omar ALDabbas. "An efficient radix trie-based semantic visual indexing model for large-scale image retrieval in cloud environment." *Software: Practice and Experience* 51, no. 3 (2021): 489–502.

26. Tayyebi, Yakuta, and D. Bhilare. "Cloud security through Intrusion Detection System (IDS): Review of existing solutions." *International Journal of Emerging Trends & Technology in Computer Science* 4, no. 6 (2015): 213–215.
27. Cai, Ken, Huazhou Chen, Wu Ai, Xuexue Miao, Qinyong Lin, and Quanxi Feng. "Feedback convolutional network for intelligent data fusion based on near-infrared collaborative IoT technology." *IEEE Transactions on Industrial Informatics*, 18, no. 2 (2021), 1200–1209.
28. Elrawy, Mohamed Faisal, Ali Ismail Awad, and Hesham FA Hamed. "Intrusion detection systems for IoT-based smart environments: a survey." *Journal of Cloud Computing* 7, no. 1 (2018): 1–20.
29. Iwendi, Celestine, Praveen Kumar Reddy Maddikunta, Thippa Reddy Gadekallu, Kuruva Lakshmanna, Ali Kashif Bashir, and Md Jalil Piran. "A metaheuristic optimization approach for energy efficiency in the IoT networks." *Software: Practice and Experience* 51, no. 12 (2021): 2558–2571.
30. Raghav, Iti, Shashi Chhikara, and Nitasha Hasteer. "Intrusion detection and prevention in cloud environment: A systematic review." *International Journal of Computer Applications* 68, no. 24 (2013).
31. Shen, Hua, Mingwu Zhang, Hao Wang, Fuchun Guo, and Willy Susilo. "A cloud-aided privacy-preserving multi-dimensional data comparison protocol." *Information Sciences* 545 (2021): 739–752.
32. Ahmadian Ramaki, Ali, Abbas Rasoolzadegan, and Abbas Javan Jafari. "A systematic review on intrusion detection based on the Hidden Markov Model." *Statistical Analysis and Data Mining: The ASA Data Science Journal* 11, no. 3 (2018): 111–134.
33. Öney, Mehmet Uğur, and Serhat Peker. "The use of artificial neural networks in network intrusion detection: A systematic review." In *2018 international conference on artificial intelligence and data processing (IDAP)*, pp. 1–6. IEEE, 2018.
34. Ganeshan, R., and T. Daniya. "A systematic review on anomaly based intrusion detection system." In *IOP Conference Series: Materials Science and Engineering*, vol. 981, no. 2, p. 022010. IOP Publishing, 2020.
35. Mishra, Nivedita, and Sharnil Pandya. "Internet of things applications, security challenges, attacks, intrusion detection, and future visions: A systematic review." *IEEE Access* 9 (2021): 59353–59377.
36. Salo, Fadi, Mohammadnoor Injadat, Ali Bou Nassif, Abdallah Shami, and Aleksander Essex. "Data mining techniques in intrusion detection systems: A systematic literature review." *IEEE Access* 6 (2018): 56046–56058.
37. Sangher, Kanti Singh, and Archana Singh. "A systematic review–intrusion detection algorithms optimisation for network forensic analysis and investigation." In *2019 international conference on automation, computational and technology management (ICACTM)*, pp. 132–136. IEEE, 2019.
38. Salo, Fadi, Mohammad Noor Injadat, Ali Bou Nassif, and Aleksander Essex. "Data mining with big data in intrusion detection systems: A systematic literature review." *arXiv preprint arXiv:2005.12267* (2020).
39. Kaur, Bhavkanwal, and Pushpendra Kumar Pateriya. "A survey on security concerns in Internet of Things." In *2018 Second International Conference on Intelligent Computing and Control Systems (ICICCS)*, pp. 27–34. IEEE, 2018.

40. Idrissi, Idriss, Mostafa Azizi, and Omar Moussaoui. "IoT security with deep learning-based intrusion detection systems: A systematic literature review." In *2020 Fourth international conference on intelligent computing in data sciences (ICDS)*, pp. 1–10. IEEE, 2020.
41. Aludhilu, Hilma, and Rafael Rodríguez Puente. "A systematic literature review on intrusion detection approaches." *Revista Cubana de Ciencias Informáticas* 14, no. 1 (2020): 58–78.
42. Ferdiana, Ridi. "A systematic literature review of intrusion detection system for network security: Research trends, datasets and methods." In *2020 4th International Conference on Informatics and Computational Sciences (ICICoS)*, pp. 1–6. IEEE, 2020.
43. Tranfield, David, David Denyer, and Palminder Smart. "Towards a methodology for developing evidence-informed management knowledge by means of systematic review." *British Journal of Management* 14, no. 3 (2003): 207–222.
44. Brereton, Pearl, Barbara A. Kitchenham, David Budgen, Mark Turner, and Mohamed Khalil. "Lessons from applying the systematic literature review process within the software engineering domain." *Journal of Systems and Software* 80, no. 4 (2007): 571–583.
45. Ghorbani, Ali A., Wei Lu, and Mahbod Tavallaee. *Network intrusion detection and prevention: concepts and techniques.* Vol. 47. Springer Science & Business Media, 2009.
46. He, Yuan, Lingcheng Dai, and Hongtao Zhang. "Multi-branch deep residual learning for clustering and beamforming in user-centric network." *IEEE Communications Letters* 24, no. 10 (2020): 2221–2225.
47. Ma, Xu, Fangguo Zhang, Xiaofeng Chen, and Jian Shen. "Privacy preserving multi-party computation delegation for deep learning in cloud computing." *Information Sciences* 459 (2018): 103–116.
48. Zhang, Yiyang, Feng Liu, Zhen Fang, Bo Yuan, Guangquan Zhang, and Jie Lu. "Learning from a complementary-label source domain: theory and algorithms." *IEEE Transactions on Neural Networks and Learning Systems* 33, no. 12 (2021): 7667–7681.
49. Sun, Gan, Yang Cong, Jiahua Dong, Yuyang Liu, Zhengming Ding, and Haibin Yu. "What and how: generalized lifelong spectral clustering via dual memory." *IEEE Transactions on Pattern Analysis and Machine Intelligence* 44, no. 7 (2021): 3895–3908.
50. Sari, Arif. "A review of anomaly detection systems in cloud networks and survey of cloud security measures in cloud storage applications." *Journal of Information Security* 6, no. 2 (2015): 142–154.
51. Aljamal, Ibraheem, Ali Tekeoğlu, Korkut Bekiroglu, and Saumendra Sengupta. "Hybrid intrusion detection system using machine learning techniques in cloud computing environments." In *2019 IEEE 17th international conference on software engineering research, management and applications (SERA)*, pp. 84–89. IEEE, 2019.
52. Ravindranath, Vaishali, Sasikala Ramasamy, Ramasubbareddy Somula, Kshira Sagar Sahoo, and Amir H. Gandomi. "Swarm intelligence based feature selection for intrusion and detection system in cloud infrastructure." In *2020 IEEE Congress on Ecomputation (CEC)*, pp. 1–6. IEEE, 2020.

53. Salman, Tara, Deval Bhamare, Aiman Erbad, Raj Jain, and Mohammed Samaka. "Machine learning for anomaly detection and categorization in multi-cloud environments." In *2017 IEEE 4th International Conference on Cyber Security and Cloud Computing (CSCloud)*, pp. 97–103. IEEE, 2017.
54. Otoum, Yazan, Dandan Liu, and Amiya Nayak. "DL-IDS: a deep learning–based intrusion detection framework for securing IoT." *Transactions on Emerging Telecommunications Technologies* 33, no. 3 (2022): e3803.
55. Verma, Abhishek, and Virender Ranga. "Machine learning based intrusion detection systems for IoT applications." *Wireless Personal Communications* 111, no. 4 (2020): 2287–2310.
56. Ge, Mengmeng, Xiping Fu, Naeem Syed, Zubair Baig, Gideon Teo, and Antonio Robles-Kelly. "Deep learning-based intrusion detection for IoT networks." In *2019 IEEE 24th Pacific Rim International Symposium on Dependable Computing (PRDC)*, pp. 256–25609. IEEE, 2019.
57. Zhang, Ying, Peisong Li, and Xinheng Wang. "Intrusion detection for IoT based on improved genetic algorithm and deep belief network." *IEEE Access* 7 (2019): 31711–31722.
58. Loukas, George, Tuan Vuong, Ryan Heartfield, Georgia Sakellari, Yongpil Yoon, and Diane Gan. "Cloud-based cyber-physical intrusion detection for vehicles using deep learning." *IEEE Access* 6 (2017): 3491–3508.
59. Luo, Jinnan, Mengling Li, Xinzhi Liu, Wenhong Tian, Shouming Zhong, and Kaibo Shi. "Stabilization analysis for fuzzy systems with a switched sampled-data control." *Journal of the Franklin Institute* 357, no. 1 (2020): 39–58.
60. Tsvetanov, Tsvetomir Iliev. "Pattern-based network defense mechanism." U.S. Patent Application 11/838,812, filed November 27, 2008.
61. Huang, Kaixing, Qi Zhang, Chunjie Zhou, Naixue Xiong, and Yuanqing Qin. "An efficient intrusion detection approach for visual sensor networks based on traffic pattern learning." *IEEE Transactions on Systems, Man, and Cybernetics: Systems* 47, no. 10 (2017): 2704–2713.
62. Aparicio-Navarro, Francisco J., Konstantinos G. Kyriakopoulos, Yu Gong, David J. Parish, and Jonathon A. Chambers. "Using pattern-of-life as contextual information for anomaly-based intrusion detection systems." *IEEE Access* 5 (2017): 22177–22193.
63. Liu, Jinping, Wuxia Zhang, Zhaohui Tang, Yongfang Xie, Tianyu Ma, Jingjing Zhang, Guoyong Zhang, and Jean Paul Niyoyita. "Adaptive intrusion detection via GA-GOGMM-based pattern learning with fuzzy rough set-based attribute selection." *Expert Systems with Applications* 139 (2020): 112845.
64. Wang, Yazi, Yingbo Liang, Huaibo Sun, and Yuankun Ma. "Retracted Article: Intrusion detection and performance simulation based on improved sequential pattern mining algorithm." *Cluster Computing* 23, no. 3 (2020): 1927–1936.
65. Rao, Chinta Someswara, and K. Butchi Raju. "Mapreduce accelerated signature-based intrusion detection mechanism (idm) with pattern matching mechanism." In *Soft Computing in Data Analytics: Proceedings of International Conference on SCDA 2018*, pp. 157–164. Springer Singapore, 2019.
66. Iqbal, Irshad M., and Ricardo A. Calix. "Analysis of a payload-based network intrusion detection system using pattern recognition processors." In *2016 International Conference on Collaboration Technologies and Systems (CTS)*, pp. 398–403. IEEE, 2016.

67. Manickam, M., and S. P. Rajagopalan. "Retracted Article: A hybrid multi-layer intrusion detection system in cloud." *Cluster Computing* 22, no. Suppl 2 (2019): 3961–3969.
68. Wu, Zhiang, Changsheng Li, Jie Cao, and Yong Ge. "On scalability of association-rule-based recommendation: A unified distributed-computing framework." *ACM Transactions on the Web (TWEB)* 14, no. 3 (2020): 1–21.
69. Kumar, Vikash, Ditipriya Sinha, Ayan Kumar Das, Subhash Chandra Pandey, and Radha Tamal Goswami. "An integrated rule based intrusion detection system: analysis on UNSW-NB15 data set and the real time online dataset." *Cluster Computing* 23 (2020): 1397–1418.
70. Papamartzivanos, Dimitrios, Félix Gómez Mármol, and Georgios Kambourakis. "Dendron: Genetic trees driven rule induction for network intrusion detection systems." *Future Generation Computer Systems* 79 (2018): 558–574.
71. Haugerud, Hårek, Huy Nhut Tran, Nadjib Aitsaadi, and Anis Yazidi. "A dynamic and scalable parallel Network Intrusion Detection System using intelligent rule ordering and Network Function Virtualization." *Future Generation Computer Systems* 124 (2021): 254–267.
72. Lou, Ping, Guantong Lu, Xuemei Jiang, Zheng Xiao, Jiwei Hu, and Junwei Yan. "Cyber intrusion detection through association rule mining on multi-source logs." *Applied Intelligence* 51 (2021): 4043–4057.
73. Riyaz, B., and S. Ganapathy. "An intelligent fuzzy rule based feature selection for effective intrusion detection." In *2018 International Conference on Recent Trends in Advance Computing (ICRTAC)*, pp. 206–211. IEEE, 2018.
74. Kshirsagar, Deepak, and Jahed Momin Shaikh. "Intrusion detection using rule-based machine learning algorithms." In *2019 5th International Conference On Computing, Communication, Control And Automation (ICCUBEA)*, pp. 1–4. IEEE, 2019.
75. Panigrahi, Ranjit, Samarjeet Borah, and Debahuti Mishra. "A Proposal of Rule-Based Hybrid Intrusion Detection System Through Analysis of Rule-Based Supervised Classifiers." In *Intelligent and Cloud Computing: Proceedings of ICICC 2019*, Vol. 2, pp. 623–633. Springer Singapore, 2021.
76. Gaffney, John E., and Jacob W. Ulvila. "Evaluation of intrusion detectors: A decision theory approach." In *Proceedings 2001 IEEE Symposium on Security and Privacy. S&P 2001*, pp. 50–61. IEEE, 2000.
77. Patel, Ahmed, Mona Taghavi, Kaveh Bakhtiyari, and Joaquim Celestino Júnior. "An intrusion detection and prevention system in cloud computing: A systematic review." *Journal of Network and Computer Applications* 36, no. 1 (2013): 25–41.
78. Alkadi, Osama, Nour Moustafa, and Benjamin Turnbull. "A review of intrusion detection and blockchain applications in the cloud: approaches, challenges and solutions." *IEEE Access* 8 (2020): 104893–104917.
79. Lv, Zhihan, Liang Qiao, Jinhua Li, and Houbing Song. "Deep-learning-enabled security issues in the internet of things." *IEEE Internet of Things Journal* 8, no. 12 (2020): 9531–9538.
80. Asharf, Javed, Nour Moustafa, Hasnat Khurshid, Essam Debie, Waqas Haider, and Abdul Wahab. "A review of intrusion detection systems using machine and deep learning in internet of things: Challenges, solutions and future directions." *Electronics* 9, no. 7 (2020): 1177.

81. Huang, Zhu, Tao Wang, Wei Liu, Luis Valencia-Cabrera, Mario J. Pérez-Jiménez, and Pengpeng Li. "A fault analysis method for three-phase induction motors based on spiking neural P systems." *Complexity* 2021, no. 1 (2021): 2087027.
82. Vermesan, Ovidiu, and Peter Friess. *Internet of things applications-from research and innovation to market deployment*. Taylor & Francis, 2014.
83. Khraisat, Ansam, and Ammar Alazab. "A critical review of intrusion detection systems in the internet of things: techniques, deployment strategy, validation strategy, attacks, public datasets and challenges." *Cybersecurity* 4 (2021): 1–27.
84. Yi, Haibo. "Secure social internet of things based on post-quantum blockchain." *IEEE Transactions on Network Science and Engineering* 9, no. 3 (2021): 950–957.
85. Li Beibei, Gaoxi Xiao, Rongxing Lu, Ruilong Deng, and Haiyong Bao. "On feasibility and limitations of detecting false data injection attacks on power grid state estimation using D-FACTS devices." *IEEE Transactions on Industrial Informatics* 16, no. 2 (2019): 854–864.
86. Hromic, Hugo, Danh Le Phuoc, Martin Serrano, Aleksandar Antonić, Ivana P. Žarko, Conor Hayes, and Stefan Decker. "Real time analysis of sensor data for the Internet of Things by means of clustering and event processing." In *2015 IEEE International conference on communications (ICC)*, pp. 685–691. IEEE, 2015.
87. Ahmed, Ejaz, Ibrar Yaqoob, Ibrahim Abaker Targio Hashem, Imran Khan, Abdelmuttlib Ibrahim Abdalla Ahmed, Muhammad Imran, and Athanasios V. Vasilakos. "The role of big data analytics in Internet of Things." *Computer Networks* 129 (2017): 459–471.
88. Lv, Zhihan, Liang Qiao, M. Shamim Hossain, and Bong Jun Choi. "Analysis of using blockchain to protect the privacy of drone big data." *IEEE Network* 35, no. 1 (2021): 44–49.
89. Dehghani, Moslem, Mohammad Ghiasi, Taher Niknam, Abdollah Kavousi-Fard, Mokhtar Shasadeghi, Noradin Ghadimi, and Farhad Taghizadeh-Hesary. "Blockchain-based securing of data exchange in a power transmission system considering congestion management and social welfare." *Sustainability* 13, no. 1 (2020): 90.
90. Dehghani, Moslem, Mohammad Ghiasi, Taher Niknam, Abdollah Kavousi-Fard, Mokhtar Shasadeghi, Noradin Ghadimi, and Farhad Taghizadeh-Hesary. "Blockchain-based securing of data exchange in a power transmission system considering congestion management and social welfare." *Sustainability* 13, no. 1 (2020): 90.
91. Gong, Jianhu, and Nima Jafari Navimipour. "An in-depth and systematic literature review on the blockchain-based approaches for cloud computing." *Cluster Computing* 25, no. 1 (2022): 383–400.
92. Lv, Zhihan, Dongliang Chen, and Qingjun Wang. "Diversified technologies in internet of vehicles under intelligent edge computing." *IEEE Transactions on Intelligent Transportation Systems* 22, no. 4 (2020): 2048–2059.
93. Benaddi, Hafsa, and Khalil Ibrahimi. "A review: Collaborative intrusion detection for IoT integrating the blockchain technologies." In *2020 8th International Conference on Wireless Networks and Mobile Communications (WINCOM)*, pp. 1–6. IEEE, 2020.

94. Lv, Zhihan, Amit Kumar Singh, and Jinhua Li. "Deep learning for security problems in 5G heterogeneous networks." *IEEE Network* 35, no. 2 (2021): 67–73.
95. Paul, Aditi, Somnath Sinha, Parveen Kumar, Shirasthi Choudhary, Krishna Samdani, and Saumya Mishra. "Overview of Cyber Security in Intelligent and Sustainable Manufacturing." In *Handbook of Intelligent and Sustainable Manufacturing*, pp. 95–118. CRC Press, 2025.
96. Lavanya, Raja, G. Bhavani, C. Santhiya, G. Vinoth Chakkaravarthy, and Parveen Kumar. "Digital Twins and IoT for Sustainable Manufacturing: A Survey of Current Practices and Future Directions." In Handbook of Intelligent and Sustainable Manufacturing, pp. 80–94. CRC Press, 2025.
97. Krishnamurthy, Bhargavi, and Parveen Kumar. "Nature-Inspired Optimization Techniques of Industry 4.0 for Sustainable Manufacturing." In *Handbook of Intelligent and Sustainable Manufacturing*, pp. 42–50. CRC Press, 2025.
98. Anand, Sourabh, Manoj Kumar Satyarthi, Pushpendra S. Bharti, Parveen Kumar, and Ajay Kumar. "Digital Twin Integration for Enhanced Control in FDM 3D Printing." In *Industry 4.0 Driven Manufacturing Technologies*, pp. 373–388. Springer Nature Switzerland, 2024.
99. Kumar, Ajay, Yang Liu, and Rakesh Kumar (Eds). *Handbook of Intelligent and Sustainable Manufacturing: Tools, Principles, and Strategies*. CRC Press, 2024.
100. Kumar, Ajay, Parveen Kumar, and Yang Liu. *Industry 4.0 Driven Manufacturing Technologies*. Springer, 2024. https://doi.org/10.1007/978-3-031-68271-1
101. Arunachalam, S. Jothi, R. Saravanan, T. Sathish, Abdullah A. Alarfaj, Jayant Giri, and Ajay Kumar. "Enhancing mechanical performance of MWCNT filler with jute/kenaf/glass composite: a statistical optimization study using RSM and ANN." *Materials Technology*, 39 no. 1 (2024), 2381156. https://doi.org/10.1080/10667857.2024.2381156
102. Giri, Jayant, T. Sathish, Taukeer Sheikh, Neeraj Sunheriya, Pallavi Giri, Rajkumar Chadge, Chetan Mahatme, and A. Parthiban. "Automatic liver segmentation using U-Net deep learning architecture for additive manufacturing." *Deleted Journal*, 245 no. 1 (2024): 90. https://doi.org/10.1007/s10751-024-01927-9
103. Narasimhamu, Katta Lakshmi, Manikandan Natarajan, Pasupuleti Thejasree, Emad Makki, Jayant Giri, Neeraj Sunheriya, Rajkumar Chadge, Chetan Mahatme, Pallavi Giri, and T. Sathish. "Development of hybrid optimization model using Grey-ANFIS-Jaya algorithm for CNC drilling of aluminium alloy." *Journal of Engineering*, 2024 no. 1 (2024), 1–12. https://doi.org/10.1155/2024/1476770
104. Kamble, Prashant D., Jayant Giri, Emad Makki, Neeraj Sunheriya, Shilpa B. Sahare, Rajkumar Chadge, Chetan Mahatme, Pallavi Giri, and Hitesh Panchal. "An application of hybrid Taguchi-ANN to predict tool wear for turning EN24 material." *AIP Advances*, 14 no. 1 (2024). https://doi.org/10.1063/5.0186432

Chapter 3

Utilizing advanced machine learning to improve network security in industries against changing cyberthreats

Roshan Chandel, Gajanand Sharma, Shekhar Kumar Swarnkar, and Ajay Kumar

3.1 INTRODUCTION

The rapid growth of digital technologies has transformed the way we conduct business, communicate, and interact with the world. However, this increasing reliance on digital infrastructure also exposes us to a wide range of cyberthreats. Cybercriminals constantly adapt their tactics, exploit vulnerabilities, and devise new attack vectors, making it challenging for traditional security measures to keep pace [1].

Traditional security approaches, such as rule-based systems and signature-based detection, are often insufficient to detect and mitigate the evolving nature of cyberthreats. Attackers utilize sophisticated techniques, including polymorphic malware, zero-day exploits, and advanced evasion tactics, to bypass traditional security defences. As a result, there is a critical need for more advanced and proactive security mechanisms to safeguard networks and protect sensitive information [2]. In Figure 3.1, the network security pathway is described, which starts with training of the employees and ends with encryption of the entire network.

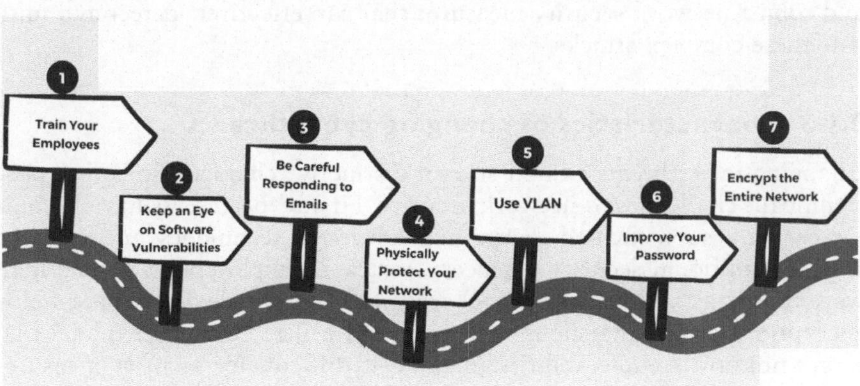

Figure 3.1 Network security pathway.

DOI: 10.1201/9781003512585-4

3.1.1 Problem statement

The dynamic nature of cyberthreats requires a proactive and adaptive approach to network security. Conventional security solutions are static and rule-driven, lacking the ability to adapt to emerging threats. Organizations struggle to detect and respond to sophisticated attacks in real time, leading to potential data breaches, financial losses, and reputational damage. A rising number of people are interested in using machine learning methods to enhance network security as a solution to this problem. In order to discover threats early and take appropriate action, machine learning algorithms have the capacity to scan massive amounts of data, spot trends, and provide precise predictions.

3.1.2 Evolution of cyberthreats

Cyberthreats have evolved significantly over the years, keeping pace with advancements in technology and the increasing interconnectedness of digital systems. In the early days of computing, threats were relatively simple, often consisting of individual hackers attempting to gain unauthorized access or disrupt systems. However, as technology became more pervasive, the motives and capabilities of cybercriminals expanded. Today, cyberthreats encompass a wide range of sophisticated techniques, including malware, phishing attacks, ransomware, advanced persistent threats (APTs), and nation-state-sponsored attacks [3]. Attackers have become more organized, leveraging advanced tactics such as social engineering, zero-day exploits, and botnets to compromise networks and steal sensitive information. Moreover, the rise of interconnected devices and the Internet of Things (IoT) has opened up new attack surfaces, making critical infrastructure, healthcare systems, and even household devices vulnerable to cyberattacks [4]. The evolution of cyberthreats underscores the urgent need for adaptive and robust network security measures that can effectively detect and mitigate these complex attacks.

3.1.3 Characteristics of changing cyberthreats

Changing cyberthreats exhibit several distinctive characteristics that pose significant challenges to network security. First, cyberthreats have become increasingly sophisticated, leveraging advanced techniques and tools to bypass traditional security defences. Attackers employ polymorphic malware that constantly changes its characteristics to evade detection, employ encryption to hide malicious activities, and utilize zero-day exploits that target unknown vulnerabilities, making it difficult for security measures to keep up. Second, cyberthreats are dynamic and constantly evolving. Attackers adapt their tactics and strategies based on the changing threat landscape, making it crucial for security systems to continuously update and respond in real time. Third, cyberthreats are often highly targeted, focusing on specific organizations or individuals to maximize their impact and gain valuable information or resources. This individualized strategy necessitates adaptive security systems that can recognize and react to targeted threats [5]. Finally, the interconnected nature of digital systems and the increasing reliance on cloud computing and IoT devices create complex attack surfaces, expanding the potential impact and reach of cyberthreats [6]. As a result, network security must encompass holistic approaches that consider the interconnected ecosystem and implement defences at multiple levels to mitigate these evolving threats.

3.1.4 Impacts of cyberattacks

Cyberattacks can have wide-ranging and significant impacts on individuals, organizations, and society as a whole. The financial repercussions of a successful cyberattack can be substantial, resulting in direct financial losses, theft of sensitive financial information, and disruption of business operations. Beyond financial implications, cyberattacks can also lead to reputational damage, eroding trust in an organization's ability to protect customer data and maintain security [7]. The loss of sensitive information through data breaches can have long-term consequences, including identity theft, fraud, and unauthorized access to personal or confidential data. Additionally, cyber assaults have the potential to impair key infrastructure, like transportation, healthcare, and power grids, endangering the security and well-being of the general population. Nation-state-sponsored attacks or cyber warfare can have severe geopolitical consequences, disrupting government operations, compromising national security, and undermining trust between nations. The impacts of cyberattacks highlight the pressing need for robust network security measures to mitigate the potential damages and safeguard individuals, organizations, and the broader digital ecosystem.

3.1.5 XGBoost

It is a boosting approach that repeatedly incorporates decision trees into the model by using gradient boosting. By adding additional trees to the model, a loss function, such as mean squared error (MSE) or log loss, is minimized. To reduce the overall error of the model, each new tree is trained using the residual errors from the preceding trees. The total of all individual trees' forecasts becomes the final prediction.

The promise of machine learning-based techniques for network intrusion detection and classification is generally well-explained in this chapter. The ultimate goal of this study is to enhance the state of the art in network intrusion detection and classification, assisting in the development of a more secure and resilient cyberspace for both businesses and people.

3.2 BACKGROUND INFORMATION

Using cutting-edge machine learning methods to improve network security in the face of growing cyberthreats is the subject of the literature review that follows. The chapter includes research, papers, and articles that emphasize the benefits, drawbacks, and practical uses of machine learning algorithms and models as they relate to network security. In order to use sophisticated machine learning for network security, the evaluation also analyses trends, obstacles, and potential future paths.

The value of machine learning in enhancing network security has been highlighted in several research studies. Musleh et al. [8] performed a thorough examination of machine learning-based intrusion detection systems (IDSs) and emphasized their superior capacity to accurately identify innovative and complex threats when compared to conventional rule-based methods. They discussed the usage of several machine learning approaches, including decision trees, support vector machines, and neural networks, as well as how effective they are in identifying various types of network intrusions [9]. Behavioural analysis work done by Liu et al. proposed a user and entity behaviour analytics (UEBA) framework that employs machine learning algorithms to detect anomalous user behaviour and identify potential insider threats. Their study showed how machine learning may be used to identify hostile actions that elude conventional security measures and detect anomalous patterns. Furthermore, the application of machine learning in malware detection and analysis has garnered significant attention. Mwanza and Kalita [10] examined convolutional neural networks (CNNs), a subset of deep learning models, as a means of malware identification. Their research demonstrated the ability of deep learning techniques to extract intricate features and classify malware samples accurately, even when confronted with polymorphic and obfuscated malware variants. A number of scholars have also carried out surveys and in-depth analyses of machine learning in network security in addition to individual investigations [3]. The performance, feature selection techniques, and difficulties of machine learning-based intrusion detection systems were evaluated by Kumar and Rose in a systematic study. In network security applications, they emphasized the need for strong training datasets, feature selection methods, and machine learning models that are easy to understand. The literature has also discussed difficulties in applying machine learning to network security [3]. A key obstacle has been recognized as the issue of data quantity and quality, especially in the context of labelled training datasets. To address the issue of data scarcity, Elmasry et al. [5] examined the drawbacks of just using publicly available datasets and suggested methods for creating synthetic data. Furthermore, it is now well acknowledged that machine learning models' susceptibility to hostile assaults is a serious issue [5]. In the context of network security, Ali and Hussain investigated adversarial assaults on machine learning models and suggested defensive methods to increase the resilience of these models [11].

Overall, the assessment of the literature shows the expanding body of research on using cutting-edge machine learning methods to improve network security against evolving cyberthreats [12]. The experiments show how well machine learning algorithms perform for network security applications, including virus detection, intrusion detection, and behavioural analysis. However, issues including poor data quality, hostile assaults, and model interpretability continue to be major sources of worry. In order to progress the subject of machine learning in network security, this chapter also highlights the need for more study in areas including federated learning, interaction with other security technologies, and training data creation [13].

3.3 RESEARCH METHODOLOGY

Previous implementations of network intrusion detection and classification have largely relied on signature-based techniques that identify known patterns of network attacks. However, these methods are limited in their ability to detect previously unseen or unknown types of attacks, leaving organizations vulnerable to emerging threats. Moreover, traditional methods often generate a high number of false positives, which can lead to a high volume of alerts and unnecessary investigations.

Machine learning-based algorithms have become a potential option for network intrusion detection and classification in order to get beyond these restrictions. These methods use algorithms that can recognize trends and abnormalities that might be signs of a network assault by learning from massive amounts of data [14–17]. The performance of several machine learning techniques, such as decision trees, random forests, and support vector machines, for network intrusion detection and classification using the UNSW-NB15 dataset is evaluated in this research, building on earlier work in the area.

The research also offers XGBoost, which is shown to perform better than the basic model in terms of accuracy, precision, recall, and F1-score, as an implementable model for network intrusion detection and classification. The use of an efficient decision tree implementation, a gradient boosting strategy that permits iterative learning from mistakes, and regularization approaches that avoid overfitting are a few of the aspects that contribute to the XGBoost model's increased performance. The implementation of the XGBoost classifier demonstrated the effectiveness of advanced machine learning techniques in network security. By leveraging the powerful capabilities of XGBoost, the classifier was able to enhance network security measures by accurately detecting and classifying evolving cyberthreats.

It is crucial to remember that the representativeness of the training dataset has a significant impact on how well the XGBoost classifier performs. Additionally, the interpretability of the XGBoost classifier should be considered, as complex ensemble models may lack transparency in explaining their decision-making process.

3.3.1 Hyperparameter optimization

Finding the optimum hyperparameters for a particular machine learning algorithm that performs the best when tested on a validation set is known as hyperparameter optimization in machine learning. The equation for hyperparameter optimization looks as follows:

$$X^* = \frac{\arg\min f(X)}{x \in X} \tag{3.1}$$

As shown in Equation (3.1), x is a collection of hyperparameters that produces the lowest score value and may take on any value in the X domain; $f(x)$ is an objective score to minimize that is assessed in the validation set. Simply put, we are looking for the model hyperparameters that provide the highest score on the validation set metric. Here, random search is used to accomplish our task since it produces better results.

3.3.2 Steps to implement XGBoost classifier

Step 1 – Initialization: The XGBoost algorithm starts by initializing the model with a single decision tree. The predicted output from this tree is denoted as $\hat{y}\ 0$.

Step 2 – Prediction: The model uses the current decision tree to make predictions on the training data. The predicted output from the ith decision tree is given by

$$yi = \hat{y}\ (i - 1) + fi \tag{3.2}$$

As shown in Equation (3.2), $\hat{y}\ (i - 1)$ is the predicted output from the previous $(i - 1)$ decision trees, and fi is the output from the current decision tree.

Step 3 – Loss function: The difference between the expected and actual outputs is calculated using the loss function. In XGBoost, the loss function is provided by

$$L(y, \hat{y}\ ) = \Sigma\ l(yi, \hat{y}i) + \Omega(f) \tag{3.3}$$

As shown in Equations (3.3)–(3.5), l is the loss function for individual samples, Σ is the sum over all samples, and Ω is the regularization term to prevent overfitting.

Step 4 – Gradient calculation: Calculated is the gradient of the loss function with respect to the expected output. The decision tree's weights are updated using this gradient. For the *i*th sample, the gradient of the loss function is given by

$$gi = \partial\, L(yi, \hat{y}i) \,/\, \partial\, \hat{y}i \tag{3.4}$$

Step 5 – Hessian calculation: The curvature of the loss function is determined from the second derivative of the loss function. This curvature is used to adjust the step size when updating the weights of the decision tree. The second derivative of the loss function for the *i*th sample is given by

$$hi = \partial 2L(yi, \hat{y}i) \,/\, \partial\, \hat{y}i\text{^}2 \tag{3.5}$$

Step 6 – Tree construction: A decision tree is built using the gradient and Hessian values for each sample. Recursively dividing the data into smaller groups depending on the values of the input characteristics creates the tree. The splits are selected to meet a regularization condition and minimize the loss function.

Step 7 – Update predictions: Based on the revised decision tree, the model revises its forecasts. To get the new expected output, the predicted output from the new decision tree is combined with the predicted output from the prior decision trees:

$$\hat{y}i = \hat{y}\ (i - 1) + \gamma fi \tag{3.6}$$

As shown in Equation (3.6), γ is the learning rate, which controls the step size when updating the model.

Step 8 – Repeat: To increase the number of decision trees in the model, steps 3–7 are performed many times. Each tree is constructed with the goal of minimizing the loss function while considering the regularization term.

Figure 3.2 describes the steps for the base model, starting from loading of dataset to evaluation of the performance of the model. Several preprocessing and feature engineering approaches were used in this research project to extract useful characteristics from the raw data. Then, using a variety of performance indicators, including accuracy, precision, recall, F1-score, and area under the ROC curve, we trained and assessed the XGBoost classifier.

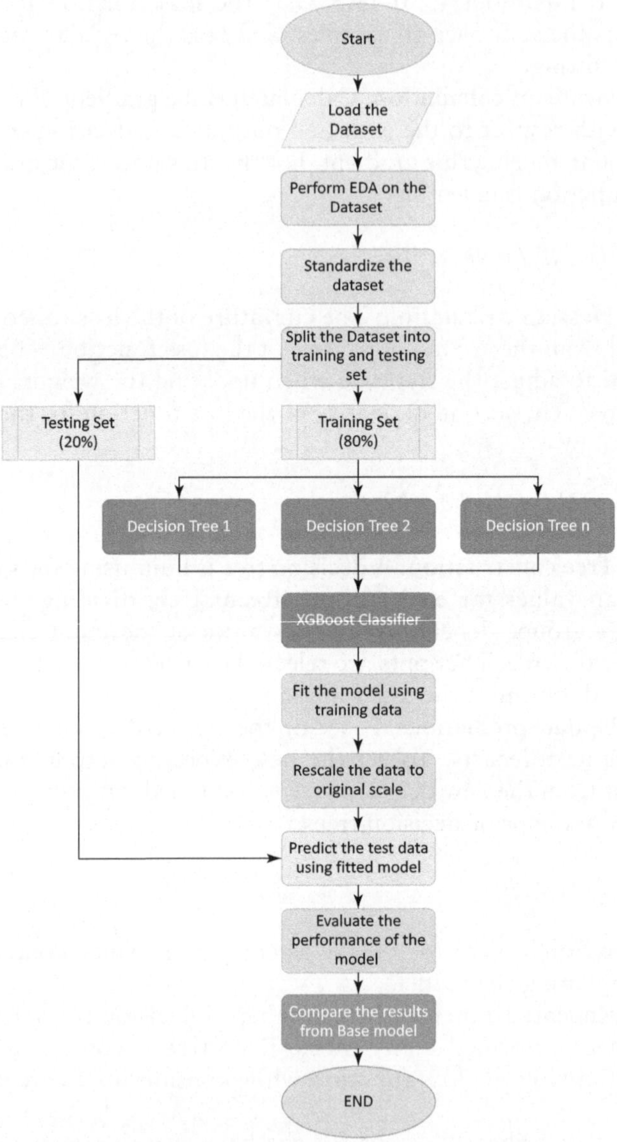

Figure 3.2 Flow diagram of proposed work.

3.4 RESULTS AND DISCUSSION

A specific table called a disarray grid is used to evaluate how well the confusion matrix computation is presented. It is often used in controlled learning and is referred to as "coordinating with matrix." Actual and predicted are the two parameters of the disarray grid. The numerical comprehension of many disarray framework limits includes exactness, affectability, explicitness, error rate, accuracy, F measurements, review, and so on.

TPR (True Positive Rate) = (TN+TP)/All or accuracy of the first class

FPR (False Positive Rate) = FP/(TN+FP) or (1 accuracy of other class)

Precision = iTP/(iFP+iTP)

iSensitivity or Recall = iTP/(iFN+iTP)

iF_measures = 2 (Precision*Recall) / (iPrecision + Recall)

The result of the count plot for the categories is shown in Figure 3.3. The XGBoost classifier performed well in predicting and categorizing network security occurrences after being initialized using hyperparameters such as the number of trees, learning rate, and maximum tree depth. Precision predictions on unseen data were made possible by the model's effective identification of patterns and correlations in the training dataset during the model training phase. The test result of XGBoost classifier is shown in Figure 3.4.

The model's performance was evaluated using a number of evaluation criteria, such as accuracy, precision, recall, F1-score, and ROC–AUC. These metrics gave us information about how well the classifier was able to categorize network security occurrences and differentiate between various cyberthreats. The classifier's capacity to accurately detect and classify security events is shown by the findings, which were very accurate and precise. Table 3.1 provides a comparison of various models based on accuracy, precision, recall, and F1-score.

Furthermore, hyperparameter tuning was conducted to optimize the classifier's performance. By fine-tuning the hyperparameters using techniques like grid search or random search, the model's performance was further improved. This step helped identify the optimal configuration of hyperparameters that maximized the classifier's accuracy and robustness. The improved result of proposed model is shown in Figure 3.5.

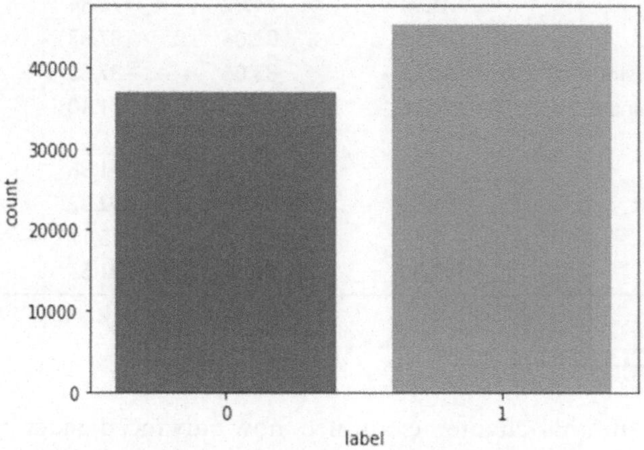

Figure 3.3 Count plot for categorical feature.

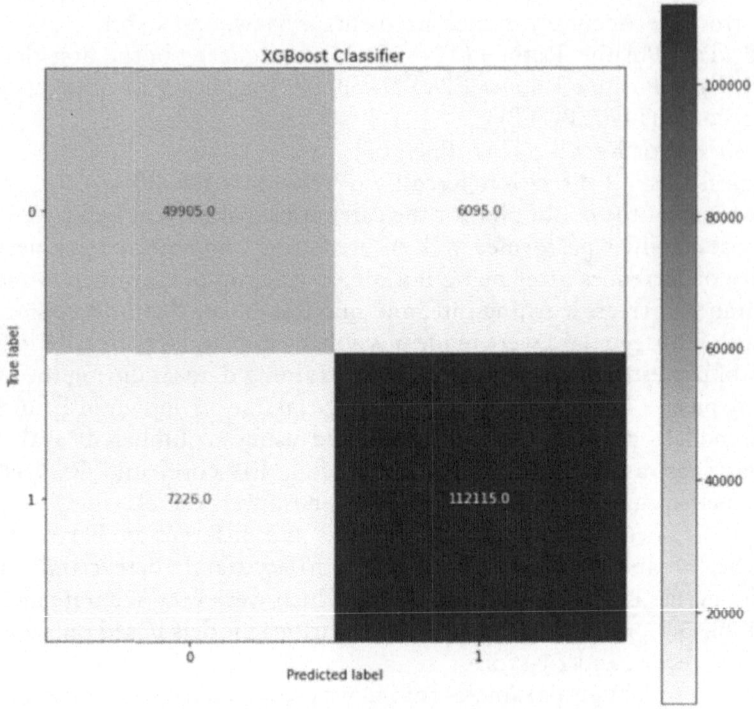

Figure 3.4 XGBoost classifier test result.

Table 3.1 Comparison value in various models

	Accuracy	*Precision*	*Recall*	*F1-score*
Naive Bayes	74.94	74.82	78.54	74.06
KNN	84.55	83.06	87.63	83.68
Logistic regression	84.55	83.06	87.63	83.68
Linear discriminant analysis	62.26	71.23	71.40	62.26
Decision tree	91.28	89.26	91.86	90.31
Random forest	91.31	89.24	92.32	90.41
AdaBoost	75.47	72.78	75.17	73.38
XGBoost	92.40	91.10	91.53	91.31

3.5 CONCLUSION

In conclusion, this chapter examined how advanced machine learning approaches, notably the XGBoost classifier, might improve network security against shifting cyberthreats. Advanced machine learning algorithms may reliably identify and categorize developing cyberthreats, improving

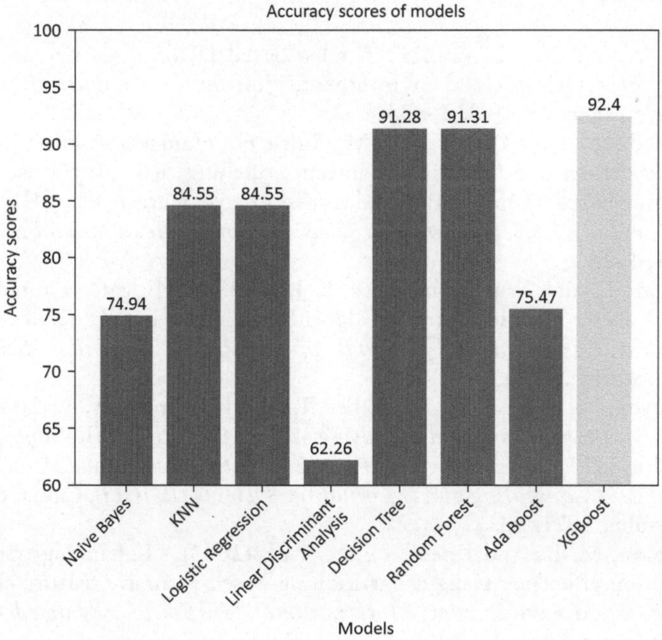

Figure 3.5 Improved result of proposed model.

network security. Data preparation, model setup, training, prediction, assessment, and hyperparameter tweaking let the XGBoost classifier detect and categorize network security occurrences. The classifier learnt patterns and correlations in the training data to accurately and precisely detect security concerns. The XGBoost classifier showed the relevance of data quality, representativeness, and diversity in training strong models. Complex ensemble model interpretability difficulties need continuous study. In conclusion, our chapter highlights the pivotal role of advanced machine learning in enhancing network security against dynamic cyber-threats. Traditional security measures fall short in the face of evolving threats, and machine learning offers a robust solution. Our study outlines the benefits of integrating machine learning algorithms to swiftly and accurately identify and respond to threats, reducing the risk of breaches and disruptions. However, successful implementation requires ongoing vigilance, ethical considerations, and adaptation to changing threat land-scapes. Collaboration and innovation in the field are essential to secure a more resilient digital future.

Future research directions may include exploring ensemble methods and combining XGBoost with other machine learning techniques to further improve network security. Additionally, the integration of XGBoost with real-time network monitoring systems can enhance the responsiveness and proactive nature of network security measures.

REFERENCES

1. Pan, Y., and Lin, L. (2021). "A trust-based DDoS discovery approach for encrypted track in cloud environment," *Journal of Computer Research and Development*, 58(4), 822–833.
2. Ahmad, M., Riaz Q., Zeeshan M., Tahir H., Haider S. A., and Khan M. S. (2021). "Intrusion detection in internet of things using supervised machine learning based on application and transport layer features using UNSW-NB15 data-set," *EURASIP Journal on Wireless Communications and Networking*, 2021, 1–23.
3. Kumar, J., and Arul Leena Rose, P. J. (2023). "Mitigate volumetric DDoS attack using machine learning algorithm in SDN based IoT network environment," *International Journal of Advanced Computer Science and Applications*, 14(1).
4. El-Sayed, R., El-Ghamry, A., Gaber, T., and Hassanien, A. E. (2021). "Zero-Day Malware Classification Using Deep Features with Support Vector Machines," In *Proceedings of the 2021 Tenth International Conference on Intelligent Computing and Information Systems (ICICIS)*, Cairo, Egypt, 5–7 December 2021, pp. 311–317.
5. Hesham, M. E., and Khedr, A. E., et al. (2023). "Enhancing the intrusion detection efficiency using a partitioning-based recursive feature elimination in big cloud environment," *International Journal of Advanced Computer Science and Applications*, 14(1).
6. Hindy, H., Bayne, E., Bures, M., Atkinson, R., Tachtatzis, C., and Bellekens, X. (2021). "Machine learning based IoT intrusion detection system: an MQTT case study," (MQTT-IoT-IDS2020 Dataset). *Lecture Notes in Networks and Systems*, 180, 73–84.
7. Bu, Z., Zhou, B., Cheng, P., Zhang, K., and Ling, Z. H. (2020). "Encrypted network traffic classification using deep and parallel network-in-network models," *IEEE Access*, 8, 132950.
8. Musleh, D., Alotaibi, M., et al. (2023). "Intrusion detection system using feature extraction with machine learning algorithms in IoT," *Journal of Sensor and Actuator Networks*, 12, 29. https://doi.org/10.3390/jsan12020029
9. Liu, J., Wang, L., et al. (2023). "Spatial-temporal feature with dual-attention mechanism for encrypted malicious traffic detection," *Hindawi Security and Communication Networks*, 2023, 7117863. https://doi.org/10.1155/2023/7117863
10. Mwanza, N. P., and Kalita, J. (2023). "Detecting DDoS attacks in software defined networks using deep learning techniques: a survey," *International Journal of Network Security*, 25(2), 360–376. https://doi.org/10.6633/IJNS.202303_25(2).19
11. Ali, Z., Hussain, F., Ghazanfar, S., Husnain, M., Zahid, S., and Shah, G. A., (2021). "A Generic Machine Learning Approach for IoT Device Identification." In *Proceedings of the 2021 International Conference on Cyber Warfare and Security (ICCWS)*, Islamabad, Pakistan, 23–25 November 2021, pp. 118–123.
12. Lin, K., Xu, X., and Gao, H., (2021). "A novel classification scheme of encrypted traffic based on flow spatiotemporal features for efficient management of IIoT," *Computer Networks*, 190.

13. Attia, A., Faezipour, M., and Abuzneid, A. (2020). "Network Intrusion Detection with XGBoost and Deep Learning Algorithms: An Evaluation Study," In *International Conference on Computational Science and Computational Intelligence (CSCI).*
14. Giri, J., Sathish, T., Sheikh, T., Sunehriya, N., Giri, P., Chadge, R., Mahatme, C., and Parthiban, A. (2024). "Automatic liver segmentation using U-Net deep learning architecture for additive manufacturing," *Deleted Journal*, 245(1). https://doi.org/10.1007/s10751-024-01927-9
15. Kumar, E. K., Patel, S. S., Panda, S. K., Patle, B. K., Makki, E., and Giri, J. (2024). "A comprehensive exploration of shape memory alloys: Fundamentals, structural reinforcements, nano-analysis, machine learning perspective, and emerging applications," *Mechanics of Advanced Materials and Structures*, 1–34. https://doi.org/10.1080/15376494.2024.2307471
16. Sharma, A., Singh, P. K., Makki, E., Giri, J., and Sathish, T. (2024). "A comprehensive review of critical analysis of biodegradable waste PCM for thermal energy storage systems using machine learning and deep learning to predict dynamic behavior," *Heliyon*, 10(3), e25800. https://doi.org/10.1016/j.heliyon.2024.e25800
17. Thakur, R., Sikka, G., Bansal, U., Giri, J., and Mallik, S. (2024). "Deadline-aware and energy efficient IoT task scheduling using fuzzy logic in fog computing," *Multimedia Tools and Applications*. https://doi.org/10.1007/s11042-024-19509-w

Chapter 4

Big data analytics and FinTech in BFSI industries

Emerging trends

Sunitha Ratnakaram and Venkamaraju Chakravaram

4.1 INTRODUCTION

The nexus of big data analytics and financial technology (FT) is helping organizations to grow stronger, both technically and financially. The digital financial sector is going through a revolutionary trend with the prevalence of big data analytics, artificial intelligence, mobile communication, and cloud computing services. The widespread popularity of intelligent computing, smartphones, and the ubiquity of mobile cloud services have revolutionized the world, especially the financial service industry, in the form of financial technologies. The financial industry must meet the global changes by actively looking for destructive innovation ideas and escalating their customer experience and service quality (Seng et al., 2018; Dalal et al., 2024). With the development of service advances and innovation integrations, the emphasis has shifted to the growth of new financial services with mobile technologies, data science, analytics, and cloud computing that diversify and modify the new functionalities available to the customers. Most worldwide financial holding corporations have faced foremost challenges from the dynamic business situation, heterogeneous data that must be marshalled in some integrated, real-time, diverse telecommunication situations, in a seamless manner (Seng et al., 2018; Kumar et al., 2024a, 2024b). FT has become an important topic because of the number of driven forces, business innovation expectations, technical development, customer demand, and cost-saving requirements (Mawandiya et al., 2024; Kumar et al., 2024c; Basu et al., 2025). This chapter attempts to identify the role of big data in the financial engineering (FE) process of three important sectors of global economies: banking, financial services, and insurance (BFSI). The main objective of the chapter is to deliberate the evolution, drivers uses, and tools of big data analytics in the BSFI sector.

 DOI: 10.1201/9781003512585-5

4.2 FINANCIAL ENGINEERING

Financial engineering (FE) uses various financial technologies in the design, development, and implementation process of innovation, and creativity of financial products. The FE is a process of solving existing financial problems of corporate by creating innovative financial models, and products using the tools, techniques, models, and knowledge of financial technologies. FE uses financial mathematics and statistics, big data, computer software, computational intelligent systems, and algorithms. It also uses data analytics, data science, blockchain technology (BCT), mobile apps, artificial intelligence (AI), Internet of Things (IoT), information and communication technologies (ICTs), machine learning tools, and InsurTech in addition to econometrics, financial management, and computer software (Joshi et al., 2024; Singh et al., 2024). Figure 4.1 shows the various steps in the financial engineering process of BFSI sector products and services (Baporikar, 2018).

The FE is widely used in the banking sector and it is in core operational activities like investment banking, derivatives trading and risk management, investment management, pension schemes, insurance policy, credit default swap, and market mechanism design (Baporikar, 2018). It is also used in forecasting, corporate strategic planning, securities trading, and financial risk management.

4.2.1 Financial technology

The FT industry is booming with the inexpensive android and other operating platforms providing financial services. These services not only help its customers to perform transactions quickly but also record and manage

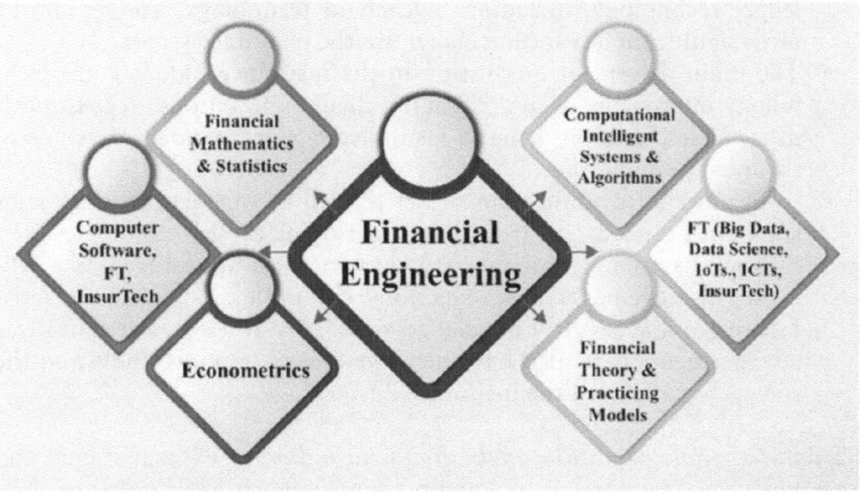

Figure 4.1 Financial engineering.

their data that can be further used in quick decision-making. It has been considered a unique taxonomy that mainly defines the sector in a wide range of operations for organizations or enterprises, which mainly talks about the enhancement of service quality by using information technology (IT) applications. The involvement of big data with FT creates the need for cybersecurity and data safety against unwanted attacks (Rahul & Banyal, 2019). The constant development of the investment has been fuelling the improvement of FT to advance technological innovations in multiple areas, such as mobile networks (Gai et al., 2016), trust management (Abawajy et al., 2016), big data (Gai et al., 2016), cloud computing (Gai et al., 2018b), mobile embedded systems (Gai et al., 2017), data analytic techniques (Lee & Kim, 2015), and image processing (Castiglione et al., 2007). For most competitive financial firms, FT is considered to be a major investment (Wigglesworth, 2016). Parag Diwan (2018) has listed the key drivers of FT that are transforming the BFSI sector such as the emergence of multiple technologies, customer pull, Government support, and availability of funds. The study has also summarized the potential BFSI technology applications: digital wallets, artificial intelligence, blockchain, big data, and virtual reality (Diwan, 2018). So, the global financial markets and multinational corporations are adopting this technology-driven ecosystem to have robust financial and investment decisions.

FT developments in the insurance industry: Innovations in the field of FT are a broad category of innovative technologies and management concepts which have the potential to transform the business insurance market.

- Advancing technology solutions, such as digital channels, IoT, telecommunications equipment, cloud computing, database management, converters, humanoid advisors, computer vision, AI, and distributed ledger technology, including blockchain technology, are by far the most significant advancements within the insurance sector.
- The main drivers of innovation in the insurance industry are technology improvements and changing client expectations. Peer-to-peer, usage-based, and on-demand insurance business models are some of the most popular ones.
- Conventional insurance businesses as well as emergent tech firms or start-ups branded "Insurtech" are developing these technologies.
- Financial systems can bring a wide variety of advantages, like efficiency improvements, cost reductions, risk evaluation, customer satisfaction, and access to financial services. Nevertheless, a few of these advancements may also have negative effects on individuals and the solvency position of the insurance business.

FT and the transformation of the financial industry: FT is a simple and uncomplicated coupling of a domain (financial) with technologies. The finance industry has been growing over time, with the very first financial institution established around 1472, followed by a plethora of many other businesses (such as security companies, insurance providers, and real estate firms). Financial institutions have frequently alluded to it as service providers and they both support primary market firms with their processes and help shape a secondary market wherein financial institutions (such as brokers, banks, and investment bankers) communicate with each other over time. Unlike the old industrial and retail enterprises, technology developed into something like a sophisticated, reciprocating, and much less unidirectional system of interlinkages. The second section of the FT acronym includes technology, which has grown increasingly essential in improving monetary operations. Technology is a method of effectively structuring things, integrating operations, and performing tasks.

FT: ecosystem, business models, investment decisions, and challenges: FT is indeed a new revolution wherein financial sector innovations are driven by digital technologies. Financial technology is now being heralded as a revolutionizing, innovative concept with the potential to impact formal economic markets. This chapter offers an overview of the history of FT as well as a look at the financial services market. Afterwards, they get above some financial technology business concepts and investment opportunities. This chapter shows how legitimate choices could be used to evaluate FT investment choices. Finally, technical and managerial issues plague both the FT start-ups and the traditional financial institutions.

The Journal of FT: The major goal of *the Journal of FT* is to have a comprehensive forum for research and analysis of problems concerning the new golden age. Finance processes, technologies, and applications are indeed the three major aspects of financial technology. They would give a venue promoting creative ideas and advancements wherein computer technology and telecommunications networks are disrupting the following old approaches:

- Legislation
- Artificial intelligence in finance
- Blockchain technology and decentralized ledgers
- Currency innovations
- Text processing
- Rob advice
- Peer-to-peer lending and crowdsourcing
- Distribution network
- Financial planning
- Monitoring
- Adherence
- Confidentiality

It would promote the debate among authorities, financiers, capitalists, academicians, including professionals in Hitech, who discover different prospects and create a networking forum for people, concepts, and issues regarding current processes which need to be resolved.

FT and innovation collection: The financial industry is being revolutionized by financial technology, a phrase that refers to the intersection of finance and technology. Traditional corporations confront the unique difficulty of adjusting to a changing financial landscape as new technologies are introduced into the financial industry by young start-ups and tech firms. FT encompasses a wide range of activities, starting with mobile payments, money transfers, and crowdfunding and progressing to the more recent worlds of blockchain, cryptocurrencies, and rob investing. The significance of technology in mortgage lending, trading, information generation, rob advising, and other areas is explored in this new collection from the Society for Financial Studies. Technology and creativity are combining to alter the financial sector, making it more accessible and competitive than ever before.

Academic production and technological emergence in finance – bibliometric study on FTs: This research intends to conduct a scientific investigation into FT businesses and related management theories. The first method reveals that 45% of papers are not blind-reviewed. The second method demonstrates a lack of focus on any journal. It denotes a lack of focus on any one journal in particular. Financial technology classifications can be based on the base, financial technology in connection to disruptive innovation theory, financial technology, and organizational or economical concepts, plus financial technology – legal and regulatory challenges are also the four different materials identified by the third approach throughout all research papers.

The rise and rise of financial technology: The good, the bad, and the verdict: This essay aims to encapsulate the FT conversation, present criticism of both the research and suggest areas for future research. A database comprising consensus publications had been created, recognized, investigated, categorized, and categorized under slightly elevated topics to be evaluated, evaluated, or understood to accomplish this. This article recommends numerous prospective avenues for additional investigation following summarizing the concept of FT inside the academia, which are grouped into the following concepts: definitions, qualities, acceptance, supervision, and rivalry.

The FT phenomenon: antecedents of financial innovation perceived by the popular press: The phrase "FT" is a hot topic in public discussion in domains like commerce, finance, and innovation, but many folks are unsure what it means. This uncertainty is experienced by both FT specialists and the intended outside users who are watching this developing industry. The FT industry's uniqueness and phenomenal sudden increase might be one cause for uncertainty. FT is indeed a broad concept that is also changing on a frequent basis as new technology innovators join the marketplace and modify it to meet social demands. On the one hand, FT could be defined as a financial service that is influenced by cutting-edge technology to meet the needs of the future: high accuracy, reduced costs, business process management, immediacy, adaptability, and originality. The chapter further talks about the key fundamentals of the emergence of FT as a new dynamic ecosystem and how it is slowly getting into the mainstream. Relevant statistical reports are given for the same to initiate better understanding.

The emergence of the global FT market: economic and technological determinants: The review looked into the technological and economic factors that motivate innovators to start businesses to revolutionize financial technology. In addition, the number of secured network servers, cell phone subscribers, and available workforce all contribute to the growth of this emerging market area. Finally, the more challenging it would be for businesses to obtain financing, the more financial start-ups there are in a particular country. Ultimately, this evidence suggests that the development of the FT industry is not to be left to happenstance and that active policy can affect its development.

Taming the beast – a scientific definition of FT: This study examines the complexity of FT and attempts a characterization based on an analysis of far more than 200 academic published works over 40 years that discuss the keyword FT. The goal of this research is to come up with a definition that is both distinct and brief in its communication while also being sufficiently broad in its scope of applicability. Because the term's roots cannot be pinpointed with certainty in either academia or practice, the definition focuses on capturing the spirit of FT using both spheres. As per syntactic research as well as the similarities of 13 consensus definitions of the term, FT is indeed a modern financial industry which employs technologies to optimize financial operations. The implications of such a concept, including its shortcomings, are discussed.

FT – the digital (r)evolution in the financial sector: The following topics were covered in the research paper, and the concerns raised were as follows:

- Internet companies have thrown down a challenge towards the financial industry. Several web- and data-based global financial goods and services are available in elements of the financial sector which consumers could get via the banks or another comparable source. As a result, a unique disruptive climate emerges. Pure economic services are being entered by non-bank, primarily technology-driven companies. Regulatory variations have a key impact.
- The less experienced finance sector and the need for readily understandable and standardizable processes are indeed the major areas impacted. These newbies' offerings currently cover everything from online payment and data technology to savings and deposit-taking, as well as contemporary internet banking, multichannel advice and securities services, simplified financing arrangements, as well as the usage of appropriate financial technology.
- Nevertheless, the importance of technological systemic reform in several business categories is routinely overlooked. Digitalization has affected no other aspects of valuation practices and business strategies,

but the systems world in general, and they must be adjusted towards the digital era's design overall.

- Commercial banks must offer a high emphasis (not just) on all digitization approaches in the long run. Despite the massive strain on certain profits, the unresolved aftermath of economic meltdown, business or service spending habits, and more stringent regulatory standards, banks must undertake a dramatic treatment of innovative treatment during the transition process. So over the medium term, it will eat up more resources.

FT is merging with IoT and AI to challenge banks: how entrenched interests can prepare: Sensing, connectivity, computing, and storage costs are plummeting, allowing start-ups to forsake the inner realm of data structure that has dominated finance for decades. Companies may build innovative disciplines of risks, information management, financial forecasting, and insurance aggregation, as well as trades which are dependent on billions of data points from external sensors and smartphones. This is a world where data drives commerce instead of the other side. That's the IoT era, and computers will progressively make human-like judgements about all types of hazards. Several financial firms, according to the researchers, are unprepared for such a transition (Giri et al., 2024; Kumar et al., 2024; Sharma et al., 2024; Thakur et al., 2024).

FT transformation: how IT-enabled innovations shape the financial sector: Financial technology, a phrase that encompasses the fields involving information technology and financial development, is gathering steam and drawing interest of professionals, financiers, and academics. The media frequently discusses FT, which shapes the public's view and conveys societal thought; nonetheless, the view of FT must be backed up with empirical evidence.

As a result, we consider the nature underlying FTs by looking at five Swiss FTs through the prism of such a framework for understanding FT as well as its aspects. As a result, we broaden our knowledge of FT and create fertile ground for future research in this area.

FT, biotech, and the future of banks: Banks seem exceptional among financial institutions such as typically producing both liquid claims (demand deposits) and loans. Although bankers could reproduce almost all of what FT firms could accomplish, FT companies profit from such an unequal field of play given that they are less controlled. Non-banking FT firms might fight banks in different types of products wherein performance is not related to what keeps banks distinctive, savings capabilities as well as the possibility for cooperation among borrowers given by deposits – due to the unequal field of play.

Though banking's reactions to FT have indeed been delayed with legacy IT systems as well as the organizational frictions that come with large, varied companies, FT's limited current products plus the absence of recognized "franchises" seem to restrict FT's capacity to replace banking institutions. BigTech enterprises, unlike FT, have had several characteristics which banks would find difficult to reproduce, and as a result, will pose a considerably stronger alternative to established banks in two main areas: customer financing and business loans.

Furthermore, both FT and BigTech are adding to a pattern wherein banks were gaining a competitive edge by possessing greater easy access to information about credit applicants. The capacity of financial institutions to fend off these risks would be based on their ability to start making efficient possibly better of the same information systems that their new competitors are using, as well as their ability to accomplish scale economies that their non-banking rivals will find tough to match.

4.2.2 Evolution of FT

Similarly, FT also has a long history of evolution concerning different periods. FT 1.0 era can be considered from the year 1866 to 1967, this was the time when industrialization and the focus was on the linkages of FT and infrastructure building for computerization. This period is known as financial globalization. Alex Graham (2017) emphasized that FT 1.0 needs to upgrade FT 2.0 for the future of banks in light of the FT era (Graham, 2017). FT 2.0 was considered from the year 1967 to 2008. In this period, more focus was on digitization and internet connectivity. In this time frame, the banking business also evolved and incorporated digitization processes. This is the era of the digital computer, mobile phone, internet banking, and quantitative risk management. The third phase of FT is known as FT 3.0 and is classified into two different parts FT 3.0 and FT 3.5. This classification is done based on the developed and developing economies. The third phase of FT includes high-speed internet connectivity, digital payment systems, start-ups, fast capital transactions and settlement procedures, and the huge advancement of technologies in the global financial market. The connectivity of people with the banking sector is overwhelming. It is the time for new entrants in the financial markets with new technological advancements over other competitors and cryptocurrencies. The increased use of technology in financial markets leads to innovative financial products and thus improved financial efficiency along with less transactional cost (Liu, Peng & Yu, 2018).

4.3 BIG DATA

Data, quality, and timely accessibility are very important in the decision-making process of any business. The data which a decision-maker uses should be harmonized information, meet needs and provision, mutual understanding of data capability, and at the same time, it should be trustworthy and credible. The quality data possess several characteristics, namely "Accurate, Precise Relevant, Complete, Consistent, Transparent and Timely" (Sarfin, 2020).

Coming to the various types of data quantity and volumes, generally small data will be in megabytes, large data will be in gigabytes, and big data will be in terabytes to petabytes. Big data means simply too large, a massive amount of complex data which cannot be dealt with, traditional database handling, and storing methods. There is no use of big data and it is worthless unless some big data analytics, irrespective of data size, type, or speed. It can create some value to that data by adding systematic big data analytics by converting it into a piece of meaningful valid information. Big data analytics mechanisms, systems, and software are highly imperative to hold and maintain efficient and effective database management. Financial big data endorses FT development where it can provide value realization for big data. The comprehensive use of big financial data promotes strong banking and capital market networking, infrastructure, and research capabilities.

The network stock exchange produces in a day the data which equals 1 terabyte. Facebook ingests 500+ terabytes of data in a day in the form of messages, videos, and images. This data is mainly in the form of uploading videos and images as well as exchanging comments and messages. A single jet engine will produce in 30 minutes of flight time 10+ terabytes of data. Just imagine with thousands of flights flying in a day, this data will be in petabytes. This big data will be in three different forms, they are structured, unstructured, and semi-structured. Under structured data, they can be stored, processed, and accessible in the form of systematic tables or fixed formats. Salary statements of employees with several working days paid, basic salary, tax deductions, the net payable, and other details come under structured data. If the data is available in the form of an unknown form full of text messages with images, and videos in between in huge volumes will be known as unstructured data. The best example is when searching for content in Google search, it will display the different file formats such as doc, pdf, ppt, images, audio, and videos. Applying strong big data analytics and mechanisms can create some value propositions for the data. Semi-structured data can be seen in both structured and unstructured forms. However, it is difficult to understand and make value creation out of that. The best example of personal information stored in an XML file comes under semi-structured data "Big data path."

4.3.1 Characteristics of big data

Big data is characterized by five different characteristics, as shown in Figure 4.2. They are volume, velocity, variety, value, and veracity (Goel, 2023).

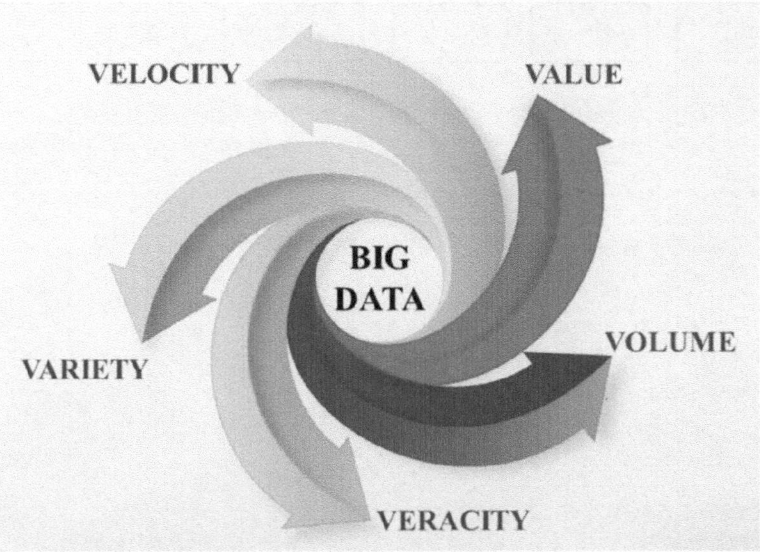

Figure 4.2 Characteristics of big data.

Variety: A combination of data types that are being dumped into the system. BFSI sectors have to process varieties of data of customers, transactions, profiles, their history, and so on, in huge numbers and various types.

Velocity: The speed of data at which it is emanating and changes occurring between datasets.

Volume: The sheer volume of data being generated every second.

Value: This can be derived from accessing and analysing big data.

Veracity: The discrepancies found in the data.

4.3.2 Big data value chain

Figure 4.3 shows the scientific flow of the value chain of the big data begins with data-generating points or sources and ends with the data analysis. In between the sourced data from various data generation points, like a business, e-commerce, social media, healthcare, financial records, and telecom, flows through the data collection, data transmission, and data processing channels. The processed data will be stored in data storage devices to apply data analytical techniques to analyse to get meaningful value output or to convert the processed data into valuable information to use various business decision-making processes (Bhadani & Jothimani, 2016).

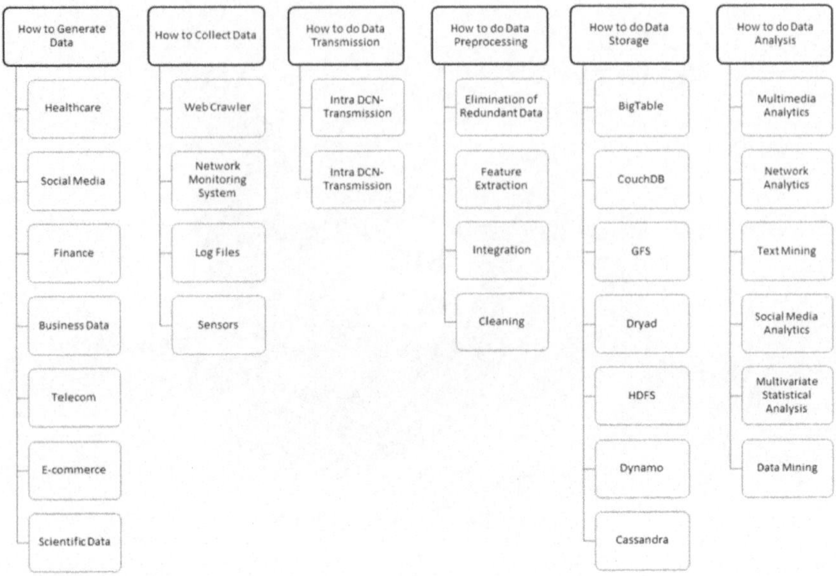

Figure 4.3 Big data value chain.

4.3.3 Big data analytics

The increased education and job opportunities in emerging markets and developing nations increase the adoption of new technologies in households. As information technology spread fast, most of the data were exchanged on the internet as well as produced digitally. According to Lyman and Varian (2003), the size of new data stored in digital media devices was more than 5 exabytes in 2002. The problem of analysing big data has not occurred suddenly but has been there for several years. The creation of data digitally is easier than finding relevant data. Even though the computer system today is more developed nowadays, large-scale data is hectic to analyse. According to Sophie Cockcroft (2018), risk, security, data visualization, predictive analytics, data management, and data quality are the main big data research areas in finance and accounting.

Data analytics is a process in which computer programs apply statistical and mathematical methods to extract previously known and new information from textual data. The vision of computing as a utility took shape in the mid-1990s after the development of critical internet technology (Buyya et al., 2009). Big data analytics involves data inspection, cleaning, transforming, and modelling of big data to generate useful information, identify trends, provide conclusions, and improve decision-making (Cao et al., 2015). It is used to generate robust computational and analytical models that can help organizations interpret the data in the appropriate form (Iqbal et al., 2020; Saggi and Jain, 2018). Apart from that, big data analytics can help in risk management, failure detection, detection of fraud, and identifying potential customers with their buying behaviours.

These days the data are not just large, but they are also composed of various types of data, including streaming data. Since big data has the unique features of "high dimensional, complex, massive, heterogeneous, unstructured, noisy, incomplete, and erroneous," it may change data analysis and statistical approaches (Ma et al., 2014). Although big data makes it feasible to collect more data, more data doesn't mean more information. It may contain more abnormal or ambiguous data. For instance, an account may be used by multiple users or multiple accounts used by a single user, which degrades the accuracy of the results. Therefore, several new issues arise in big data like security, storage, privacy, quality of big data, and fault tolerance (Katal et al., 2013).

Big data is impacting remarkably quantitative finance, from business process modelling to back-end statistics (Fang & Zhang, 2016). The main trend of big data processing in FT is migrating from simple statistics work and structured data to unstructured data and complex computation work (Gai et al., 2018a). Big data applications are widely used in the design and development of BFSI products or services. The role of big data in the design development and execution of innovative insurance products is narrated by Chakravaram et al. (2019).

Big data phase 1 can be the period from 1970–2000, big data phase 2 from 2000–2010, and big data phase 3 from 2010 to the present period.

4.4 BIG DATA APPLICATIONS

For decades, the BFSI industries have been using conventional models in the design, development, operations, and facilitation of their services and products. Nowadays, the financial markets are becoming more sophisticated and competitive. Customization and digitalization are increasing in the sector worldwide. To exploit the dynamically changing financial markets, and to face day-to-day technological challenges and changes, particularly in the said sectors, adoption of the financial engineering process is highly imperative.

4.4.1 Big data tools

Big data requires a set of tools to create value for that data and mine that data to convert it into useful information for the users. Tools are very important in this process to mine and analyse the big data using various software algorithms: "A Good Tool improves the way you work. A Great Tool improves the way you think" by Coriolis Jeff Duntemann. Table 4.1

Table 4.1 Big data analytical tools and their utilization in BFSI institutions

Big data analytical tool	*Basic concerns of this tool in BFSI institutions*
Apache Spark (AS)	The main focus of AS is on filling the space of Hadoop considering data processing. Spark can be handled easily with real-time and batch data
Hadoop	Across clusters of computers, Hadoop distributed the processing of large business datasets enormously. Hadoop is a robust ecosystem highly suitable to the needs and requirements of the developer and it runs on a cloud infrastructure
Cassandra	Cassandra processes the structured datasets. Also, it enables the distribution of easy data across all data centres, since Cassandra is used for managing a large set of data across the servers
Storm	For the unbounded data stream, Storm offers a fault-tolerant processing system. Also, it is a free open-source source-distributed real-time framework
RapidMiner (RM)	RM is the best platform for statistical modelling, evaluation, deployment, and so on. This platform is also for data science activities and it provides an integrated environment for data preparation, machine learning, text mining, and so on. RM is the best platform for data visualization, predictive analysis, application development, prototyping, and model development. RM offers suitable products to develop new data mining processes
Apache SAMOA (AS)	AS is used for clustering, classification, regression, programming abstractions for new algorithms, and so on, like machine learning tasks. This is also useful for big data mining through its distributed streaming algorithms
MongoDB	MongoDB, in a cloud structure, can easily partition the data across the servers. It is a simple and flexible cloud-based infrastructure. This helps the business tycoons to make instant, fastest, and most accurate decisions in their day-to-day operational and strategic areas
HPCC	HPCC maintains code and data encapsulation. It nicely supports the management of the end-to-end big data workflow of businesses and corporates. HPCC offers high redundancy and availability. One can use HPCC for both complex data processing and Thor cluster data processing
R computing tool (RCT)	RCT is a well inside SQL server that helps to run Windows and Linux. It can be used for data analysis, data modelling, and statistics
Neo4j	Neo4j widely supports a query language known as Cypher to resolve queries on graphs. This was used in the big data industry to maintain graph databases and it can be integrated easily with other databases. Neo4j maintains a key-value pattern in data storage

shows the available tools for the big data analytics market worldwide. These data analytical, processing, and open tools are helping the core stakeholders value for creative process of their structured, unstructured, and semi-structured data. The top ten big data tools in the data analytics market worldwide by 2020 are provided in Table 4.1.

4.4.2 Big data applications in the BFSI sector

According to UpGrad, the following are the significant financially engineered operational areas broadly, where big data applications are widely used in the sector (UpGrad, 2019).

Customer interaction and insights have been increased: The customers of BFSI industries are happy on digital platforms in terms of great insights, engagement, products, services, and handling queries. This helps the companies to keep their customers in a friendly environment with a great digital experience, and also to stay ahead in the competitive market at the same time. For example, big data tools are used for customer segmentation to classify customers based on demographic profiles and behavioural profiles. This helps the banking sectors to execute customer retention strategies and to know their spending patterns (Ray, 2017). Big data analytics can identify spending patterns of each customer season-wise, and age-wise.

Fraud detection and prevention capabilities have been increased: Big data applications and tools can analyse massive data to predict and identify fraudulent transactions and entries in the segregation process of systematic datasets. This helps the BFSI sector to a great extent to prevent losses and data hacking threats. Due to the increase of online and digitalization in banking transactions, fraud has increased. If banker knows the spending patterns of a customer, they can easily be alert to unusual transactions done by hackers and fraud people.

Market trading analysis: The improved and sophisticated big data analysis algorithms are helping marketers in their trading. The massive data available in the market channels based on geographical location, in various markets, are being analysed on time by the algorithms. Hybrid datasets offer a real-time 360-degree market picture and related accurate information covering various aspects like the latest tax policies, pricing structure, risk management, and profit and loss estimation.

Risk management: Big data analytics are analysing financial data to reduce operational and integrated risk under the enhanced risk management task. Banks can prevent bad loans if they detect them in an early stage by using big data analytical tools to observe the customers' behaviours and their day-to-day lifestyle profiles.

It helps in the prevention of fraudulent practices in loan management. Big data tools also strengthen and boost the predictive power of risk models for enhancing the response, and time to offer more extensive risk coverages in addition to optimum cost-saving opportunities.

Employee engagement: By leveraging big data techniques, organizations can also estimate employee involvement and engagement to analyse the performance appraisal. Organizations can easily track and monitor the employees to identify the best-performing employees, happy employees, and unhappy employees at their given tasks. These metrics will help the management in the recruitment, training, and development process of their man force.

Customized and personalized services: Offering special customized and personalized services to the customers is also one of the output big data analytics achievements. The industry collects huge data from e-commerce websites to analyse and estimate the buying behaviour and buying habits of customers. This valid information will help the bakers to offer their products and services to their targeted customers based on their interests and habits.

Statutory compliances and requirements: To address various statutory compliances to government regulating agencies, bankers are using big data tools and techniques. To show future launching services, and products, or to show some predictions in the documents submitted to the regulating agencies, big data analytics will serve the valid data for the bankers to fulfil and submit the compliances from time to time.

4.4.3 Insurance sector

The following are the key and financially engineered operational areas of the insurance sector where big data applications are used. Generally, companies are collecting their required data from telematics, sensors, various government agencies, social media sites, and direct customer interaction (Exastax, 2017).

Risk assessment and calculation of premiums: Whenever the insurance companies are fixing the pricing of the policies, they need accurate information to assess the exact risk in the form of authenticated documents in their files. These companies always rely on a principle called "The law of large numbers" to predict and make statistically insurable events. Companies started extracting the actionable behaviour and habits of the customers to price their insurance requirements and assess the risk. For this purpose, companies started collecting transaction data or big data by installing a box in the vehicle or by downloading an app on the mobile phone of the customer.

Underwriting process considering medical data: During the process of underwriting the insurance policies and premiums, big data provides huge data in the form of medical and diagnosis data information of the prospective customers to the companies.

Fraud detection: Insurance companies are using big data and big data analytics widely to identify criminal natured activities of the people through data management and predictive modelling techniques. Companies match the variables of past claims with present claim applications and also match with the behaviour of applicants at various social sites to understand the genuine and mental status of the customer.

Insurance claim settlement: Big data helps insurance companies in the claim settlement process. They can calculate an accurate claim settlement amount after thorough verification of the applicant's various transactions extracted from various sources. For example, if a bike person claims an accidental claim amount, insurance companies can check the data from different sources: several accidents he/she made, fines he paid, his driving skills, vehicle servicing habits, driving habits, and his psychological status through his circles and social sites.

Smarter data and finance: By extracting real-time big data from the customer through some advanced digital wearables, insurance companies can charge premiums according to the risk assessment from time to time by applying advanced underwriting techniques and algorithms.

Companies can also adjust premium rates daily based on the big data supplied by the equipment.

Automation and big data: Companies have started using automation techniques in operational activities. By using big data analysis, they can avoid asking repeated questions to the customers at the time of new purchase and at the time of availing various services on the existing policies. This helps the companies save valuable time, and paperless work also helps to assess the exact risk of releasing the product or service in a shorter time.

Right care at the right time: Big data analytics helps health insurance insurers and their customers in selecting the right medical policies. Big data analytics suggests appropriate coverage be required for the proposer after analysing his/her existing coverages, habits, age, and health records (Sullivan, 2023). The predictive analytical concepts are greatly helping fraudulent claims by analysing social media posts and unstructured data of the applicant. Based on applicants' profile data, health insurers are migrating from a product-centric strategy to a customer-centric strategy. By introducing wearable bands kind of IoT-enabled devices, health insurers are encouraging the health habits of the applicants by sharing time-to-time and day-to-day health-related information.

4.5 RESULTS AND CONCLUSIONS

The potential of the sector is so strong that huge big tech multinational enterprises are providing various financial services along with data resources. An expectation of using FT has caused a great challenge in its acceptance and planning, due to the complicated integrated system, under crossed realms, and distinctive demands. Therefore, an up-to-date and accurate awareness of FT is an urgent requirement for both professionals and academics (Gai et al., 2018a). In the FT era, various disruptive innovations are coming very rapidly such as P2P lending, blockchain, and cryptocurrency. The regulation of FT innovation is still a challenging task for governments around the world. With the value proposition, big data has brought some big challenges to the industry. One more challenge is the corporate need to advance new breed high-volume data storing, capturing systems, and servers in the operations same time. Researchers can identify the role of big data in the design and development of financial services and products in the BFSI sector. Also, how these tools are effectively used in the financially engineered operational areas of the targeted sectors.

4.6 SCOPE FOR FURTHER RESEARCH

There is a rising trend in the financially engineered applications in the products and services of the BFSI sector along with big data applications in the designing and execution process: particularly for fraud detection, insurance claim settlements, risk assessment and risk management, and other important operational areas of the BFSI sector. Big data tools are used effectively in the designing process of customized financially engineered products. They are also used in the development and product launching process of FE products and services in the BFSI sector as well as in the operational areas of the sector, wherever required to get better results. Companies in the BFSI sector are trying to use effectively new and innovative big data tools from time to time to achieve operational efficiency, compete with global trends, and also come out from traditional operational methods. At present, some of the big data applications are in their initial stages. As evident, they are at an inflexion point, and sooner than later, their impact would be seen in the BFSI sector. At present, hundreds of new FT start-ups are working on big data applications and solutions for the BFSI sector to take FT to another plane.

REFERENCES

Abawajy, J., Wang, G., Yang, L. T., & Javadi, B. (2016). Trust, security, and privacy in emerging distributed systems. *Future Generation Computer Systems*, *55*, 224–226. https://doi.org/10.1016/j.future.2015.11.008

Baporikar, N. (2018). *Understanding Financial Engineering – Researchgate*. ResearchGate. Retrieved from: https://www.researchgate.net/profile/Neeta-Baporikar/publication/328630424_Understanding_Financial_Engineering/links/5bd964caa6fdcc3a8db2f9a5/Understanding-Financial-Engineering.pdf

Basu, R., Das, M. C., Kumar, A., & Sarkar, B. (2025). Identifying and Prioritizing Quality 4.0 Practices in Sustainable Manufacturing Using Rough Number-Based AHP-MABAC. In *Handbook of Intelligent and Sustainable Manufacturing* (pp. 135–165). CRC Press.

Bhadani, A. K., & Jothimani, D. (2016). Big data: challenges, opportunities, and realities. *Effective big data management and opportunities for implementation*, 1–24.

Buyya, R., Yeo, C. S., Venugopal, S., Broberg, J., & Brandic, I. (2009). Cloud computing and emerging IT platforms: Vision, hype, and reality for delivering computing as the 5th utility. *Future Generation Computer Systems*, *25*(6), 599–616. https://doi.org/10.1016/j.future.2008.12.001

Cao, M., Chychyla, R., & Stewart, T. (2015). Big data analytics in financial statement audits. *Accounting horizons*, *29*(2), 423–429. https://doi.org/10.2308/acch-51068

Castiglione, A., De Santis, A., & Soriente, C. (2007). Taking advantages of a disadvantage: Digital forensics and steganography using document metadata. *Journal of Systems and Software*, *80*(5), 750–764. https://doi.org/10.1016/j.jss.2006.07.006

Chakravaram, V., Srinivas, J., & Ratnakaram, S. (2019). The role of big data, data science and data analytics in financial engineering. In *Proceedings of the 2019 International Conference on Big Data Engineering* (pp. 44–50).

Cockcroft, S., & Russell, M. (2018). Big data opportunities for accounting and finance practice and research. *Australian Accounting Review*, *28*(3), 323–333.

Dalal, S., Kumar, A., Lilhore, U. K., Dahiya, N., Jaglan, V., & Rani, U. (2024). Optimizing cloud service provider selection with firefly-guided fuzzy decision support system for smart cities. *Measurement: Sensors*, *35*, 101294.

Diwan, P. (2018, August 30). *Fintech: Transforming the BFSI sector*. Medium. Retrieved from: https://medium.datadriveninvestor.com/fintech-transforming-the-bfsi-sector-4fd7d59e38e9

Exastax, B. (2017, May 9). *Top 7 big data use cases in the insurance industry*. Medium. Retrieved from: https://medium.com/@exastax/top-7-big-data-use-cases-in-insurance-industry-a0da62484f16

Fang, B., & Zhang, P. (2016). *Big Data in Finance*. 391–412. https://doi.org/10.1007/978-3-319-27763-9

Gai, K., Qiu, L., Chen, M., Zhao, H., & Qiu, M. (2017). Sa-East. *ACM Transactions on Embedded Computing Systems*, *16*(2), 1–22. https://doi.org/10.1145/2979677

Gai, K., Qiu, M., & Sun, X. (2018a). A survey on FT. *Journal of Network and Computer Applications*, *103*, 262–273. https://doi.org/10.1016/j.jnca.2017.10.011

Gai, K., Qui, M., Tao, L., & Zhu, Y. (2016). Efficient and secure multi-functional searchable. October 2015, 34–42. https://doi.org/10.1002/sec

Gai, K., Qiu, M., & Zhao, H. (2018b). Energy-aware task assignment for mobile cyber-enabled applications in heterogeneous cloud computing. *Journal of Parallel and Distributed Computing*, *111*, 126–135. https://doi.org/10.1016/j.jpdc.2017.08.001

Giri, J., Sathish, T., Sheikh, T., Sunehriya, N., Giri, P., Chadge, R., Mahatme, C., & Parthiban, A. (2024). Automatic liver segmentation using U-Net deep learning architecture for additive manufacturing. *Deleted Journal*, *245*(1). https://doi.org/10.1007/s10751-024-01927-9

Goel, K. (2023, October 3). *Introduction to big data*. Intellipaat. Retrieved from: https://intellipaat.com/blog/tutorial/hadoop-tutorial/big- data overview/

Graham, A. (2017). *Fintech and Banks: How Can the Banking Industry Respond to the Threat of Disruption*. Toptal Finance Blog. Retrieved from: https://www.total.com/finance/investmentbanking-freelancer/fintech-and-banks (accessed on 3rd October, 2021).

Iqbal, R., Doctor, F., More, B., Mahmud, S., & Yousuf, U. (2020). Big data analytics and computational intelligence for cyber-physical systems: Recent trends and state of the art applications. *Future Generation Computer Systems*, *105*, 766–778.

Joshi, V. D., Sharma, M., Kumar, A., Cepova, L., Kumar, R., & Dogra, N. (2024). Navigating uncertain distribution problem: a new approach for resolution optimization of transportation with several objectives under uncertainty. *Frontiers in Mechanical Engineering, 10*, 1389791.

Katal, A., Wazid, M., & Goudar, R. H. (2013). Big data: Issues, challenges, tools, and Good practices. In *2013 6th International Conference on Contemporary Computing, IC3 2013* (pp. 404–409). https://doi.org/10.1109/IC3.2013.6612229

Kumar, E. K., Patel, S. S., Panda, S. K., Patle, B. K., Makki, E., & Giri, J. (2024). A comprehensive exploration of shape memory alloys: Fundamentals, structural reinforcements, nano-analysis, machine learning perspective, and emerging applications. *Mechanics of Advanced Materials and Structures*, 1–34. https://doi.org/10.1080/15376494.2024.2307471

Kumar, S., Kumar, A., & Kumar, A. (2024a). Improvising the Quality of Manifold Production Using Six-Sigma Technique for Implementation in Automobile Manufacturing Industries: A Case Study. In *Industry 4.0 Driven Manufacturing Technologies* (pp. 355–372). Cham: Springer Nature Switzerland.

Kumar, R., Kumar, R., & Kumar, A. (2024b). Exploring the Challenges of Integrating Lean Green Practices in Industry 4.0 Manufacturing Frameworks: An Empirical Study. In *Industry 4.0 Driven Manufacturing Technologies* (pp. 277–292). Cham: Springer Nature Switzerland.

Kumar, S., Kumar, R., Phanden, R. K., Kumar, A., Bala, J., Bharath Kumar, S., ... & Vishwanatha, H. M.. (2024c). A study on critical failure factors for implementation of sustainable lean six sigma from Indian manufacturing industries perspective using BWM technique. *Frontiers in Mechanical Engineering, 10*, 1451568.

Lee, T. H., & Kim, H. W. (2015, August). An exploratory study on FinTech industry in Korea: Crowdfunding case. In *2nd Int'l conf. on Innovative Engineering Technologies* (pp. 58-64).

Liu, Y., Peng, J., & Yu, Z. (2018), Big Data Platform Architecture under The Background of Financial Technology: In *The Insurance Industry As An Example* (pp. 31–35). Association for Computing Machinery, New York. https://doi.org/10.1145/3297730.3297743

Lyman, P., & Varian, H. (2003). How much information 2003? Tech. Rep, 2004. Retrieved from: http://www2.sims.berkeley. edu/research/projects/how-much-info-2003/printable_report.pdf

Ma, C., Zhang, H. H., & Wang, X. (2014). Machine learning for Big Data analytics in plants. *Trends in Plant Science, 19*(12), 798–808. https://doi.org/10.1016/j.tplants.2014.08.004

Mawandiya, B. K., Choksi, A., Lokapur, A., Kumar, A., Makhesana, M. A., Patel, K. M., & Gopal, P. R. C. (2024). Environment-Friendly Practices for Integrating Green Business with Green Supply Chain Management: Industry 4.0 Perspectives and Beyond. *Industry 4.0 Driven Manufacturing Technologies*, 207–231.

Rahul, K., & Banyal, R. K. (2019, March). Data cleaning mechanism for big data and cloud computing. In *2019 6th International Conference on computing for Sustainable Global Development (INDIACom)*, New Delhi, India (pp. 195–198). IEEE.

Ray, T. (2017, December 3). *Scopes of Big Data & Data Science in the Banking & Finance (Fintech) sector*. Stoodnt. Retrieved from: https://stoodnt.com/blog/scopes-of-big-data-data-science-in-the-banking-finance-fintech-sector/

Saggi, M. K., & Jain, S. (2018). A survey towards integration of big data analytics to big insights for value-creation. *Information Processing & Management*, *54*(5), 758–790. https://doi.org/10.1016/j.ipm.2018.01.010

Sarfin, R. L. (2020, May 15). 5 Characteristics of Data Quality [Blog Post]. Retrieved from: https://www.precisely.com/blog/data-quality/5-characteristics-of-data-quality

Seng, J.-L., Chiang, Y.-M., Chang, P.-R., Wu, F.-S., Yen, Y.-S., & Tsai, T.-C. (2018). Big data and FT. *Big Data in Computational Social Science and Humanities*, 139–163. https://doi.org/10.1007/978-3-319-95465-3_6

Sharma, A., Singh, P. K., Makki, E., Giri, J., & Sathish, T. (2024). A comprehensive review of critical analysis of biodegradable waste PCM for thermal energy storage systems using machine learning and deep learning to predict dynamic behavior. *Heliyon*, *10*(3), e25800. https://doi.org/10.1016/j.heliyon.2024.e25800

Singh, P., Das, L., Jha, R., Kumar, A., & Rani, S. (2024). Revolutionizing Sales and Marketing: The Power of AI in Action. In *Security and Risk Analysis for Intelligent Cloud Computing* (pp. 297–315). CRC Press.

Sullivan, J. (2023, March 8). *Big Data & You: A Guide to data analytics in the health insurance industry*. Hitachi Solutions. Retrieved from: https://global.hitachi-solutions.com/blog/big-data-in-health-insurance/

Thakur, R., Sikka, G., Bansal, U., Giri, J., & Mallik, S. (2024). Deadline-aware and energy efficient IoT task scheduling using fuzzy logic in fog computing. *Multimedia Tools and Applications*. https://doi.org/10.1007/s11042-024-19509-w

UpGrad (2019, September 11). *Top 5 big data applications in Banking & Insurance*. upGrad blog. Retrieved from: https://www.upgrad.com/blog/big-data-applications-in-banking-insurance/

Wigglesworth, R. (2016, January 20). *FT: Search for a super-algo*. The Financial Times Limited, pp. 1–7. Retrieved from: https://www.ft.com/content/5eb91614-bee5-11e5-846f-79b0e3d20eaf

Chapter 5

Internet of Things products

Management of youth's purchase intention

Priyanka Malik, Shalini Gautam, and Astha Gupta

5.1 INTRODUCTION

In today's scenario, it has been widely accepted that for the growth of a business, innovation in technologies is necessary [1]. Survival without the internet is impossible in the 21st century. It is one of the most widely accepted and used inventions for human beings. It has enabled to spread knowledge and made communication easy at the global level. The internet is currently used for almost everything, whether it is searching for information, sending an email, talking to someone, or at a company level. The Internet of Things has emerged in recent years and is heading to a new industrial revolution. It would change the way we are living in this world today.

The Internet of Things is a system of interconnected digital and mechanical machines, computing devices, people, animals, or objects that allow data transfer over a network without the need of a human being. A system or a persona can directly interact with the computer [2–6].

Researchers are continuously innovating in the field of IoT to exploit new opportunities. There are lots of issues which are technical in nature but other issues also exist related to acceptability of IoT products by the users and citizens [7–9].

This study aims to determine the four characteristics of IoT products that influence the emotional and functional experience of the customer's intent to buy. This study would help the marketers to identify market opportunities and the scope for IoT products in a specific area. The youth is fascinated toward these latest products, that is why this study has considered youth. The findings of this study would help the companies to make their strategies targeting youth. Companies can also improve their customer services, profit, and reputations if they are aware about the importance of emotional and functional experience for a customer.

5.1.1 What is IoT and how does it work?

The IoT devices are also known as "smart systems" which can be considered from simple smart home devices to complex industrial machines and

 DOI: 10.1201/9781003512585-6

GPS-enabled transportation systems. IoT technologies are able to envision "smart cities," which the Indian government is trying to create in future. IoT enables these smart devices to communicate with each other with the help of the internet. Smart phones, smart watches, smart TV, or smart washing machines are few home appliances which are commonly used recently. IoT consists of sensors, processors, and communication devices to collect, process, and deliver the data which is received from the environment [10]. These smart devices may be paired to send/receive data from each other. The major benefit of IoT-enabled devices is that humans are not required to intervene with the devices, though human beings can also interact with these devices to get the data. IoT has emerged as one of the most significant digital transformation trends in business organizations and economies. Our way of life and the way we work have changed and improved because of the development of IoT technology as well as its integration into the commercial economy [11]. According to a report by McKinsey [12], the potential value of the IoT ecosystem can reach $12.6 by 2030.

An IoT platform manages connectivity of devices which can be software or cloud service [13]. The purpose of an IoT platform is to monitor software, hardware, and processing abilities. IoT sensors fetch data from the real world so that device can interpret and share. Different types of sensors fetch different types of data from an environment. For example, motion sensors detect the movement of ultrasonic waves and temperature sensors detect temperature and convert that into data. Because IoT is communicating with a large number of devices and users within a large network, they create unique identifiers to communicate with other devices. The identifiers can be in the form of patterns (numeric or alphanumeric string). The internet should be active for connecting IoT devices. IoT devices use natural language processing to take input from users and interact with other devices. Amazon Alexa is the one example which takes the input from users in their language and process accordingly. The Internet of Things generates massive volumes of data from millions of devices. IoT machine learning is powered by data and generates insight from it. Machine learning serves as a toolkit for developing and evaluating algorithms facilitating prediction, pattern recognition, and classification [14]. Machine learning uses past behavior to identify patterns and builds models that help predict future behavior and events.

5.1.2 Importance of IoT

The IoT industry is growing continuously and estimated to be a $12.5 trillion industry by 2030 [12]. It includes customers of IoT products and services. The IoT devices are making people's lives easy and they are able to work smarter. Many home appliances are smart now, which are optimizing the processing time and increasing the efficiency in terms of output. That way we are able to save a lot of money at home. Similarly, industries are also able to cut the cost by automating the process, improving the efficiency and productivity [15]. IoT devices generate a lot of data which can be used for making informed decisions [16]. New business models and strategies can be made after resolving all the issues. IoT devices gather data about customer behavior for creating personalized services and engaging experience for them.

5.1.3 IoT applications

IoT products and services are revolutionizing human life by making it simpler and efficient. IoT products are being used in the real world at home and in industries. It is used in many sectors like telecommunication, automobile, manufacturing, and energy. For example, in the consumer sector, smart homes are having smart home appliances like smart lights, smart AC, smart TV, and smart camera which can be controlled remotely through smartphones and computers. This technology is finding a way into our day-to-day life and the potential of this technology appears to be limitless [17].

In the healthcare industry, patients can be monitored remotely and real-time data can be collected through IoT devices. The person can monitor his/her blood pressure, heart rate, oxygen level, and so on. The pattern can be recorded by the sensors and potential health issues can be identified by the doctors. These records can be very beneficial for the doctors/ patients. Many wearable devices are coming into the market which are monitoring health patterns of a person [18].

Manufacturing sector can implement industrial IoT devices to monitor the performance of machines, detect machine failure, and for optimizing the process of production. IoT devices can keep track of the humidity and environmental conditions to monitor the appropriate condition for producing the products [19]. IoT devices can keep track of the inventory, quality checks of the finished products, and manage the supply chain.

Retail industry is another industry where IoT devices are used. IoT devices keep track of the customer behavior and analyze the foot traffic in a retail outlet. The retail outlook can analyze the data and optimize the position of products that would lead to improve the customer experience.

In the agriculture sector, IoT devices can monitor the quality of soil, crop growth, and weather patterns to produce more. For example, sensors can monitor the moisture in soil, water level, and irrigation time of a crop. IoT devices can automate the process of seeding and plucking fruits/vegetables.

Smart cities are the future of countries so that resources can be optimized. Smart traffic lights, smart street lights, smart sensitization, smart meters, smart homes, and smart transportation would reduce the traffic, solve environmental problems, save energy, and increase sustainability. Smart cities are the need of the hour.

In a smart city, IoT sensors and deployments, such as smart street lights and smart meters, can help reduce traffic, save energy, monitor and solve environmental problems, and improve sanitation [20].

This study aims to understand the new growing Internet of Things technology. This study examines the purchasing behavior of IoT devices for making better marketing strategies to increase the demand. We have considered intelligence, interactivity, privacy, and social interaction impact on functional and emotional experience which further lead to measuring the purchase intention of youth. The youth is tech savvy and understand the importance of IoT devices which can bring change in the world. These devices can optimize the allocation and use of resources which are required for the existence of human beings in future.

5.2 LITERATURE REVIEW AND HYPOTHESIS DEVELOPMENT

5.2.1 Intelligence

The rapid progress of IoT has facilitated the connection between many smart objects and helped to collect more data to analyze and appreciate. Predicting the interconnectedness of these elements is automation in all areas of work, and human intervention [13, 21, 22]. Intelligence indicates the degree of automated tasks depending on the operation of an IoT product. The intelligent IoT product integrates sensing, memory, processing of data, and communication capabilities. The efficiency of the IoT product will be reduced in the absence of these characteristics and the consumer would consider it to be a low-quality product [7]. The smarter IoT products are not concerned with any task, service, or activity because they can do other activities. Intelligence is an important characteristic of the IoT products, and it is important to examine the influence of intelligence on the functional and emotional experience of buying intentions [23]. Thus, the following hypothesis is proposed:

H1a: Intelligence has a positive influence on emotional experience.
H1b: Intelligence has a positive influence on functional experience.

5.2.2 Interactivity

The normal daily activity can be related to the set of instructions using Internet of Things. For example, using soap from the grocery basket can become a reminder to buy online to replace it [24]. Using IoT to add articles will enhance the consumer opportunities and at the same time, the way of interaction of these goods with the environment also changes [25, 24]. The consumer satisfaction can be increased by good consumer–producer engagement [7]. As interactivity helps in describing IoT of products, we propose to examine its influence on the functional and emotional experience of buying intent:

H2a: Interactivity has a positive influence on emotional experience.
H2b: Interactivity has a positive influence on functional experience.

5.2.3 Privacy

An individual would always try to increase the confidentiality of their sensitive data and protect the same. The privacy concern has two parts. First, an individual would be sensitive to his personal information available in the public domain and, second, he would be concerned regarding what happens to that information [26]. The cross-border commercial activities and use of connected things are limited due to privacy concerns [27, 28]. The IoT will be opposed if people believe that it will pose a serious threat to their privacy [29]. The lack of an individual motivation due to privacy concerns acts as a major barrier to adopt IoT-based products [26]. Trust plays an important role during online transactions. It also impacts privacy and security. To reduce these barriers, online stores create a circle of trust with consumers, which makes loyalty important [30]. Privacy is an important characteristic that describes the Internet of Things and allows the characteristics of "IoT" products to influence the functional and emotional experience of intent to be investigated [31]. Thus, the following hypothesis is proposed:

H3a: Privacy has a positive influence on emotional experience.
H3b: Privacy has a positive influence on functional experience.

5.2.4 Social interaction

The daily lives of the people are stimulated through social interaction [32, 33]. The decision-making process and the customer behavior get impacted by social interaction, which acts as an external factor [34]. Social interaction is single or multiple communication between single or multiple companies or customers in an online or offline mode [31, 35]. In Ref.[36], the authors argue that social interaction is a "stimulus" to IoT use because sharing information with others helps to understand its benefits. The Internet of Things is also seen as a disruptive innovation as it dramatically changes the way social interactions take place. However, these social interactions are limited in the ability to collect sufficient information about the use of IoT technologies [37]. Thus, the following hypothesis is proposed:

H4a: Social interaction has a positive influence on emotional experience.
H4b: Social interaction has a positive influence on functional experience.

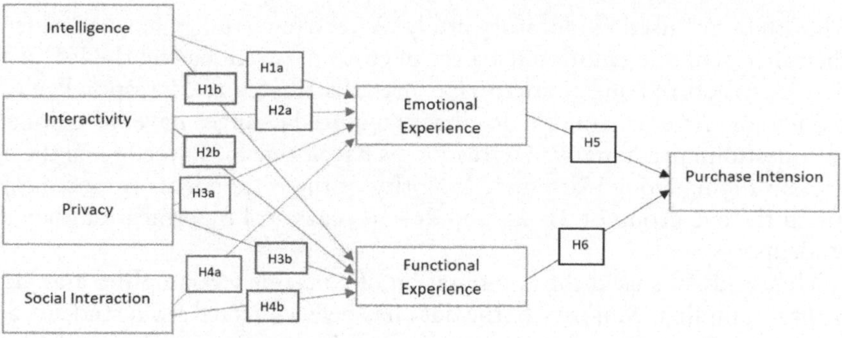

Figure 5.1 Conceptual model.

5.2.5 Functional and emotional experience

It is easier for consumers to establish a product relationship that gives them emotional and functional experiences. The customer satisfaction towards the product is dependent on the consumer perception of the product and his expectation from the product. The satisfaction arises due to perceived usefulness of IoT technology that defines the experience of the customer [38].

The companies tend to strengthen emotional bonds with their customers to enhance the emotional experience. Increasingly, companies are presenting their products with the help of IoT technologies [19, 32]. They are trying to develop customers' relationships that reflect cultural characteristics such as customer values by creating feelings of trust, respect, and closeness [38]. The functional experience arises if the product answers any economic issues and gives the utilitarian satisfaction to the customer [39, 40]. Thus, the proposed hypotheses are as follows:

H5: Emotional experience has a positive influence on purchase intention.
H6: Functional experience has a positive influence on purchase intention.

Based upon the above-mentioned hypothesis, the authors proposed the conceptual model shown in Figure 5.1.

5.3 RESEARCH METHODOLOGY AND DATA ANALYSIS

A structured questionnaire has been created for the respondents to test hypotheses. In the next section, we would discuss the data gathering process and various items to measure the constructs.

5.3.1 Sample and data collection procedure

The study has used exploratory study. A very few studies have considered the functional and emotional aspect of customers purchasing the IoT products. A structured questionnaire has been distributed to 50 respondents for pilot study. After getting the feedback, required changes have been done in the questionnaire. Finally, 270 responses have been considered from the age group ranging from 15 to >35. Majority of the respondents in our sample are in the age group of 21–25 and 26–30 years and by profession they are students.

This study has used the combination of convenience sampling and judgmental sampling. Majority of the data has been collected from students and it was easy for us to collect the data from colleges and universities. We have considered only those students who are purchasing IoT devices. The questionnaire had 25 questions from the standard scale, including demographic factors like age, education, occupation, and income. The other variables which have been considered to make the questionnaire were intelligence, interactivity, privacy, social interaction, functional experience, emotional experience, and purchase intention. The Likert scale has been used in the questionnaire. Responses of respondents are symbolized as follows: 1 – strongly agree; 5 – strongly disagree. Cronbach alpha to check reliability of data and regression analysis in SPSS software has been used to analyze and interpret the data toward purchase intention for IoT devices.

5.3.2 Validity and reliability

Validity and reliability is very crucial in every research study. Reliability is often considered as to what extent results are consistent over a time and an accurate representation of the entire population in study. Cronbach alpha shows the internal consistency of the items which have been tested in the study. The lower limit of Cronbach alpha is 0.70 [41]. The reliability coefficients of seven categories is more than 0.70 which is acceptable (Figure 5.2).

5.3.3 Multiple regression

The multiple regression was conducted on various factors. The results show that there is a linear positive significant relationship between intelligence, interactivity, privacy, and social interaction with emotional experience. Hence, hypotheses H1a, H2a, H3a, and H4a are accepted. The model is significant with R^2 value of 0.707. Further, it was shown that there is a significant relationship between interactivity, privacy, and social interaction with functional experience (FE); therefore, H2b, H3b, and H4b are accepted. Intelligence does not show a significant relation with FE, hence H1b is rejected. The value of R^2 is 0.597 and is significant [42–45]. The emotional and functional experience also show a positive significant

Category	Cronbach's Alpha
Intelligence (INT)	.873
Interactivity (IN)	.835
Privacy (P)	.889
Social Interaction (SI)	.871
Functional Experience (FE)	.833
Emotional Experience (EE)	.850
Purchase Intention (PI)	.843

Figure 5.2 Reliability coefficients.

Factors	Beta	t-value	Significance	Hypothesis	Hypothesis Accepted/ Rejected
Intelligence	.130	2.156	.032	H1a	Accepted
	.094	1.334	.183	H1b	Rejected
Interactivity	.206	3.242	.001	H2a	Accepted
	.287	3.989	.000	H2b	Accepted
Privacy	.158	1.998	.047	H3a	Accepted
	.301	3.455	.001	H3b	Accepted
Social Interaction	.413	6.255	.000	H4a	Accepted
	.235	3.071	.002	H4b	Accepted
Emotional Experience	.595	9.098	.000	H5	Accepted
Functional Experience	.206	3.148	.002	H6	Accepted

Figure 5.3 Hypotheses summary.

relationship with purchase intention, therefore, H5 and H6 are accepted. The value of R^2 is 0.593 and is significant. The summary of the hypotheses is shown in Figure 5.3.

5.4 CONCLUSION

This study has tried to analyze the influence of various attributes of the IoT products on the functional and emotional experiences and consumers' purchase intention, especially youth. Our model showed that most of the constructs have significant association with independent constructs.

The study shows that intelligence factor is not significantly associated with the functional experience of the consumer to purchase IoT products. It means consumers may be intelligent but they are not bothered about the functionality of the products like security and privacy [46]. Indian consumers purchase the products based on reviews or if they heard from someone about the product. Other constructs like interactivity, intelligence, privacy, and social interaction have a significant effect on the emotional experience. It means consumers are connected with their devices emotionally.

The constructs like interactivity, intelligence, privacy, and social interaction have a significant effect on the functional experience. Functional and emotional experience have significant effects on purchase intention of IoT products. Our study shows that emotional experience has a greater influence on purchase intention in comparison to functional experience. The study proves that a youth is emotionally connected to their devices and they don't want to stay away from these technologies.

This study can be very useful for the market players who are manufacturing IoT products. They can make their strategies better so that consumers can purchase their products/services. It is very important for the players to understand what factors are important to handle the emotional and functional experience of a consumer. The industries are growing and the usage of IoT devices are also increasing day by day. In this scenario, industries need to understand and consider various factors which can attract more and more consumers. IoT devices are the future of an economy for optimizing the process, resources, and sustainability.

5.5 LIMITATIONS AND FUTURE SCOPE

We have considered a few important constructs. Other constructs like price of the products, privacy and security, performance, and convenience can be considered for the future study. Majority of our respondents were from urban areas, future study can consider respondents from tier II or tier III cities. We have considered youth for this study, whereas in future the comparison can be done between different age groups. Different generations can also be considered for future studies. The IoT industry is growing rapidly and understanding this industry is very important for making informed decisions.

REFERENCES

1. Steegen, An. "Technology innovation in an IoT Era." In *2015 Symposium on VLSI Circuits (VLSI Circuits)*, IEEE (2015): C170–C171.
2. Atzori, Luigi, Antonio Iera, and Giacomo Morabito. "The internet of things: A survey." *Computer Networks* 54, no. 15 (2010): 2787–2805.
3. Paul, Aditi, Somnath Sinha, Parveen Kumar, Shirasthi Choudhary, Krishna Samdani, and Saumya Mishra. "Overview of Cyber Security in Intelligent and Sustainable Manufacturing." In *Handbook of Intelligent and Sustainable Manufacturing*, pp. 95–118. CRC Press, 2025.

4. Krishnamurthy, Bhargavi, and Parveen Kumar. "Nature-Inspired Optimization Techniques of Industry 4.0 for Sustainable Manufacturing." In *Handbook of Intelligent and Sustainable Manufacturing*, pp. 42–50. CRC Press, 2025.
5. Bhatia, Vineet, Ajay Kumar, Sumati Sidharth, Sanjeev Kumar Khare, Surendra Chandrakant Ghorpade, Parveen Kumar, and Gaydaa AlZohbi. "Industry 4.0 in Aircraft Manufacturing: Innovative Use Cases and Patent Landscape." In *Industry 4.0 Driven Manufacturing Technologies*, pp. 103–137. Cham: Springer Nature Switzerland, 2024.
6. Lavanya, Raja, G. Bhavani, C. Santhiya, G. Vinoth Chakkaravarthy, and Parveen Kumar. "Digital Twins and IoT for Sustainable Manufacturing: A Survey of Current Practices and Future Directions." In *Handbook of Intelligent and Sustainable Manufacturing*, pp. 80–94. CRC Press, 2025.
7. YaPing Chang, XueBing Dong, and Wei Sun. "Influence of characteristics of the Internet of Things on consumer purchase intention." *Social Behavior and Personality: An International Journal* 42, no. 2 (2014): 321–330.
8. Kumar, Ajay, Parveen Kumar, and Yang Liu. "Industry 4.0 Driven Manufacturing Technologies." (2024). https://doi.org/10.1007/978-3-031-68271-1
9. Kumar, Ajay, Yang Liu, and Rakesh Kumar, eds. *Handbook of Intelligent and Sustainable Manufacturing: Tools, Principles, and Strategies.* CRC Press, 2024. https://doi.org/10.1201/9781003405870
10. Ashton, Kevin. "That 'internet of things' thing." *RFID Journal* 22, no. 7 (2009): 97–114.
11. Mahajan, N., N. Kaur, V. Singh, A. Gupta, and N. Garg. "Green Trade: Transition Towards a Sustainable Economy." In *IOP Conference Series: Earth and Environmental Science*, vol. 1279, no. 1, p. 012023. IOP Publishing, 2023.
12. McKinsey. "What is the Internet of Things", May 26, 2024.
13. Bąk, S., R. Czarnecki, and S. Deniziak. "Synthesis of real-time cloud applications for Internet of Things." *Turkish Journal of Electrical Engineering and Computer Sciences* 23 (2015): 913–929.
14. Thakur, Ayush, Sanskar Chauhan, Astha Gupta, Amrish Kumar Choubey, and Chitra Krishnan. "Enhancing Breast Cancer Detection via Optimized Machine Learning." In *2024 4th International Conference on Innovative Practices in Technology and Management (ICIPTM)*, IEEE (2024): 1–5.
15. Vermesan O. and Friess P. *Internet of Things–From Research and Innovation to Market Deployment.* River Publisher Series in Communication, 2014.
16. Chen, Zhikui, Feng Xia, Tao Huang, Fanyu Bu, and Haozhe Wang. "A localization method for the internet of things." *The Journal of Supercomputing* 63 (2013): 657–674.
17. Maurya, Alka, and J. Mark Munoz. 2023. "Introduction: Disruptive technologies and their implications for international business." In *Disruptive Technologies in International Business: Challenges and Opportunities for Emerging Markets*, by J. Mark Munoz Alka Maurya, 1–4. Walter de Gruyter.
18. Chalfen, Richard. "'Your panopticon or mine?' Incorporating wearable technology's Glass and GoPro into visual social science." *Visual Studies* 29, no. 3 (2014): 299–310.

19. Bian, Jiang, Kenji Yoshigoe, Amanda Hicks, Jiawei Yuan, Zhe He, Mengjun Xie, Yi Guo, Mattia Prosperi, Ramzi Salloum, and François Modave. "Mining Twitter to assess the public perception of the "Internet of Things"." *PloS One* 11, no. 7 (2016): e0158450.
20. Kamel Boulos, Maged N., and Najeeb M. Al-Shorbaji. "On the Internet of things, smart cities and the WHO Healthy Cities." *International Journal of Health Geographics* 13 (2014): 1–6
21. Eisenberg, Michael B., and Sean P. Fullerton. "ED and INFO 2052: Oh, the places you'll go!." *Information Services & Use* 32, no. 3–4 (2012): 103–115.
22. Hu, Jun, Bram van der Vlist, Gerrit Niezen, Willem Willemsen, Don Willems, and Loe Feijs. "Designing the internet of things for learning environmentally responsible behaviour." *Interactive Learning Environments* 21, no. 2 (2013): 211–226
23. Pinochet, Luis Hernan Contreras, Evandro Luiz Lopes, Caio Henrique Fernandes Srulzon, and Luciana Massaro Onusic. "The influence of the attributes of "Internet of Things" products on functional and emotional experiences of purchase intention." *Innovation & Management Review* 15, no. 3 (2018): 303–320.
24. Manches, Andrew, Pauline Duncan, Lydia Plowman, and Shari Sabeti. "Three questions about the internet of things and children." *TechTrends* 59 (2015): 76–83.
25. Christophe, Benoit, Mathieu Boussard, Monique Lu, Alain Pastor, and Vincent Toubiana. "The web of things vision: Things as a service and interaction patterns." *Bell Labs Technical Journal* 16, no. 1 (2011): 55–61.
26. Weber, Rolf H. "Internet of Things–New security and privacy challenges." *Computer Law & Security Review* 26, no. 1 (2010): 23–30.
27. Chang, Ya Ping, Xuebing Dong and Wei Sun. "Influence of characteristics of the internet of things on consumer purchase intention." *Social Behavior and Personality* 42 (2014): 321–330.
28. Wu, Hui, Haiting Han, Xiao Wang, and Shengli Sun. "Research on artificial intelligence enhancing internet of things security: A survey." *IEEE Access* 8 (2020): 153826–153848.
29. Emami-Naeini, Pardis, Henry Dixon, Yuvraj Agarwal, and Lorrie Faith Cranor. "Exploring how privacy and security factor into IoT device purchase behavior." In *Proceedings of the 2019 CHI Conference on Human Factors in Computing Systems*, (2019): 1–12.
30. Chang, Che-Chang. "What Otaku consumers care about: The factors influential to online purchase intention." In *11th International Conference of Numerical Analysis and Applied Mathematics 2013: ICNAAM 2013*, vol. 1558, no. 1, (2013): 450–454.
31. Ho-Sam-Sooi, Nick, Wolter Pieters, and Maarten Kroesen. "Investigating the effect of security and privacy on IoT device purchase behaviour." *Computers & Security* 102 (2021): 102–132.
32. Behr, Michael, and Martina Becker. "Scales for experiencing emotions: Awareness, appraisal and regulation of one's own emotions." *Hellenic Journal of Psychology* (2012).
33. Sicari, Sabrina, Alessandra Rizzardi, Luigi Alfredo Grieco, and Alberto Coen-Porisini. "Security, privacy and trust in Internet of Things: The road ahead." *Computer Networks* 76 (2015): 146–164.

34. Said, O., and M. Masud. "Towards internet of things: Survey and future vision." *International Journal of Computer Networks* 5, no. 1 (2013): 1–17.
35. Lieberman, Alicea, and Juliana Schroeder. "Two social lives: How differences between online and offline interaction influence social outcomes." *Current Opinion in Psychology* 31 (2020): 16–21.
36. Uckelmann, Dieter, Mark Harrison, and Florian Michahelles. "An architectural approach towards the future internet of things." In *Architecting the internet of things*, Berlin, Heidelberg: Springer Berlin Heidelberg, (2011): 1–24.
37. Scuotto, Veronica, Alberto Ferraris, and Stefano Bresciani. "Internet of Things: Applications and challenges in smart cities: a case study of IBM smart city projects." *Business Process Management Journal* 22, no. 2 (2016): 357–367.
38. Pinochet, Luis Hernan Contreras, Evandro Luiz Lopes, Caio Henrique Fernandes Srulzon, and Luciana Massaro Onusic. "The influence of the attributes of "Internet of Things" products on functional and emotional experiences of purchase intention." *Innovation & Management Review* 15, no. 3 (2018): 303–320.
39. Chandran, Sucharita, and Vicki G. Morwitz. "Effects of participative pricing on consumers' cognitions and actions: A goal theoretic perspective." *Journal of Consumer Research* 32, no. 2 (2005): 249–259.
40. Caputo, Francesco, Veronica Scuotto, Elias Carayannis, and Valentina Cillo. "Intertwining the internet of things and consumers' behaviour science: Future promises for businesses." *Technological Forecasting and Social Change* 136 (2018): 277–284.
41. Hair, Joseph F., William C. Black, Barry J. Babin, Rolp E. Anderson, and Ronald L. Tatham. *Multivariate data analysis*. 6th Edition. 2006. Springer Science & Business Media.
42. Kamble, P. D., J. Giri, E. Makki, N. Sunheriya, S. B. Sahare, R. Chadge, C. Mahatme, P. Giri, and H. Panchal. (2024). An application of hybrid Taguchi-ANN to predict tool wear for turning EN24 material. *AIP Advances*, 14(1). https://doi.org/10.1063/5.0186432
43. Narasimhamu, K. L., M. Natarajan, P. Thejasree, E. Makki, J. Giri, N. Sunheriya, R. Chadge, C. Mahatme, P. Giri, and T. Sathish. "Development of hybrid optimization model using Grey-ANFIS-Jaya algorithm for CNC drilling of Aluminium Alloy." *Journal of Engineering*, 2024 (2024), 1–12. https://doi.org/10.1155/2024/1476770
44. Natarajan, M., Pasupuleti, T., Abdullah, M. M. S., Mohammad, F., Giri, J., Chadge, R., Sunheriya, N., Mahatme, C., Giri, P., & Soleiman, A. A. (2023). Assessment of Machining of Hastelloy Using WEDM by a Multi-Objective Approach. Sustainability, 15(13), 10105. https://doi.org/10.3390/su151310105
45. Natarajan, M., T. Pasupuleti, J. Giri, N. Sunheriya, L. N. Katta, R. Chadge, C. Mahatme, P. Giri, S. Mallik, and K. Ray. "Machinability of titanium grade 5 alloy for wire electrical discharge machining using a hybrid learning algorithm." *Information*, 14, no. 8 (2023): 439. https://doi.org/10.3390/info14080439
46. Gautam, Shalini, and Priyanka Malik. "Importance of perceived security, perceived privacy and website design of active online investors: an Indian market perspective." *International Journal of Electronic Finance* 11, no. 1 (2022): 30–48.

Chapter 6

Artificial intelligence and its futuristic approach in industries

Pranjal Rawat, Suruchi Sharma, Rajesh Tiwari, and Smitha Kothari

6.1 INTRODUCTION

Change is the only constant, and in this context, continual innovation and development form the basic pillars for the successful expansion of any business organization Aggarwal and Goyal (2022). Most firms have been motivated to incorporate the latest techniques and processes to remain relevant in the tough market competition. Frank et al. (2019) proposed that invention is the backbone and lifeline of any organization. Besides, flexibility is another key component in the rapidly changing business spectrum, and disruptive changes are being introduced in all core functional areas. These help the companies navigate toward strengthening their core competencies and employability skills. Artificial intelligence (AI) has the potential to remodel any business, spanning from manufacturing process to last-mile customer service. Once integrated into the organizational processes, it is easy to implement. Finding its usage in multiple industries (Kumar et al., 2023), AI has shown tremendous growth over the last few years. AI has immense benefits to offer for the small business segment as well (Yang et al., 2019). Artificial intelligence is a display of human intelligence by machines. To put it another way, the dimensions of human intelligence like reasoning, learning, and self-correction are demonstrated by AI. Since its inception, AI has shown exponential growth in multiple domains over the period, but in spite of this, there are pertinent issues related to its optimum usage and future direction. Artificial intelligence works on the concept of machine reactions toward certain activities, lemmatization of memory, theory related to human cognitive response, and their interrelationships. Psychologists have often considered human intelligence as a basket of multiple characteristics like knowledge, logical reasoning, troubleshooting, insight or sensitivity, and the ability to use language. This is the reason why research studies related to artificial intelligence have focused on these building blocks. The disruptive domain of AI is still in the early evolutionary phase, where there has been not much consensus on its definition. AI has been considered as a term, encompassing expanding business through high-tech and innovative tools and techniques, including the approach of transformation, conversion,

 DOI: 10.1201/9781003512585-7

smartness, futuristic approach, and long-term sustainability (Frank, 2019). With the passage of time, AI concept has been revolutionized and it is now considered to be one of the key factors that promotes and fosters economic development (Kumar et al., 2023). Artificial intelligence is suddenly taking up the important role of one-stop business solution dealing with image recognition, document classification, and resolving interdisciplinary problems. The indispensability of AI routes to economic and social prosperity has been well acknowledged by most countries, which is evident through their huge investment in related research. The IT sector has been growing by leaps and bounds even in the 21st century. The pandemic phase has only accelerated and expanded it.

6.2 LITERATURE REVIEW

The domain of artificial intelligence has been the focal point of recent research. The plethora of research works by eminent experts on a global scale has enriched the repository of AI and related uses. Artificial intelligence has deeply impacted the organic and inorganic aspects of doing business (Kumar et al., 2024b). It affects not only the profitability aspects but also the working culture of the business firm Singh et al. (2022). Artificial intelligence and its extent of use are dependent on factors like need, intellectual ability, affordability of the firm, manpower availability, and technical optimality (Singh et al., 2022). Mirlean et al., 2021, discussed the effects of AI and its related uses on the financial knowledge, attitude, behavior, and the changes in the life of students. He conducted a detailed analysis on the multiple dimensions of AI as a key for any business organization in today's world. Innovation and adoption of high technological interface are mandated for gaining competitive edge among all the organizations in the world. A research study on 225 women entrepreneurs in Malaysia, focusing on the underlying motivations for choosing this field, also threw light on the relevance of technical aspects and expertise for success. Business organizations are keenly interested in gaining financial edge through optimum use of technological interfaces (Perez et al., 2022)). Another study showed the interrelationship between culture and innovation and their future coexistence. The surge in government spending on artificial intelligence and related dimensions, thereby suggesting that the regulatory bodies are taking due cognizance of the importance and critical nature of the investment. This study was done after analyzing the AI strategic plans of many governments across the globe and highlights the need for a planned investment on the part of the governments, along with providing solution for the failure of such plans. Dhaliwal et al. (2024) focused on AI and its impact on customer loyalty as the key to survival. Pilipczuk (2024) provided certain numerical measures to understand the financial aspect of AI investment.

The impact of AI in the medical field and discussed the current trends as well as the future possibilities in this field. Passetti et al. (2023) shared their research on fuzzy expert systems that help in the prediction and detection of fraudulent cases in the healthcare insurance industry, besides focusing on its futuristic approach. Mirlean et al., (2021) in his research proposed the benefits of boosting innovation by artificial intelligence and how it can benefit the society. Benítez-Buenache et al. (2020) explains the importance of AI and the ways in which it can be beneficial in NLP, biomedical, and malware classification.

6.3 ARTIFICIAL INTELLIGENCE AND DEVELOPED ECONOMIES

The developed nations have been able to make use of the disruptive technology of artificial intelligence quite extensively and in a wide array of sectors (Acemoglu and Restrepo, 2020).

The healthcare industry is a critical area with respect to human health and survival (Kumar et al., 2024a). The extent of services aided by AI-based methods include the detection of diseases and drugs, analyzing the medical imaging results, charting the customized medical care for the patients, and performing the routine (Kumar et al., 2024c) and repetitive work related to general administration. This helps in enhanced blanket care for patients, and adequate treatment planning and scheduling on the basis of precise diagnosis. It also helps in preventive medication aided by regular monitoring of the patients.

In the financial sector, the stakes are very high just like the healthcare industry. The novel technology is essentially helpful in troubleshooting fraudulent activities in financial transactions, assessing the risk related to security markets, assisting in algorithmic trading, providing customer service and grievance redressal through the use of chatbots, and assisting in preparation of a customized financial plan for the investors based on the risk assessment and options available. The financial data, which is a huge volume, can be easily analyzed and used for making intelligent and real-time decisions by the stakeholders (Jayaprakash et al., 2024; Sathish et al., 2024a, 2024b; Pooja et al., 2023).

The manufacturing sector is rapidly transforming with AI assistance in precision manufacturing and enhancing the performance of the product to the highest level. The processes are metamorphosed through the use of automation, and human error has been brought to negligible levels. Tasks like predicting the maintenance of machines, controlling the quality parameters, and optimizing the logistics and supply chain along with inventory management have led to an enormous reduction in the breakdown, maintenance, and overall cost overheads, simultaneously improving the quality of products to unprecedented quality standards.

Another area that has been greatly transformed by the use of AI is the transportation sector. The driverless vehicles, selection of optimum route, and managing the traffic system flawlessly are some of the benefits in this sector (Brynjolfsson and McAfee, 2014). AI has extended to developed countries. Safety of people, reduced traffic congestion, and optimal use of time and fuel are some of the many benefits that AI has made possible in this sector. The tourism and hospitality industry is aligned with this industry and the benefits percolate down to the end user with great ease with the help of AI-based services.

AI and related uses have also greatly advanced the performance of other sectors like retail merchandising, education, and government services. The retail sector has seen a huge boom in the last few decades. The AI revolution has immensely changed it for the better, increasing the proximity of the retail zone and the potential customers. Consumer behavior has been completely revolutionized with the use of AI platforms in the retail sector. Personalized selling and product or service recommendation using AI-based applications have transformed consumer experiences. The retailers are now better positioned to showcase their offerings to a sensitized and aware consumer. The overall shopping experience has been tremendously enhanced by the use of AI. The tutoring facilities powered by the AI platforms, coupled with the modern-day learning management systems, and customized content creation tools, are in the front line of the education sector. The masses are now ready for the state-of-the-art content delivery and educational experiences in developed nations. The government sector has been greatly relieved from the performance pressure in addressing public grievances timely and effectively. The AI and its relevant platforms come to the rescue of government agencies (Frank et al., 2019), thereby increasing the overall efficiency of the governance function. The areas of public monitoring and safety, crime prevention, detection and troubleshooting, management of scarce resources, forecasting the calamities and weather vagaries for the benefit of the general population at large, monitoring the ecological balance and pollution control, and other routine management tasks are flawlessly performed by using AI-led interfaces in public administration (Shiller, 2019). In summation, AI has resulted in streamlining administrative procedures, effective decision-making, and improving citizen service delivery.

The use of AI has ushered in a new dawn in the developed nations, which are at the forefront of AI adoption and usage. This has allowed the developed world to leverage AI in enhancing innovation, efficiency, overcoming challenges, and solving problems in multiple domains.

Countries like the United States and the European Union are leading the world in institutionalizing the regulation related to AI and its use. The recently enacted AI Act in the European Union and the Executive Order released by the Highest Office of US President on AI have signaled transformative work toward making AI development safe and ethical for all the stakeholders. According to a government statistic, the AI-related regulations in the United States increased by 56.3%, indicating the proactive approach of the country to leverage the AI potential with responsibility. Developed countries are making great strides by investing magnanimously in AI research and development. The United Kingdom has become a major player in this area, with predominant investments in this field and setting up of a flourishing ecosystem of AI-based start-ups. Besides the UK, Canada has also progressed in this field by leaps and bounds in AI-related investments and responsible adoption. In the developed nations, Europe, the UK, and Germany are ahead in providing relevant degree programs in AI and related spectrum to the young populace. This has ensured that the skilled population is ready to take up the AI storm easily. Countries like France have adopted a strategy to leverage ethical AI by design. On the other hand, Japan is investing heavily in establishing infrastructure and start-ups for AI.

6.4 ARTIFICIAL INTELLIGENCE AND DEVELOPING ECONOMIES

The developing economies have shown great promise in the advancement of AI and its related usage, but are still in the growing and evolving phase, although the use has been spread across multiple sectors of the economy. Generally, the developing economies are characterized by the social disparities in terms of income and status. The resources are also skewed in favor of the powerful segments of the population. (Ahmed and Waheed, 2020). The administrative lacuna coupled with the unfavorable geography has rendered many untouched by the regular governance systems. Therefore, in the case of developing economies, the access to optimum services among all in a fair and equitable manner is a big challenge. In this context, artificial intelligence has come to the rescue in an unprecedented way.

In the healthcare sector, AI has aided many through the adequate use of facilities like telemedicine and diagnostics, through which the underprivileged and masses can avail best medical aid without incurring much cost. These AI-enabled tools and techniques form a serviceable method of reaching the last-mile citizen and developing adequate treatment plans (Kumar and Gupta, 2019), including preventive medication, and prevention of epidemics and other dreaded diseases. The database related to the health profiles can be easily managed using AI tools.

The developing economies are also characterized by a prominent share of primary or agriculture sectors in the GDP. Therefore, AI also comes to the rescue of agricultural problems related to yields, irrigation, etc. Methods like precision farming, forecasting of weather and natural disasters, smart irrigation systems, continuous monitoring of the soil health, and remote guidance by experts for different agriculture-related problems have been made possible by AI. The resource crunch can be adequately addressed by the AI-powered processes.

The mass education system has been greatly improved by the use of AI and other smart technological tools. Remote support by the best teachers in the sector, domestic or international, can be easily accessed by the population in developing economies. Besides, personalized learning, based on the background of the beneficiary, can be availed using this tool.

As far as the financial sector is concerned, access to financial services by one and all without any discrimination (financial inclusion) can be made possible only by the use of AI (Chen and Zhang, 2021). This technology is capable of providing real-time data pertaining to banking services, credit records, and fraudulent activities, and bringing unbanked masses within the ambit of the modern banking system. Businesses are also benefited by AI-based services for better customer management, financial remedies related to business, and other umbrella benefits.

The overall administration of developing economies can be improved by adequate infrastructure development and bridging the gap between rural and urban areas (Perez and Torres, 2022). The "smart cities" concept encompasses improved infrastructure development, optimum traffic movement and management, ensuring public safety through monitoring and real-time service, forecasting natural disasters and timely preventive interventions.

Last but not least, the governance services related to developing nations, especially e-Governance and related policy development mandate intelligent and ethical use of AI tools. AI can improve public administration and governance through transparent systems of revenue collection and fund disbursement to the last beneficiary, taking informed decisions based on proper management of citizen database.

Illustrating the cases of developing economies like India, the national AI strategy formulated lays strict emphasis on citizen healthcare, agriculture, education, and infrastructure development through smart city projects. Kenya, on the other hand, has used AI in various sectors like finance through the use of mobile banking, diagnosis of diseases using AI applications, and improving agriculture produced by AI-based farm management systems. Brazil has been using the smart agriculture system to optimize crop production and resource utilization, promoting public safety and improving overall civil administration.

The developing economies have been assisted by the developed ones through international collaborations for sharing knowledge and real-time information bytes for better management of scarce resources.

6.5 CHALLENGES IN THE USE OF ARTIFICIAL INTELLIGENCE

The major concerns regarding the proliferated use of AI and its enabling technologies are pertaining to the data privacy and security lapses in data

Table 6.1 Comparative analysis between developed and developing nations

Point of comparison	*Developed nations*	*Developing nations*
Innovation	Very hi-tech and innovative in approach	Developing nations like India are now adapting new and innovative e-techniques
Risk factor	Are always ready to take risk	Are not frequent risk-takers as developed nations
Labor efficiency	Labors are more efficient and effective in terms of skills	Labors lack skilled efficiency
Adaptation	Can adapt new and innovative techniques very rapidly	Take time in implementation and adaptation of new techniques
Smartness	Workers are smarter than those from developing nations	Workers are not as smart as those from developed nations
Technical skills	Workers are having more technical knowledge and skills in comparison to those from developing nations	Workers are having less technical knowledge in comparison to those from developed nations
Competitiveness	Are always ready to face any sort of competition	Are also ready, but take time to face severe competition
Entrepreneurship	Entrepreneurs from nations like the United States are using AI in a most rapid and frequent way	Entrepreneurs from nations like India now have started using AI in their businesses
Dynamics	Are most adaptive and dynamic in nature	Are trying to be more adaptive and dynamic in nature in order to manage their business
Finance modeling	Are using the finance modeling technique to manage AI	Are now trying to be more focused in their finance modeling so that they can manage their business effectively
Marketing and promotions	America, Russia, and most of the developed nations are using AI in high capacity to manage their marketing and promotions	If we talk about developing nations, they have now started taking risks to promote their business through AI in their firms or organizations

Source: Author's self-observation.

management services. The infrastructural limitations related to technology and its widespread reach among the masses stand tall. The differentiated social strata in terms of the use of tech-enabled services and their working knowledge play a barrier in this case. There is a huge skill gap that needs to be addressed by the governance system, and the shortage of such manpower prohibits the optimum use of AI technology among the masses. The manpower should be highly skilled to leverage the benefits of AI and its related services to develop and manage the AI infrastructure. Lastly, ethical considerations need to be adequately considered for keeping the future prospects of the technology growing. Unethical practices need to be curbed through proper legislation and protocols, with sincere monitoring and redressal mechanisms. In the case of the companies that are en route to an extensive use of AI have to incur the heavy cost of laying down the AI infrastructure and its management systems. This may lead to a sharp increase in the cost of production, resulting in price hike at the consumer's end. There are chances of increased unemployment in the economy, as AI systems replace human hands and minds, simultaneously providing higher efficiency and time management (Table 6.1).

6.6 CONCLUSION AND FUTURE SCOPE

The study concludes that in the context of the adoption and efficient usage of AI tools and techniques, there is a need for continual innovation and technological upgrades among all the countries, especially the developing ones. This is the only key to growth and survival for the countries and the business organizations. The countries will have to think in terms of human-machine synergy as the rate of unemployment may rise steeply due to AI adoption. So, it totally depends on the firm to implement the techniques leading to growth as well as employability. Though the idea of creating AI is aimed at making human life easier, there is still much debate about the overall benefits and drawbacks. Artificial intelligence has a wider future in terms of quality, innovation, time management, advancement, increasing productivity, sustainability, and most importantly, adaptation, which will be the key to the progress of any nation. However, it will also depend on how we use it.

REFERENCES

Acemoglu, D., and P. Restrepo. (2020). "Artificial Intelligence, Automation, and Work." In *The Economics of Artificial Intelligence: An Agenda*, edited by A. Agrawal, J. Gans, and A. Goldfarb, 197–236. University of Chicago Press.

Aggarwal, Kriti, and Gulshan Goyal. (2022). "IoT-Based Innovative Technological Solutions for Smart Cities and Villages." In *Proceedings of International Conference on Innovative Technologies for Clean and Sustainable Development (ICITCSD – 2021)*, 21–32. Cham: Springer International Publishing.

Ahmed, S., and A. Waheed. (2020). "The Impact of Artificial Intelligence on Economic Growth in Developing Countries." *Journal of Economic Perspectives 34*(2): 45–62.

Benítez-Buenache, Alexander, Lorena Álvarez-Pérez V. John Mathews, and Aníbal R. Figueiras-Vidal. (2020). "Corrigendum to 'Likelihood Ratio Equivalence and Imbalanced Binary Classification' [Expert Systems with Applications, Volume 130 (2019), Pages 84–96]." *Expert Systems with Applications 146*: 113299. https://doi.org/10.1016/j.eswa.2020.113299.

Brynjolfsson, E., and A. MacAfee. (2014). *The Second Machine Age: Work, Progress, and Prosperity in a Time of Brilliant Technologies*. W.W. Norton & Company.

Chen, L., and Y. Zhang. (2021). "AI and the Digital Divide: The Case of Developing Countries." *Technology in Society 67*.

Dhaliwal, Salwinder Singh, Vivek Sharma, Arvind Kumar Shukla, Jadgish Singh, Vibha Verma, Manmeet Kaur, Prabhjot Singh, and Jagbir Rehal. (2024). "Corrigendum to 'Assessment of Optimum Mineral Zinc Fertilizer Rate for Quantitative and Qualitative Production of Sugarcane in North-Western India'. [Journal of Trace Elements and Minerals 2C (2022) 100021]." *Journal of Trace Elements and Minerals 8*: 100143. https://doi.org/10.1016/j.jtemin.2024.100143.

Frank, Morgan R., David Autor, James E. Bessen, Erik Brynjolfsson, Manuel Cebrian, David J. Deming, Maryann Feldman, et al. (2019). "Toward Understanding the Impact of Artificial Intelligence on Labor." *Proceedings of the National Academy of Sciences of the United States of America 116*(14): 6531–6539. https://doi.org/10.1073/pnas.1900949116.

Jayaprakash, R., Natarajan, K., Daniel, J. A., Chinnappan, C. V., Giri, J., Qin, H., & Mallik, S. (2024). Heuristic machine learning approaches for identifying phishing threats across web and email platforms. *Frontiers in Artificial Intelligence* 7: 1414122. https://doi.org/10.3389/frai.2024.1414122

Kumar, A., D. K. Rajak, P. Kumar, and A. Kumar. (Eds.). (2024a). *Advances in Sustainable Biomaterials: Bioprocessing 4.0, Characterizations, and Applications* (1st ed.). CRC Press. https://doi.org/10.1201/9781003434313.

Kumar, Ajay, Parveen Kumar, and Yang Liu. (2024b). *Industry 4.0 Driven Manufacturing Technologies*. Springer. https://doi.org/10.1007/978-3-031-68271-1

Kumar, Ajay, Sangeeta Rani, Sarita Rathee, and Surbhi Bhatia, (Eds). (2023). *Security and Risk Analysis for Intelligent Cloud Computing: Methods, Applications, and Preventions*. CRC Press.

Kumar, Ajay, Virendra Kumar Shrivastava, Parveen Kumar, Ashwini Kumar, and Vishal Gulati. (2024c). "Predictive and experimental analysis of forces in die-less forming using artificial intelligence techniques." *Proceedings of the Institution of Mechanical Engineers, Part E: Journal of Process Mechanical Engineering* 09544089241235473.

Kumar, R., and S. Gupta. (2019). "Artificial Intelligence in Emerging Economies: Opportunities, Challenges, and Policy Implications." *International Journal of Innovation and Technology Management 16*(5).

Mirlean, Nicolai, Lauro J. Calliari, and Karen Johannesson. (2021). "Reply to MPB-D-20-01629. Carlos Alberto Eiras Garcia Heitor Evangelista Osmar Olinto Möller Jr.Comments on 'Dredging in an Estuary Causes Contamination by Fluid Mud on a Tourist Oceanbeach. Evidence via REE Ratios' by N. Mirlean, L. Calliari, and K. Johannesson in Marine Pollution Bulletin 159

(2020) 111495. https://doi.org/10.1016/j.marpolbul.2020.111495." *Marine Pollution Bulletin 165*: 112161. https://doi.org/10.1016/j.marpolbul.2021.112161.

Passetti, Lora, Jennifer Smith Ramey, Brooke Hunter, and Mark Godley. (2023). "Predicting Response to Services for Homeless Adolescents and Transition Age Youth (TAY) with Substance Use and/or Mental Health Disorders: Implications for Youth Treatment and Recovery." *Journal of Evidence-Based Social Work 20*(5): 668–685. https://doi.org/10.1080/26408066.2023.2202663.

Perez, M., and A. Torres. (2022). "Artificial Intelligence and Labor Markets in Developing Economies: A Systematic Review." *Journal of Development Economics 149*.

Pilipczuk, Olga. (2024). "Overview of Edge Computing Applications in Energy Sector." *Przegląd Elektrotechniczny 1*(5): 242–245. https://doi.org/10.15199/48.2024.05.45.

Pooja, M. R., Giri, J., Chadge, R., Sunheriya, N., Mahatme, C., Mallik, S., Alsubaie, N., Alqahtani, M. S., Abbas, M., & Soufiene, B. O. (2023). An Optimized Machine Learning Model for the Early Prognosis of Chronic Respiratory Diseases with specific focus On Asthma. *Research Square (Research Square)*. https://doi.org/10.21203/rs.3.rs-3075448/v1

Sathish, T., S. Lalu Prasad, S. Patil, A. A. Ibrahim, S. Khan, R. Saravanan, and J. Giri. (2024a). Testing the auto-regressive integrated moving average approach vs the support vector machines-based model for materials forecasting to reduce inventory. *AIP Advances 14*(5). https://doi.org/10.1063/5.0208049

Sathish, T., D. Sai Kumar, S. Patil, R. Saravanan, J. Giri, and A. A. Aly. (2024b). Exponential smoothing method against the gradient boosting machine learning algorithm-based model for materials forecasting to minimize inventory. *AIP Advances 14*(6). https://doi.org/10.1063/5.0208491

Shiller, R. J. (2019). *Narrative Economics: How Stories Go Viral and Drive Major Economic Events*. Princeton University Press.

Singh, Simarjeet, Manika Sharma, and Sandeepa Kaur. (2022). "Eswar S. Prasad (2021): The Future of Money: How the Digital Revolution Is Transforming Currencies and Finance: Belknap Press of Harvard University Press, Cambridge Massachusetts, LCCN 2021008025." *Journal of Evolutionary Economics 32*(4): 1391–1394. https://doi.org/10.1007/s00191-022-00774-7.

Yang, Tuo, Yang Sun, Leilei Mao, Meijuan Zhang, Qianqian Li, Lili Zhang, Yejie Shi, Rehana K. Leak, Jun Chen, and Feng Zhang. (2019). "Corrigendum to 'Brain Ischemic Preconditioning Protects against Ischemic Injury and Preserves the Blood-Brain Barrier via Oxidative Signaling and Nrf2 Activation' [Redox Biol. 17 (2018) 323–337]." *Redox Biology 22*: 101067. https://doi.org/10.1016/j.redox.2018.101067.

Chapter 7

Analysis of a resource-based population model with Caputo fractional derivative

Swati Domaji Kharabe, Beena Bundela, Manish Kumar Bansal, and R. P. Chauhan

7.1 INTRODUCTION

A species that has been introduced into a habitat is referred to as an exotic species; these species are also known as introduced, nonnative, nonindigenous, or alien species. One of the key factors contributing to a general decrease in the allocation of native species is the replacement of native species by insidious species. There are several causes involving direct and indirect interspecific competition, habitat alteration, and the spread of fresh species. Native species frequently experience a reduction in the presence of imported species [1]. The issue of invasive, nonnative species has spread throughout the planet in large numbers, because of humans unwittingly introducing new species. The ensuing dynamics must be taken into account while creating effective plans for the eradication or control of invasive species, see also Ref. [2]. The Allee effect can lead to a significant increase in the complexity of the spatiotemporal dynamics of invasive species and the manifestation of invasion regimes [3]. Many autonomous systems have been employed specifically to ascertain the stable and periodic dynamic within the predator-prey system [4]. This finding has attracted a lot of attention and created a significant problem for ecologists and researchers. The Lotka–Volterra model served as the foundation for studies on predator-prey interactions in 1927. More problems developed and extensive studies were carried out, such as those found in Refs [5, 6], which separated susceptible and diseased populations of the prey (see also Refs [7–10]).

Arbitrary-order calculus (AOC) is a powerful tool for mathematical modeling of ecological phenomena. It demonstrates superior efficiency, practicality, and authenticity compared to traditional integer-order frameworks in representing memories and phenomena, as evidenced by Refs [9–16]. In this study, a fractional-order model is under examination due to the prevalence of memory effects, such as prey populations retaining memory and gaining insights from their encounters with escape and defense against potential threats, which are frequently observed in predator-prey interactions.

Several authors have proposed a large number of mathematical models such as Riemann-Liouville (RL), Caputo, Hadamard, Hilfer, and many

 DOI: 10.1201/9781003512585-8

more [17] studies. Most scientific fields are currently paying close attention to the use of fractional calculus. This fractional calculus (FC) on dynamical systems was important and fascinating, as numerous studies [18–24, 28] have just recently found. The presence of equilibrium solutions is unaffected by changes in the fractional derivative order, but stability is affected. Numerous conceptions of fractional-order derivatives have been put forth by renowned researchers. Among these, the meanings associated with Caputo are the most widely used.

The α -order CFD is mathematically defined as follows:

$$ {}^{c}D_t^{\alpha} f(t) = \frac{1}{``(n-\alpha)} \int_c^t (t-s)^{n-\alpha-1} f^n(s)ds, \qquad n-1<\alpha\le n \tag{7.1} $$

and α lies in the range $f:[c,\infty)\to\mathbb{R}$, n - 1< $\alpha \le n$ for $n \in n \in \mathbb{N}$ $f^n(s)=\left(\left(d^n f(s)\right)/ds^n\right)$. The time constant at which the system's state is comprehended and memory effects appear is given by c, where $\Gamma(.)$ is the gamma function. The order derivative's value provides how significant the information is to the model's long-lasting (continuous) memory. Additionally, it is considered the case where c = 0 and the memory effects are considered to include all knowledge acquired since the beginning of time when n = 1 and (0,1], respectively. As a result, Equation (7.1) turns into a gamma function, and c represents the time constant at which memory effects and an understanding of the system's state are both present. Additionally, it considers the case where c = 0 and the memory effects are considered to include all knowledge acquired since a long time when n = 1 and (0,1], respectively. Therefore, Equation (7.1) becomes

$$ D_t^{\alpha} f(t) = \frac{1}{``(1-\alpha)} \int_0^t (t-s)^{-\alpha} f'(s)ds, \qquad 0<\alpha\le 1 \tag{7.2} $$

This chapter is organized as follows: Section 7.2 introduces the model using the CFD Equation (7.2), which is described in Ref. [25, 26]. Boundedness and balance points are covered in Section 7.3. Boundary and equilibrium points are illustrated in Section 7.4. Numerical simulations are provided in Section 7.5 to show the practical relevance of our theoretical findings. The fractional Adams–Bashforth–Moulton (ABM) under examination is numerically presented in Section 7.7. The conclusion and upcoming projects are finally mentioned in Section 7.8.

7.2 MATHEMATICAL MODEL

This model presents the effect of exotic species on resource-dependent system, and the predator in disease that was introduced in Ref. [26]. See also Ref. [27],

$$\left.\begin{aligned} \frac{dR}{dt} &= \left[r\left(1-\frac{R}{K}\right)-a_1N_1\right]R - a_2R_1^*N_2 \\ \frac{dN_1}{dt} &= \left[a_1\varepsilon_1 R - m_1 - b_1N_1\right]N_1 \\ \frac{dN_2}{dt} &= \left[a_2\varepsilon_2 R_1^* - m_2 - b_2N_2\right]N_2 \end{aligned}\right] \tag{7.3}$$

with nonnegative initial conditions

$R(t) = R_0 > 0, N_1(t) = N_{10} > 0, N_2(t) = N_{20} > 0$

where $r, k, a_1, a_2, m_1, m_2, b_1, b_2, \varepsilon_1, \varepsilon_2$ are positive constants.

Table 7.1 represents definitions of model parameters. Now reducing the mathematical model as represented in Equation (7.3) having CFD of order α with range varying between $0 < \alpha \leq 1$, Equation (7.2) becomes

$$\left.\begin{aligned} {}^{c}D_t^{\alpha} R &= \left[r\left(1-\frac{R}{K}\right)-a_1N_1\right]R - a_2R_1^*N_2 \\ {}^{c}D_t^{\alpha} N_1 &= \left[a_1\varepsilon_1 R - m_1 - b_1N_1\right]N_1 \\ {}^{c}D_t^{\alpha} N_2 &= \left[a_2\varepsilon_2 R_1^* - m_2 - b_2N_2\right]N_2 \end{aligned}\right] \tag{7.4}$$

7.3 ANALYSIS

7.3.1 Boundedness and equilibrium points

Equation (7.4) also finds same equilibrium points as the integer system given in Equation (7.3):

(i) The trivial equilibrium $E_0(0,0,0)$
(ii) Axial equilibrium $E_1(K,0,0)$
(iii) The absence of $E_2\left(R^{\#}, 0, N_2^{\#}\right)$, where

Table 7.1 Definitions of model parameters [27]

Parameters	*Definition*
R	Intrinsic rate of biological resource
K	Carrying capacity of resource
a_1, a_2	Per capita consumption rate
b_1, b_2	Intra-specific coefficient of competition
m_1, m_2	

$$R^{\#} = \frac{K}{2r}\left[r + \sqrt{\frac{b_2 r^2 - 4a_2 rK^2\left(a_2\varepsilon_2 K - m_2\right)}{b_2}}\right],\ N_2{}^{\#} = \frac{a_2 \in_1 K - m_2}{b_2},$$

E_3 exists if $b_2 r > 4a_2K^2\left(a_2 \in_2 K - m_2\right)$ and $a_2 \in_2 K > m_2$

(iv) Boundary equilibrium $E_3\left(R^*, N_1^*, N_2^*\right) = \left(R^*, N_1^*, 0\right)$, where

$$R^*{}' = \frac{K\left(a_1 m_1 + b_1 r\right)}{a_1^2 \in_1 K + b_1 r},\quad N_1^* = \frac{r\left(a_1 \in_1 K - m_1\right)}{a_1^2 \in_1 K + b_1 r}$$

The positive interior equilibrium E_4 exists if $a_1, \in_1 K > m_1$

(v) Boundary equilibrium $E_4\left(\widetilde{R_1^*}, \widetilde{N_1^*}, \widetilde{N_2^*}\right) = \left(\widetilde{R_1^*}, \widetilde{N_1^*}, \widetilde{N_2^*}\right)$, *where*

$$\widetilde{R_1^*}{}' = \frac{b_2 K\left(a_1 m_1 + b_1 r\right) + \sqrt{\begin{array}{l} b_2^2 K^2\left(a_1 m_1 + b_1 r\right)^2 - \\ 4ka_2 b_1 b_2 \tilde{R}\left(a_1^2 \varepsilon_1 K + b_1 r\right)\left(a_2\varepsilon_2\tilde{R} - m_2\right)\end{array}}}{2b_2\left(a_1^2 \in_1 K + b_1 r\right)}$$

$$\widetilde{N_1^*} = \frac{a_1 \in_1 \widetilde{R_1^*}{}' - m_1}{b_1},\ \widetilde{N_2^*} = \frac{a_2 \in_2 R_1^*{}' - m_2}{b_2}$$

The positive interior equilibrium E_5 exists if $a_1 \in_1 \widetilde{R_1^*}{}' > m_1$, $a_2 \in_2 R_1^* > m_2$, and $b_2K\left(a_1 m_1 + b_1 r\right) > 4a_2 b_1\left(a_2\varepsilon_2 \widetilde{R_1^*} - m_2\right)$ show that $R_1^* > \widetilde{R_1^*}$ where $R_1^* = \dfrac{K\left(a_1 m_1 + b_1 r\right)}{a_1^2 \in_1 K + b_1 r}$

7.3.2 Existence and uniqueness

For Lipschitz properties in $G(M)$,

$$\Phi = \{\left(R, N_1, N_2\right) \in \mathbb{R}^3 : \max\{|R|, |N_1|, |N_2| \le \mu\}$$

denote $\bar{M} = \left(\bar{R}, \overline{N_1}, \overline{N_2}\right)$; for any M, $\bar{M} \in \Phi$ it follows that

$$G(M) - G(\bar{M}) = \left|G_1(M) - G_1(\bar{M})\right| + \left|G_2(M) - G_2(\bar{M})\right| + \left|G_3(M) - G_3(\bar{M})\right|$$

$$\left|\left[r\left(1-\frac{R}{K}\right)-a_1N_1\right]R-a_2R_1^*N_2-\left[r\left(1-\frac{\bar{R}}{K}\right)-a_1\overline{N_1}\right]\bar{R}+a_2R_1^*\overline{N_2}\right| =$$

$$\left|\left[a_1\varepsilon_1R-m_1-b_1N_1\right]N_1-\left[a_1\varepsilon_1\bar{R}-m_1-b_1\overline{N_1}\right]\overline{N_1}\right| +$$

$$\left|\left[a_2\varepsilon_2R_1^*-m_2-b_2N_2\right]N_2-\left[a_2\varepsilon_2R_1^*-m_2-b_2\overline{N_2}\right]\overline{N_2}\right| +$$

$$\left|r\left(R-\bar{R}\right)-\frac{R}{K}\left[\left(R-\bar{R}\right)\left(R+\bar{R}\right)\right]-a_1\left(N_1R-\overline{N_1R}\right)-a_2R_1^*\left(N_2-\overline{N_2}\right)\right| =$$

$$\left|a_1\varepsilon_1\left(RN_1-\overline{RN_1}\right)-m_1\left(N_1-\overline{N_1}\right)-b_1\left(N_1+\overline{N_1}\right)\left(N_1-\overline{N_1}\right)\right| +$$

$$\left|a_2\varepsilon_2R_1^*\left(N_2-\overline{N_2}\right)-m_2\left(N_2-\overline{N_2}\right)-b_2\left(N_2+\overline{N_2}\right)\left(N_2-\overline{N_2}\right)\right| +$$

$$\{|R|,|N_1|,|N_2|\}\le\delta$$

Now applying triangle inequality and noticing that max, it can be shown that

$$G(M)-G(\bar{M})\le(r-1+2\delta)\left|R-\bar{R}\right|+(a_1\varepsilon_1-a_1)\left|RN_1-\overline{RN_1}\right|$$

$$+(-m_1-2b_1\delta)\left|N_1-\overline{N_1}\right|+\left(a_2\varepsilon_2R_1^*-m_2-2b_2\delta-a_2R_1^*\right)\left|N_2-\overline{N_2}\right|$$

$$\le\left(r-1+\delta(2+a_1\varepsilon_1-a_1)\right)\left|R-\bar{R}\right|+\left(\delta\left(a_1\varepsilon_1-a_1-2b_1\right)-m_1\right)\left|N_1-\overline{N_1}\right|$$

$$+\left(a_2\varepsilon_2R_1^*-m_2-2b_2\delta-a_2R_1^*\right)\left|N_2-\overline{N_2}\right| \tag{7.5}$$

$$\text{where } T=max\begin{Bmatrix}r-1+\delta(2+a_1\varepsilon_1-a_1),\left(\delta\left(a_1\varepsilon_1-a_1-2b_1\right)-m_1\right),\\ \left(a_2\varepsilon_2R_1^*-m_2-2b_2\delta-a_2R_1^*\right)\end{Bmatrix}$$

G(M) thus satisfies the Lipschitz criterion in Equation (7.5).

7.4 BOUNDEDNESS OF THE SYSTEM

Lemma 1: The solutions of the Equation (7.3) lie in the region

$$B_1 = \left\{ (R, N_1, N_2) \in \mathbb{R}_+^3 : 0 \le R \le K, 0 \le N_1 \le \frac{a_1 \epsilon_1 K b_1}{b_1}, 0 \le N_2 \le \frac{a_2 \epsilon_2 R_1^*}{b_2} \right\} \text{ as}$$

$t \to \infty$ for all positive initial values $(R_0, N_{10}, N_{20}) \in \mathbb{R}_+^3$

Proof : From Equation (7.1), get as $t \to \infty$

$$\frac{dR}{dt} \le \left[r\left(1 - \frac{R}{K}\right) \right] R$$

Using comparison theorem, it is found that $R(t) \le K = R_{\max}$. Similarly from Equation (7.3) it is found that $t \to \infty$

$$\frac{dN_1}{dt} \le [a_1 \varepsilon_1 R - b_1 N_1] N_1$$

using comparison theorem $N_1(t) \le \dfrac{a_1 \in_1 K}{b_1} = (N_1)_{\max}$

Similarly, from Equation (7.3) $t \to \infty$

$$N_2(t) \le \frac{a_2 \in_2 R_1^*}{b_2} = (N_2)_{\max}$$

Lemma 2: The order of fractions that all solutions start in $\mathbb{R}_+^3$ of Equation (7.4) are uniformly bounded and continue positively. This is a resource-based native population model with the effect of exotic species.

7.5 DYNAMICAL BEHAVIOR OF THE SYSTEMS

This section aims to study the dynamical behavior of Equation (4) about all five feasible equilibria:

(1) Equilibrium point E_1 is unstable.
(2) Equilibrium point E_2 is linearly asymptotically stable if $a_1 \in_1 K < m_1$ and $a_2 \in_2 K < m_2$ are satisfied.
(3) Equilibrium point E_3 is linearly asymptotically stable if $a_1 \in_1 R^* < m_1$ is satisfied.
(4) Equilibrium point E_4 is linearly asymptotically stable if $m_2 \left(a_1^2 \in_1 K + b_1 r\right) > a_2 \in_2 K\,(a_1 m_1 + b_1 r)$ is satisfied.
(5) Equilibrium point E_5 is linearly asymptotically stable if

$\frac{2rR^*}{K}+(a_1+b_1)N_1^* > r$ and $b_1\left(\frac{2rR^*}{K}+a_1N_1^*\right)+a_1^2 \in_1 R^* > b_1 r$ are satisfied.

The Equation (7.4) Jacobian matrix, as measured at $\left(R^*, N_1^*, N_2^*\right)$, is

$$J=\begin{bmatrix} r-\frac{2R^*}{K}-a_1N_1^* & -a_1R^* & -a_2R_1^* \\ a_1 \in_1 N_1^* & a_1 \in_1 R^* - m_1 - 2b_1N_1^* & 0 \\ 0 & 0 & a_2 \in_2 R_1^* - m_2 - 2b_2N_2^* \end{bmatrix} \tag{7.6}$$

Now take $r = .05$; $a_1 = .12$; $a_{(2=.1)}$ $\in_1 = 1$; $\in_2 = 1$; $m_1 = m_2 = .025$; $K = 1$; $b_1 = .1 = b_2$; $R_0 = 0.1$; $N_{10} = 0.1$ $N_{20} = 0.035$ in the above equation, then:
the fixed point and the eigenvalues are

E_1= (0,0,0): $\lambda_1 = 0.0500; \lambda_2 = 0.0950; \lambda_3 = -0.0209$

E_2 = (K,0,0): $\lambda_1 = r-2; \lambda_2 = a_1 \in_1 K - m_1; \lambda_3 = a_2 \in_2 K - m_2$

Thus, the eigen values exist these conditions $r < 2$, $a_1 \in_1 K\, m_1$ and $a_2 \in_2 K < m_2$, then the equilibrium point E_1 is asymptotically stable, otherwise unstable.

Similarly, find the stability of equilibrium points E_3, E_4, and E_5.

7.6 CAPUTO DERIVATIVE NUMERICAL TECHNIQUE

The following is a brief introduction to the predictor-corrector type of numerical method. Let the general differential equation be as follows:

$$\begin{cases} {}_0^C D_t^\alpha \gamma(t) = f(t,\gamma(t)), 0 \le t \le T, \\ \gamma^{(k)}(0) = \gamma_0^{(k)}, k = 0,1,2,\ldots, m-1 \end{cases} \tag{7.7}$$

The Volterra integral equation is identical to the aforementioned DEs.

$$\gamma(t) = \sum_{k=0}^{[\alpha]-1} \gamma_0^{(k)} \frac{t^k}{k!} + \frac{1}{\text{“}(\alpha)} \int_0^t \frac{f(\xi,\gamma(\xi))}{(t-\xi)^{1-\alpha}} d\xi \tag{7.8}$$

Now, set $h = \frac{T}{N}, t_n = nh, (n = 0,1,2,\ldots, N)$. Thus, the above equation can be discretized as follows:

$$\gamma_h\left(t_{n+1}\right)=\sum_{k=0}^{[\alpha]-1}\gamma_0^{(k)}\frac{t_{n+1}^k}{k!}+\frac{h^\beta 1}{``(\alpha+2)}\left[f\left(t_{n+1},\gamma_h^p\left(t_{n+1}\right)\right)+\sum_{j=0}^{n}a_{j,n+1}f(t_j,\gamma_h\left(t_j\right))\right] \tag{7.9}$$

where the predicted value $\gamma_h^p\left(t_{n+1}\right)$ calculated by the arbitrary Adams–Bashforth technique is

$$\gamma_h^p\left(t_{n+1}\right)=\sum_{k=0}^{[\alpha]-1}\gamma_0^{(k)}\frac{t_{n+1}^k}{k!}+\frac{1}{``(\alpha)}\sum_{j=0}^{n}{}''\ {}_{j,n+1}f(t_j,\gamma_h\left(t_j\right)) \tag{7.10}$$

where

$$a_{j,n+1}=\begin{cases} n^{\alpha+1}-(n-\alpha)(n+1)^\alpha & j=0\\ (n-j+2)^{\alpha+2}+(n-j)^{\alpha+1}-2(n-j+1)^{\alpha+1} & 1\le j\le n\\ 1 & j=n+1\end{cases}$$

and

$$\Delta_{j,n+1}=\frac{h^\alpha}{\alpha}(n+1-j)^\alpha-(n-j)^\alpha,0\le j\le n,\ i=1,2,3.$$

7.7 PREDICTOR-CORRECTOR SCHEME ON PROPOSED RESOURCE-BASED NATIVE POPULATION MODEL

The aforementioned proposed predictor–corrector numerical method is used for the fractional operator to the numerical solution of the recommended variable-order resource-based native population system with arbitrary-order Caputo derivative. According to the suggested numerical scheme, the model Equations (7.4) through (7.3) can be discretized as follows:

$$R_{n+1}=R(0)+\frac{h^\alpha 1}{``(\alpha+2)}\left[\begin{array}{l}F_1\left(t_{n+1},R_{n+1}^p,(N_1)_{n+1}^p,(N_2)_{n+1}^p\right)+\\ \sum_{j=0}^{n}a_{j,n+1}F_1(t_j,R_j,(N_1)_j,(N_2)_j)\end{array}\right] \tag{7.11}$$

$$(N_1)_{n+1} = N_1(0) + \frac{h^{\alpha}1}{``(\alpha+2)}\begin{bmatrix} F_2\left(t_{n+1}, R^p_{n+1}, (N_1)^p_{n+1}, (N_2)^p_{n+1}\right) + \\ \sum_{j=0}^{n} a_{j,n+1} F_2(t_j, R_j, (N_1)_j, (N_2)_j) \end{bmatrix} \tag{7.12}$$

$$(N_2)_{n+1} = N_2(0) + \frac{h^{\alpha}1}{``(\alpha+2)}\begin{bmatrix} F_3\left(t_{n+1}, R^p_{n+1}, (N_1)^p_{n+1}, (N_2)^p_{n+1}\right) + \\ \sum_{j=0}^{n} a_{j,n+1} F_3(t_j, R_j, (N_1)_j, (N_2)_j) \end{bmatrix} \tag{7.13}$$

The predator form $R^p_{n+1}, (N_1)^p_{n+1}, (N_2)^p_{n+1}$ are suggested as follows:

$$R^p_{n+1} = R(0) + \frac{1}{``\alpha}\sum_{j=0}^{n}{}''\,{}_{j,n+1} F_1(t_j, R_j, (N_1)_j, (N_2)_j \tag{7.14}$$

$$(N_1)^p_{n+1} = N_1(0) + \frac{1}{``\alpha}\sum_{j=0}^{n}{}''\,{}_{j,n+1} F_2(t_j, R_j, (N_1)_j, (N_2)_j \tag{7.15}$$

$$(N_2)^p_{n+1} = N_2(0) + \frac{1}{``\alpha}\sum_{j=0}^{n}{}''\,{}_{j,n+1} F_3(t_j, R_j, (N_1)_j, (N_2)_j \tag{7.16}$$

where

$$F_1(t, R, N_1, N_2) = \left[r\left(1 - \frac{R}{K}\right) - a_1 N_1\right] R - a_2 R_1^* N_2$$

$$F_2(t, R, N_1, N_2) = [a_1 \varepsilon_1 R - m_1 - b_1 N_1] N_1$$

$$F_3(t, R, N_1, N_2) = \left[a_2 \varepsilon_2 R_1^* - m_2 - b_2 N_2\right] N_2$$

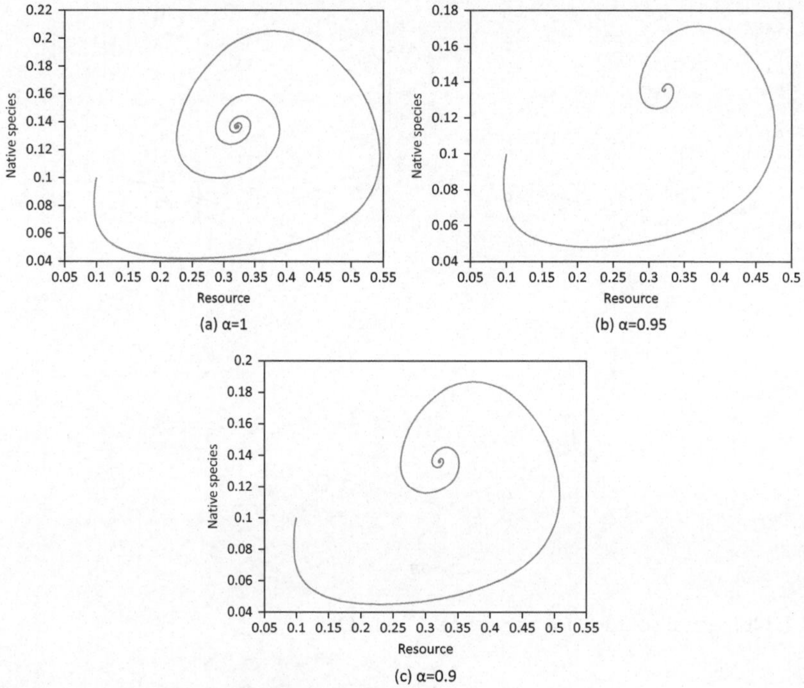

Figure 7.1 Numerical solution for the model (3) in 2D.

7.8 NUMERICAL SIMULATION

This section describes how to use the suggested computational methods to find numerical approximations for the state variables $R(t)$, $N_1(t)$, and $N_2(t)$ of the specified system under the Caputo fractional derivative (CFD) operator. The simulation is run using the parameter values provided (Figures 7.1–7.3).

7.9 CONCLUSION

With the aid of ABM, the relevant derivative was taken in the Caputo sense, and this study comes up with some suitable conditions for the existence and individuality of equilibrium points to the resource-based native population model with fractional-order derivative if the death rate is lower than the rate of generations. However, if the stability of E_3 were to be reversed, the local population would continue to exist, while the alien population would go extinct. The coexistence of resources and both species has been inferred from the stability of the positive point; however, the equilibrium resource densities are further reduced by the presence of exotic species. ABM was

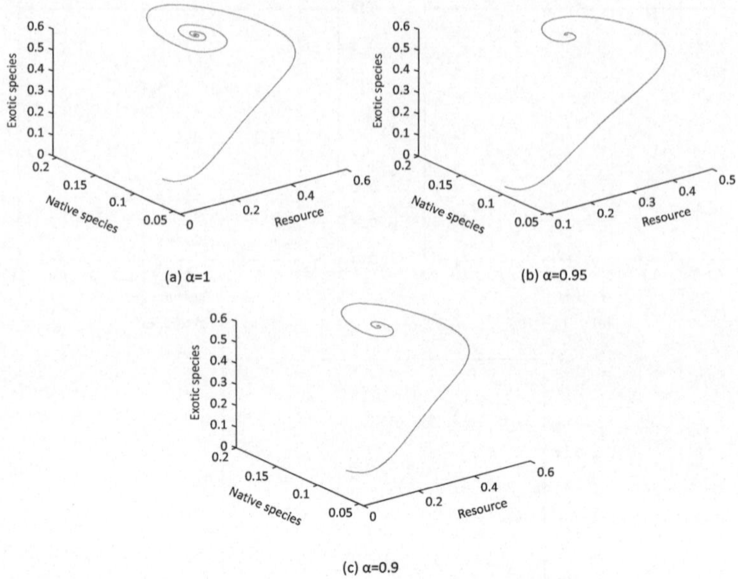

Figure 7.2 Numerical solution for the model (3) in 3D.

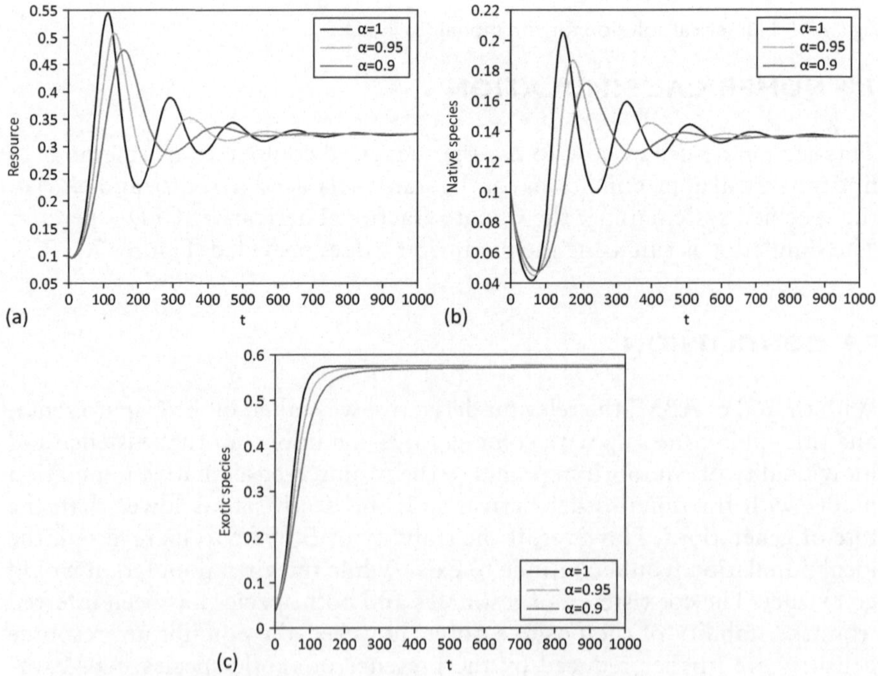

Figure 7.3 Time series for the model (3) at different values of α.

used for the representation of numerical simulations in the approximations. Through simulations, it is concluded that fractional-order systems exhibit more complex dynamics than integer-order systems.

7.10 FUTURE SCOPE

This study with fractional-order derivatives has provided valuable insights into the dynamics of resource-dependent native populations. However, several avenues remain open for future exploration to deepen our understanding and enhance the applicability of this model. This work can be expanded to include multiple native species and can be studied using fractional-order derivatives such as Atangana–Baleanu, Caputo–Fabriz, generalized Caputo, and others.

REFERENCES

1. Coblentz, Bruce E. "Exotic organisms: a dilemma for conservation biology." *Conservation Biology* 4, no. 3 (1990): 261–265.
2. Petrovskii, Sergei, Andrew Morozov, and Bai-Lian Li. "Regimes of biological invasion in a predator-prey system with the Allee effect." *Bulletin of Mathematical Biology* 67 (2005): 637–661.
3. Shigesada, Nanako, and Kohkichi Kawasaki. *Biological invasions: Theory and practice*. Oxford University Press, UK, 1997.
4. James D. Murray, *Mathematical Biology: I. An Introduction*, Vol. 17. Springer Science & Business Media, Berlin, Germany, 2007.
5. Anderson, R. M., and Robert M. May. "Regulation and stability of host-parasite population interactions." *Journal of Animal Ecology* 47, no. 1 (1978): 219–247.
6. Aljahdaly, Noufe H., and Manar A. Alqudah. "Analytical Solutions of a Modified Predator-Prey Model through a New Ecological Interaction." *Computational and Mathematical Methods in Medicine* 2019, no. 1 (2019): 4849393.
7. Fay, Temple H., Johanna C. Greeff, Benjamin E. Eisenberg, and Hendrik T. Groeneveld. "Testing the model for one predator and two mutualistic prey species." *Ecological Modelling* 196, no. 1–2 (2006): 245–255.
8. Kar, T. K., and K. S. Chaudhuri. "Harvesting in a two-prey one-predator fishery: a bioeconomic model." *The ANZIAM Journal* 45, no. 3 (2004): 443–456.
9. Křivan, Vlastimil, and Jan Eisner. "The effect of the Holling type II functional response on apparent competition." *Theoretical Population Biology* 70, no. 4 (2006): 421–430.
10. Al Themairi, Asma, and Manar A. Alqudah. "Predator-prey model of Holling-type II with harvesting and predator in disease." *Italian Journal of Pure and Applied Mathematics* 43 (2020): 744–753.

11. Abdeljawad, Thabet, Iyad Suwan, Fahd Jarad, and Ammar Qarariyah. "More properties of fractional proportional differences." *Journal of Mathematical Analysis and Modeling* 2, no. 1 (2021): 72–90.
12. Benchohra, Mouffak, and Mehdi Slimane. "Fractional differential inclusions with non instantaneous impulses in Banach spaces." *Results in Nonlinear Analysis* 2, no. 1 (2019): 36–47.
13. Chalishajar, Dimple Kumar, Duraisamy Senthil Raja, Kulandhaivel Karthikeyan, and Ponnusamy Sundararajan. "Existence results for nonautonomous impulsive fractional evolution equations." *Results in Nonlinear Analysis* 1, no. 3 (2018): 133–147.
14. Leela D., Kapula Rajendra Prasad and Mahammad Khuddush, "Existence and uniqueness of positive solutions for system of (p, q, r)laplacian fractional order boundary value problems," *Advances in the Theory of Nonlinear Analysis and its Application* 5, no. 1 (2021): 138–157.
15. Tate, Shivaji, and H. T. Dinde. "Existence and uniqueness results for nonlinear implicit fractional differential equations with non local conditions." *Palestine Journal of Mathematics* 9, no. 1 (2020): 212–219.
16. Wahash, Hanan A., and Satish K. Panchal. "Positive solutions for generalized Caputo fractional differential equations using lower and upper solutions method." *Journal of Fractional Calculus and Nonlinear Systems* 1, no. 1 (2020): 1–12.
17. Podlubny, Igor. *Fractional differential equations: an introduction to fractional derivatives, fractional differential equations, to methods of their solution and some of their applications*. Elsevier, 1998.
18. Ali, M. Syed, M. Hymavathi, Bandana Priya, Syeda Asma Kauser, and Ganesh Kumar Thakur. "Stability analysis of stochastic fractional-order competitive neural networks with leakage delay." *AIMS Mathematics* 6, no. 4 (2021): 3205–3241.
19. Ali, M. Syed, M. Hymavathi, Sibel Senan, Vineet Shekher, and Sabri Arik. "Global asymptotic synchronization of impulsive fractional-order complex-valued memristor-based neural networks with time varying delays." *Communications in Nonlinear Science and Numerical Simulation* 78 (2019): 104869.
20. Ali, M. Syed, Govindasamy Narayanan, Saeid Nahavandi, Jin-Liang Wang, and Jinde Cao. "Global dissipativity analysis and stability analysis for fractional-order quaternion-valued neural networks with time delays." *IEEE Transactions on Systems, Man, and Cybernetics: Systems* 52, no. 7 (2021): 4046–4056.
21. Syed Ali, M., Govindasamy Narayanan, Zeynep Orman, Vineet Shekher, and Sabri Arik. "Finite time stability analysis of fractional-order complex-valued memristive neural networks with proportional delays." *Neural Processing Letters* 51, no. 1 (2020): 407–426.
22. Arslan, Emel, Govindasamy Narayanan, M. Syed Ali, Sabri Arik, and Sumit Saroha. "Controller design for finite-time and fixed-time stabilization of fractional-order memristive complex-valued BAM neural networks with uncertain parameters and time-varying delays." *Neural Networks* 130 (2020): 60–74.
23. Hilfer, Rudolf, ed. *Applications of fractional calculus in physics*. World Scientific, 2000.

24. Kilbas, Anatoliĭ Aleksandrovich, Hari M. Srivastava, and Juan J. Trujillo. *Theory and applications of fractional differential equations*. Vol. 204. Elsevier, 2006.
25. Magin, Richard. "Fractional calculus in bioengineering, part 1." *Critical Reviews™ in Biomedical Engineering* 32, no. 1 (2004).
26. Alqudah, Manar A., Thabet Abdeljawad, Eiman, Kamal Shah, Fahd Jarad, and Qasem Al-Mdallal. "Existence theory and approximate solution to prey–predator coupled system involving nonsingular kernel type derivative." *Advances in Difference Equations* 2020 (2020): 1–10.
27. Om P. Mishra and Pramod Kushwah. "Effect of exotic population on a resource-based native population: A Model." In *Proceedings - The Mathematical Society, BHU*. Vol. 26, 2010.
28. Al Themairi, A., and Manar A. Alqudah. "Existence and Uniqueness of Caputo Fractional Predator-Prey Model of Holling-Type II with Numerical Simulations." *Mathematical Problems in Engineering* 2021, no. 1 (2021): 2990958.

Chapter 8

Selection of teaching pedagogy for engineering classroom learning from instructor's viewpoints using Fuzzy TOPSIS approach

A case study of selected higher education institutions in northern India

Parveen Sharma, Sonia Sharma, Ashwini Jagannath Mahamuni, Rakesh Kumar Phanden, and Ajay Kumar

8.1 INTRODUCTION

Throughout history, teaching pedagogy has evolved in response to societal needs, cultural shifts, and advancements in education theory and practice. Ancient civilizations, such as those in Mesopotamia and Egypt, employed oral traditions and apprenticeship models to pass down knowledge and skills from one generation to the next. The philosophy of education in ancient Greece, notably attributed to philosophers like Socrates, Plato, and Aristotle, laid the groundwork for systematic teaching methods, emphasizing critical thinking, inquiry-based learning, and dialogue. In the medieval period, religious institutions played a central role in education, with monastic schools and cathedral schools serving as hubs of learning. During the Renaissance, humanist ideals propelled educational reform, advocating for a broader curriculum that included literature, science, and the arts. The Industrial Revolution sparked new approaches to education, with the rise of compulsory schooling and the implementation of standardized curricula. In the 20th century, educational psychology and pedagogical theories, such as behaviorism, constructivism, and sociocultural theory, reshaped teaching practices, emphasizing student-centered learning, differentiated instruction, and the use of technology. Today, teaching pedagogy continues to evolve in response to diverse learner needs, technological advancements, and emerging educational paradigms.

Teaching pedagogy has evolved across civilizations, from ancient oral traditions to medieval cathedral schools and the Renaissance emphasis on critical thinking (Compayré & Payne, 2015; Kabulova, 2023). The

 DOI: 10.1201/9781003512585-9

Industrial Revolution standardized education, while 20th-century theorists like Dewey championed student-centered learning (Salvatori, 2003; Olson, 2014). Today, pedagogy adapts to technology and global shifts, maintaining a focus on innovation and inclusivity.

Educators grapple with a plethora of teaching methodologies, each offering unique advantages and challenges. From traditional lecture-based approaches to innovative experiential and project-based learning paradigms, the spectrum of pedagogies reflects diverse philosophies and methodologies (Ekpok & Dukku, 2021). However, amidst this diversity, the quest for identifying the most effective pedagogy remains elusive.

The selection of teaching pedagogy holds immense significance in shaping the learning experiences and outcomes of students. First, effective pedagogy plays a pivotal role in engaging learners and fostering their active participation in the learning process. By incorporating diverse instructional strategies, such as active learning, collaborative activities, and experiential learning, educators can cater to different learning styles and preferences, thus enhancing student engagement and motivation. Second, the choice of pedagogy influences the acquisition and retention of knowledge and skills. A well-designed pedagogical approach ensures that content is delivered in a manner that is accessible, comprehensible, and conducive to deep learning. Moreover, it facilitates the development of critical thinking, problem-solving, and other higher-order cognitive skills essential for academic success and lifelong learning. Additionally, teaching pedagogy plays a crucial role in promoting inclusivity and equity in education. By adopting inclusive pedagogical practices and accommodations, educators can create learning environments that are accessible to all students, regardless of their diverse backgrounds, abilities, or learning needs. Furthermore, the selection of pedagogy impacts the overall effectiveness and efficiency of teaching and learning processes. By aligning instructional methods with learning objectives, educators can optimize instructional time, resources, and efforts, thus maximizing learning outcomes. Overall, the careful selection of teaching pedagogy is essential for creating engaging, effective, and inclusive learning experiences that empower students to achieve their full potential (Karthik et al., 2024; Thakur et al., 2024b; Swain et al., 2024; Amale et al., 2024).

In the dynamic realm of education, the selection of appropriate teaching pedagogy is a complex and pivotal decision for educators (Dalton, 1998; Ahmadi & Moosav, 2024). The challenge is not only to engage students effectively but also to create environments that foster meaningful learning outcomes. This study delves into the intricate process of choosing the optimal instructional approach for classroom learning, recognizing the multifaceted nature of educational endeavors.

Selecting the right teaching pedagogy faces several challenges. These include meeting the diverse needs of students, integrating technology effectively, keeping up with educational advancements, and working within constraints like limited resources and institutional policies. Additionally, assessing pedagogy effectiveness can be tricky. Overcoming these challenges requires collaboration and support for evidence-based decision-making and professional development.

The importance of teaching pedagogy for engineering students cannot be overstated in the realm of education. Pedagogy serves as the cornerstone of effective teaching practices, guiding educators in the creation of engaging and meaningful learning experiences for students (Entz, 2007). By employing diverse pedagogical approaches tailored to the unique needs and learning styles of learners, educators can foster deep understanding, critical thinking skills, and lifelong learning habits. Moreover, pedagogy plays a pivotal role in cultivating an inclusive and supportive learning environment where all students feel valued and empowered to participate actively in their own education (Lupton & Hempel, 2012; Loughran, 2013a). Through thoughtful pedagogical design and implementation, educators have the opportunity to inspire curiosity, ignite a passion for learning, and equip students with the knowledge and skills they need to succeed in an ever-evolving world (Kabulova, 2023; Loughran, 2013b; Corbin & Bugden, 2018). Thus, teaching pedagogy stands as a fundamental pillar in the pursuit of educational excellence and the empowerment of future generations.

The need for the current study lies in addressing the pressing challenge faced by educators worldwide: the selection of the most effective teaching pedagogy for classroom learning in higher education in engineering. In today's diverse educational landscape, where students possess varied learning styles, interests, and abilities, there is a growing recognition of the importance of tailored instructional approaches. However, the abundance of pedagogical methodologies, coupled with the lack of systematic evaluation frameworks, often leaves educators grappling with uncertainty when choosing the most suitable approach. By undertaking this research, the current study aims to fill this critical gap in educational practice by employing rigorous decision-making methodologies to evaluate and compare five distinct teaching pedagogies. The insights gleaned from this study will not only empower educators with evidence-based guidance but also contribute to the ongoing discourse on instructional design and pedagogical effectiveness. Ultimately, the findings of this research hold the potential to enhance classroom learning experiences, optimize student engagement, and foster academic success for learners across diverse educational contexts.

Multiple criteria decision-making (MCDM) techniques play a crucial role in various fields, facilitating complex decision-making processes by considering multiple conflicting criteria. Among these techniques, Fuzzy Technique for Order of Preference by Similarity to Ideal Solution (TOPSIS) stands out for its effectiveness in handling imprecise and uncertain data. In today's dynamic and uncertain environments, where decision-makers often face ambiguity and vagueness, MCDM techniques offer valuable insights by incorporating diverse perspectives and priorities. Fuzzy TOPSIS, in particular, excels in scenarios where decision factors involve linguistic variables or subjective assessments, providing a systematic framework for evaluating alternatives and identifying the most favorable solutions.

Fuzzy TOPSIS addresses the inherent uncertainties in decision-making by employing fuzzy set theory to capture the imprecision and vagueness inherent in real-world data. By representing decision factors and alternatives as linguistic variables with fuzzy membership functions, Fuzzy TOPSIS allows decision-makers to express their preferences in a more intuitive and flexible manner. This approach enables the consideration of qualitative factors alongside quantitative ones, resulting in more comprehensive and robust decision outcomes. As organizations and decision-makers increasingly recognize the importance of considering multiple criteria in decision-making, MCDM techniques like Fuzzy TOPSIS continue to gain prominence for their ability to handle complexity and uncertainty effectively.

The selection of Fuzzy TOPSIS for this study is driven by several compelling reasons. First, Fuzzy TOPSIS is well-suited for decision-making scenarios characterized by imprecise, uncertain, or qualitative data. In the context of selecting the optimal teaching pedagogy for classroom learning, where factors such as student engagement strategies and alignment with educational goals may involve subjective assessments or linguistic variables, Fuzzy TOPSIS offers a structured approach to handle such complexities effectively.

Moreover, Fuzzy TOPSIS allows for the integration of both qualitative and quantitative criteria, making it suitable for evaluating teaching pedagogies, which often involve a combination of tangible and intangible factors. By employing fuzzy set theory to model uncertainty and vagueness, Fuzzy TOPSIS enables decision-makers to capture the nuances of real-world decision contexts more accurately.

While other MCDM approaches exist, such as analytic hierarchy process (AHP) or simple additive weighting (SAW), Fuzzy TOPSIS was chosen for its ability to accommodate the inherent uncertainties and subjectivity involved in pedagogy selection. Unlike AHP, which requires pairwise comparisons and a hierarchical structure, or SAW, which relies solely on numerical ratings, Fuzzy TOPSIS offers a more flexible and intuitive framework for decision-making in complex and uncertain environments.

Research objectives:

RQ1: How do engineering instructors in Northern India perceive the effectiveness of various teaching pedagogies?

RQ2: How do the rankings of teaching pedagogies, derived from the Fuzzy TOPSIS-based multiple attribute decision-making approach, correlate with factors influencing pedagogical effectiveness?

The chapter is organized as follows: in the first section, an introduction to the research problem and the significance of the study is presented. Section 8.2 provides a detailed description of the selected factors influencing teaching pedagogy, including alignment with educational goals, student engagement strategies, accommodation of diverse learning styles, integration of technology, assessment methodologies, and potential scalability. An overview of the utilized methodology for evaluating and comparing teaching pedagogies is outlined in Section 8.3, followed by a mathematical description of the proposed method in Section 8.4. Section 8.5 presents a comprehensive discussion of the results obtained from the evaluation and ranking process. Finally, the conclusions drawn from the study and implications for educational practice are provided in Section 8.6.

8.2 LITERATURE REVIEW

Exploring the diverse landscape of teaching pedagogy, this literature survey delves into various instructional approaches and methodologies employed in classroom settings.

Key term for literature search?

For a literature search on teaching pedagogy selection, key terms include teaching pedagogy, classroom instruction, instructional approaches, educational methodologies, pedagogical strategies, student engagement techniques, learning outcomes, MCDM, Fuzzy TOPSIS, modified digital logic (MDL), fuzzy logic, AHP, decision-making in education, comparative analysis of teaching methods, educational technology integration, assessment methodologies, curriculum design, learning styles, teacher effectiveness, and educational innovation.

The methodology for literature review in the study on teaching pedagogy selection follows a structured approach. The process begins by identifying relevant databases such as Scopus, Web of Science, and others to extract pertinent articles. Keywords like "teaching pedagogy," "educational strategies," "student engagement," and "learning outcomes," combined with Boolean operators, are systematically used to retrieve articles aligned with the research objectives. More than 75 research papers were considered initially and nearly 55 considered. Preference is given to articles published within the last decade and indexed in reputable databases. Non-English articles, unpublished works, and those not directly related to the research focus are excluded. Following rigorous screening based on titles, abstracts, and full texts, the selection is narrowed down to a subset of articles conducive to the study's thematic exploration of effective teaching pedagogies for classroom learning.

Yakovleva and Yakovlev (2014) focus on modern education's shift toward student autonomy, interactive methods, and practical training. They survey prevalent teaching approaches like case studies, peer feedback, and storytelling, highlighting their efficacy in fostering professional competencies and motivation. Barnett and Hallam (1999) explored innovative pedagogical approaches emphasizing student autonomy and interactive methods to prepare learners for complex professional environments. Their work underscores the importance of adapting education to meet the challenges of contemporary society. Bidabadi et al. (2016) investigate effective teaching methods in higher education, identifying requirements and barriers. Their study offers insights into optimizing teaching approaches amidst challenges faced in the higher education landscape.

Seatter and Ceulemans (2017) delve into pedagogical styles conducive to teaching sustainability in higher education. Their research underscores the significance of adopting effective teaching approaches to instill sustainability principles among students. Kersten (2018) examines engineering pedagogy approaches to enhance teaching quality in engineering education, discussed in "Vocational Teacher Education in Central Asia." The paper emphasizes strategies to elevate teaching standards and foster success in engineering learning environments within the Central Asian context. Aziz and Islam (2021) explore the impact of mixed pedagogy on engineering education. Their study investigates the effectiveness of integrating various teaching methods to enhance learning outcomes and student engagement in engineering curricula.

Law et al. (2005) delves into the significance of studying educational innovations, particularly amidst global education reforms. It highlights the scarcity of comparative studies in this field and discusses the challenges faced by the SITES M2 study, an international comparative endeavor involving 28 educational systems. The paper scrutinizes the methodological approaches used by three research teams, emphasizing the importance of comparing innovations across various dimensions to glean insights into technology-supported education advancements. Law, N. (2014) underscores the perpetual need for innovation in education, driven by the evolving demands of the knowledge society. It highlights the global context of intensifying globalization, rapid knowledge turnover, and economic competitiveness, emphasizing the necessity for new educational goals and processes to meet these challenges effectively. Dai (2014) investigates the effectiveness of discussion versus lecture pedagogy in teaching oral communication within a business course context. The paper offers insights into optimal instructional methods for enhancing students' oral communication skills in business education.

Many researchers have dedicated their efforts to studying various teaching pedagogies, often comparing them with conventional methods such as student discussions and classroom surveys. However, there remains a notable gap in the literature regarding the exploration of pedagogical practices specific to higher education in engineering. Despite the crucial role of effective teaching methods in engineering education, only a limited number of studies have delved into this area. This underscores the need for further research and analysis to better understand and optimize pedagogical approaches within the engineering discipline, ultimately enhancing the learning experience for engineering students.

Drawing inspiration from decision-making methodologies, particularly the Fuzzy TOPSIS (Esfandiari, & Rizvandi, 2014), this research embarks on a systematic exploration of pedagogy selection. By leveraging the principles of decision science, the current study endeavors to establish a structured framework for evaluating and comparing teaching pedagogies. Central to this endeavor are key factors that underpin effective pedagogy selection. These factors encompass alignment with educational objectives, strategies for student engagement, accommodation of diverse learning styles, integration of technology, assessment methodologies, and scalability. Recognizing the inherent complexities and nuances within each factor, the authors employ the MDL technique (Sharma & Singhal, 2016) to assign weights, thus capturing their relative importance accurately.

Subsequently, the current study deploys a Fuzzy TOPSIS-based multiple attribute decision-making approach to assess and rank five distinct teaching pedagogies: Lecture-Based Pedagogy (LBP), Active Learning (AL), Experiential Learning (EL), Project-Based Learning (PBL), and the Flipped Classroom (FC). Each pedagogy represents a unique approach to instructional delivery, offering varied opportunities for student engagement and learning.

Through meticulous analysis and comparison, this study aims to elucidate the strengths and limitations of each teaching pedagogy within the context of classroom learning. By identifying the most effective pedagogy, the authors aspire to provide valuable insights for educators, instructors, and instructional designers, facilitating informed decision-making and ultimately enhancing classroom learning experiences.

The objective of the current study is to systematically evaluate and compare five distinct teaching pedagogies: LBP, AL, EL, PBL, and the FC. Key factors influencing pedagogical effectiveness, such as alignment with educational goals, student engagement strategies, accommodation of diverse learning styles, integration of technology, assessment methodologies, and potential scalability, will be identified and assessed. Utilizing the MDL technique (Torrez, et al., 2012; Lutfina, & Nuraisha, 2020), weights will be assigned to these factors based on their relative importance. Subsequently, a Fuzzy TOPSIS-based multiple attribute decision-making approach will be applied to evaluate and rank the teaching pedagogies (Behzadian, et al., 2012; Lai, et al., 1994; Olson, 2004: Papathanasiou, et al., 2018). The study aims to determine the most effective pedagogy for enhancing classroom learning experiences and provide valuable insights for educators and instructional designers, facilitating informed decision-making and optimization of teaching practices in diverse educational settings.

8.2.1 Teaching pedagogies evaluation factors

The current study delineates a comprehensive set of crucial factors essential for assessing and comparing various teaching pedagogies (Malunda et al., 2016; Somekh, 2008). These factors, selected in collaboration with senior experts from both the education and industry sectors, embody the diverse dimensions of effective pedagogy. They include alignment with educational goals (AEG), strategies for student engagement (SSE), accommodation of diverse learning styles (ADLS), integration of technology (IoT), assessment methodologies (AM), potential scalability (PS), and instructor support (IS). Each factor is represented using a short name for ease of reference, and they are meticulously organized in Table 8.1, providing a structured framework for the systematic evaluation and comparison of teaching pedagogies.

Each selected factor in the evaluation of teaching pedagogies holds significant importance in shaping the overall effectiveness of instructional design. Alignment with educational goals serves as the guiding principle, ensuring that the pedagogy aligns seamlessly with defined educational

Table 8.1 Factors of teaching pedagogy and their description

Factors	*Description*	*References*
Assessment methodologies (AM)	The methods utilized to assess student learning and progress include formative and summative assessment strategies, feedback mechanisms, and grading criteria	Beatty and Gerace (2009), Baxter and Lederman (1999), Carvalho et al. (2021)
Alignment with educational goals (AEG)	The extent to which the pedagogy aligns with the defined educational objectives and learning outcomes	Krajcik et al. (2008), Larsen et al. (2018)
Strategies for student engagement (SSE)	The variety and effectiveness of methods employed to actively engage students in the learning process	Parsons and Taylor (2011), Franklin and Harrington (2019)
Accommodation of diverse learning styles (ADLS)	The ability of the pedagogy to cater to different learning preferences, including visual, auditory, kinesthetic, and tactile styles	Honigsfeld and Schiering (2004), Geleta et al. (2022)
Integration of technology	The extent to which technology is seamlessly integrated into the teaching approach to enhance learning experiences and outcomes	Okojie et al. (2006), Ching and Roberts (2020), Wentworth et al. (2008)
Potential scalability (PS)	The adaptability and scalability of the pedagogical approach to accommodate varying class sizes, educational contexts, and future growth	Mioduser et al. (2004), Niederhauser et al. (2018)
Instructor support (IS)	The availability and effectiveness of support mechanisms provided to instructors to facilitate the successful implementation of the pedagogy	Koepke and O'Brien (2012), Werth and Williams (2023)

objectives and learning outcomes. Strategies for student engagement play a vital role in fostering active participation and deep learning experiences among students, while ADLS ensures inclusivity and accessibility for learners with varying preferences. Integration of technology facilitates enhanced learning experiences through the seamless incorporation of digital tools and resources. Assessment methodologies provide valuable insights into student progress and learning outcomes, informing instructional strategies and interventions. Potential scalability enables adaptability to varying class sizes and educational contexts, ensuring sustainability and long-term effectiveness. Finally, instructor support is essential for providing educators with the necessary resources and guidance to successfully implement the chosen pedagogy, ultimately impacting the quality and success of the teaching-learning process. AM should be minimal and the instructor should be able to judge the students with the minimum number of assessments. On the other hand, the remaining factors should be toward the higher side.

8.3 METHODOLOGY USED

8.3.1 Modified digital logic

In this study, authors recognize that not all factors or attributes have an equal impact on the selection of an appropriate teaching pedagogy, and assigning equal weights to them would not accurately reflect their relative importance. Therefore, it is essential to determine the priority of each factor. The authors employ MDL, a well-established technique for estimating the weights of factors in such situations (Torrez, et al., 2012; Lutfina, & Nuraisha, 2020). Initially, each factor is assigned a priority rating of 1 for less importance, 2 for equal importance, and 3 for greater importance, based on expert opinion.

In this research, the authors reached out to 11 experts from 2 institutions located in Punjab India, representing core engineering disciplines, including Mechanical, Civil, Electrical, Electronics, and Computer Engineering. Among them, eight experts generously agreed to share their insights on teaching pedagogy within their respective fields (each expert had more than ten years of experience). Through video conferencing sessions, these experts engaged in fruitful discussions, enabling the collaborative exploration of relevant topics and the formulation of informed decisions regarding pedagogical strategies. This collaborative effort facilitated a comprehensive understanding of effective teaching methods tailored to the unique demands of engineering education.

Following the assignment of priorities, a decision matrix is constructed through pairwise comparisons. Prior to matrix formation, the number of possible positive decisions (N) is estimated using the formula $N = n(n - 1)/2$, where n represents the number of attributes or parameters under consideration. The summation of all positive decisions (P) for a particular parameter is then normalized to determine its final weight (Wj)

$$W_j = \frac{P_j}{\sum_{j=1}^{n} P_j} \tag{8.1}$$

8.3.2 Fuzzy logic

In this study, authors utilize fuzzy logic to address issues characterized by imprecision and ambiguity, where conventional binary logic may fall short (Zadeh, 1965). Fuzzy logic offers a framework for handling situations where distinctions between non-member and member objects within a set are challenging to discern. This approach has been successfully applied in multiple criteria decision-making contexts by Belleman and Zadeh (Bellman & Zadeh, (1970).

Fuzzy sets are based on fuzzy set theory, which uses a scale from 0 to 1 to show how much something belongs to a set. The authors start with words to compare things, and then give these words fuzzy numbers to turn them into numbers that can be compared more easily. This helps to make things clearer and less confusing. In this study, trapezoidal fuzzy numbers (a_1, a_2, a_3, a_4) for ($a_1, a_2, a_3, a_4 \in R; a_1 \le a_2 \le a_3 \le a_4$) are used. Figure 8.1 graphically illustrates the trapezoidal fuzzy number. The membership function $\mu a(x)$ of trapezoidal fuzzy number is defined as follows:

$$\mu a(x) = \begin{cases} \dfrac{x - a_1}{a_2 - a_1}, x \in [a_1, a_2] \\ 1, x \in [a_2, a_3] \\ \dfrac{a_4 - x}{a_4 - a_3}, x \in [a_3, a_4] \\ 0, \text{otherwise} \end{cases} \tag{8.2}$$

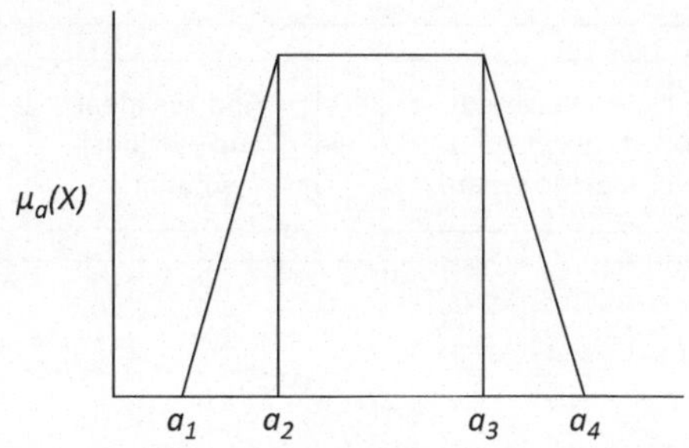

Figure 8.1 Trapezoidal fuzzy number.

8.3.3 TOPSIS method

In this study, authors employ the TOPSIS method, proposed by Hwang and Yoon in 1981 (Deng, et al.,2000; Phanden, et al., 2019). TOPSIS operates on the principle that a decision matrix comprising *n* properties and *m* alternatives can be conceptualized as a problem in an *n*-dimensional hyperplane with *m* points, each representing an alternative and defined by its attribute values. It assesses the Euclidean distance between each alternative and both the positive ideal solution (representing the most desirable attributes) and the negative ideal solution (representing the least desirable attributes). The alternative with the shortest distance from the positive ideal solution is deemed the most favorable, while those with greater distances from the negative ideal solution are considered less desirable.

8.4 PROPOSED APPROACH

In this section, the authors present a proposed methodology for selecting the optimal teaching pedagogy for classroom learning, termed the Fuzzy-TOPSIS approach. The methodology is designed to identify the most suitable pedagogical approach by integrating fuzzy logic principles with the TOPSIS. After conducting the required literate survey and considering experts' opinion, the following steps (Figure 8.2 demonstrates these steps) have been implemented:

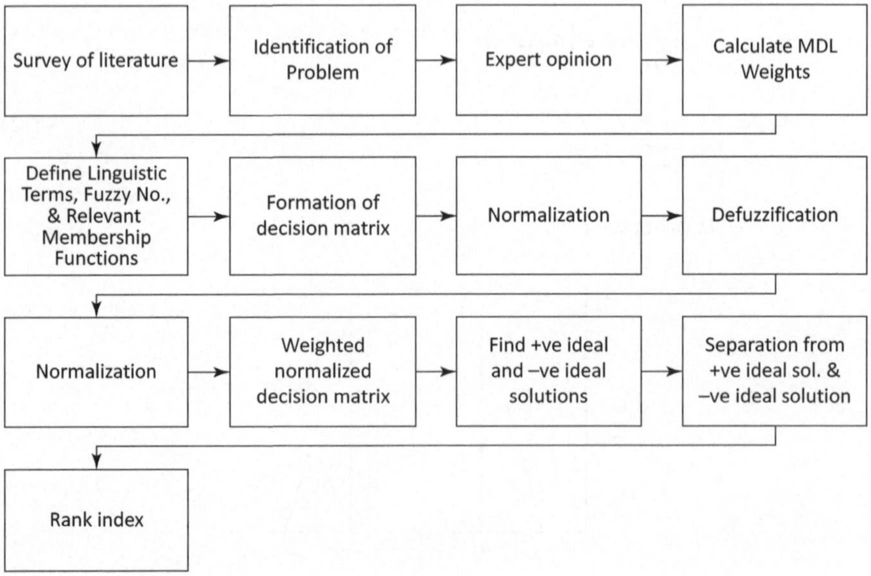

Figure 8.2 Steps of proposed approach.

Step 1: At this stage, calculate MDL weights.

First, figure out the MDL weights (*Wj*) for each factor. The details related to this step are discussed in Section 8.3.1.

Step 2: At this stage, define linguistic terms, corresponding fuzzy numbers, and relevant membership functions.

To make it easier to compare all the options for each factor, define a bunch of words to describe them. These words are decided by the person making the decision and are really important for comparing options within each factor.

Step 3: At this stage, formulate a decision matrix.

The subsequent step involves determining the decision matrix. In the proposed model, if there are *m* alternatives and *n* factors, and *k* decision-makers are considered, the aggregated fuzzy rating for C_j factors is as follows:

xijk = {*xijk*1, *xijk*2, *xijk*3, *xijk*4},

for *i* = 1,2,..., *m*; *j* = 1,2,..., *k*, *xijk* is calculated as follows (Mehrjerdi, 2013)]:

$$\begin{cases} x_{ij1} = \min_k \left\{ a_{ijk1} \right\} \\ x_{ij2} = \frac{1}{k} \sum a_{ijk2} \\ x_{ij3} = \frac{1}{k} \sum a_{ijk3} \\ x_{ij4} = \max_k \left\{ a_{ijk4} \right. \end{cases} \tag{8.3}$$

The obtained decision matrix (*D*) is as follows:

$$D = \begin{bmatrix} x_{11} & x_{12} & \dots & x_{1n} \\ x_{21} & x_{22} & \dots & x_{2n} \\ \vdots & \vdots & \ddots & \vdots \\ x_{m1} & x_{m2} & \ddots & x_{mn} \end{bmatrix}$$

Step 4: At this stage, perform normalization.

Normalization is performed as follows:

$$\mu_{ij} = \left(\frac{x_{ij1}}{x_{ij1}^{+}}, \frac{x_{ij2}}{x_{ij2}^{+}}, \frac{x_{ij3}}{x_{ij3}^{+}}, \frac{x_{ij4}}{x_{ij4}^{+}} \right), j \in \tag{8.4}$$

$$\mu_{ij} = \left(\frac{x_{\bar{ij}1}}{x_{ij1}}, \frac{x_{\bar{ij}2}}{x_{ij2}}, \frac{x_{\bar{ij}3}}{x_{ij3}}, \frac{x_{\bar{ij}4}}{x_{ij4}} \right), j \in \tag{8.5}$$

where $x_{ij4}^{+} = \max\left(x_{ij4}, j \in J_1\right.$, and $x_{\bar{ij}1} = \min\left(x_{ij1}, j \in J_2\right.$

J_1 corresponds to the higher best value criterion and J_2 corresponds to the lower best value criterion.

Step 5: At this step, perform defuzzification.

The authors do this to get clear numbers for each factor for every option. Defuzzification gives us specific numbers for the words we have been using, so as to understand them better. Using Equation (8.6), we get these clear numbers:

$$f_{ij} = Defuzz(x_{ij}) = \frac{\int \mu(x).xdx}{\int \mu(x).dx}$$

$$= \frac{\int_{x_{ij1}}^{x_{ij2}} \left(\frac{(x - x_{ij1})}{(x_{ij2} - x_{ij1})} \right).xdx + \int_{x_{ij2}}^{x_{ij3}} xdx + \int_{x_{ij3}}^{x_{ij4}} \left(\frac{(x_{ij4} - x)}{(x_{ij4} - x_{ij3})} \right).xdx}{\int_{x_{ij1}}^{x_{ij2}} \left(\frac{(x - x_{ij1})}{(x_{ij2} - x_{ij1})} \right) dx + \int_{x_{ij2}}^{x_{ij3}} dx + \int_{x_{ij3}}^{x_{ij4}} \left(\frac{(x_{ij4} - x)}{(x_{ij4} - x_{ij3})} \right) dx}$$

$$= \frac{-x_{ij1}x_{ij2} + x_{ij3}x_{ij4} + \left(\frac{1}{3}\right)(x_{ij4} - x_{ij3})^2 + \left(\frac{1}{3}\right)(x_{ij2} - x_{ij1})^2}{-x_{ij1} - x_{ij2} + x_{ij3} + x_{ij4}} \tag{8.6}$$

The numbers obtained from those equations, along with the MDL weights, are utilized to determine the final ranking using the TOPSIS method. This process entails:

Step 6: At this stage, normalize the matrix.

The data obtained in the preceding step must undergo normalization, necessitating the creation of a normalized decision matrix. This involves converting all factors into unique and standardized numerical values.

$$r_{ij} = \frac{f_{ij}}{\sqrt{\sum_{i=1}^{m} (f_{ij})^2}}; \forall_j \tag{8.7}$$

Step 7: At this stage, find weighted normalized decision matrix.

In the subsequent step, construct the weighted normalized decision matrix:

$$V_{ij} = [r_{ij}]_{m \times n} \times [W_j]_{n \times m}^{\text{diagonal}} \tag{8.8}$$

Step 8: At this step, find positive ideal and negative ideal solutions.

Two important solutions, namely, the positive ideal solution and the negative ideal solution, play pivotal roles in the decision-making process. The objective is to steer clear of the negative ideal solution while striving to approach the positive ideal solution as closely as feasible.

The positive ideal solution V_j^+ and negative ideal solution V_j^- are as follows:

$$V_j^+ = \left\{\left(\max V_{ij}, j \in J_1\right), \left(\min V_{ij}, j \in J_2\right), i = 1,2,3,\ldots,m\right\}; \forall j \tag{8.9}$$

$$V_j^- = \left\{\left(\min V_{ij}, j \in J_1\right), \left(\max V_{ij}, j \in J_2\right), i = 1,2,3,\ldots,m\right\}; \forall j \tag{8.10}$$

where J_1 and J_2 represent higher best and lower best criteria, respectively.

Step 9: Calculate the separation d_i^+ and d_i^- from the positive ideal solution and negative ideal solution, respectively.

$$d_i^+ = \left[\sum_{j=1}^{n}\left(V_{ij} - V_j^+\right)^2\right]^{0.5} ,i = 1,2,3,\ldots,m \tag{8.11}$$

$$d_i^- = \left[\sum_{j=1}^{n}\left(V_{ij} - V_j^-\right)^2\right]^{0.5} ,i = 1,2,3,\ldots,m \tag{8.12}$$

In Equation (8.11), d_i^+ is the separation of alternative i from the positive ideal solution and in Equation (8.12) d_i^- is the separation of alternative i from the negative ideal solution.

Step 10: At this step, find the rank index.

The relative closeness to the ideal solution for each alternative is calculated as follows:

$$C_i^+ = \frac{d_i^-}{d_i^- + d_i^+}$$

Alternative with higher rank index $\left(C_i^+\right)$ is preferred.

8.5 RESULTS AND DISCUSSION

In the last part, the authors looked at how to choose the best teaching methods and the important factors to consider. Throughout this research, the authors have been getting input from experts in the field to help us make decisions. There are different methods for picking the right teaching style, and used Fuzzy TOPSIS to rank them. This method seems to fit well with the need for our specific situation. The study starts by figuring out how important each factor is using a method called MDL. Then, the authors use fuzzy logic to give scores to different factors based on how they match up with our goals. After that, the authors use the TOPSIS method to find the best solutions that are closest to what they want and farthest from what don't want. Figure 8.3 illustrates the process of determining the optimal teaching style for classrooms. At Level 0, the main goal is to pick the best teaching style from the options, which can be seen at Level 2. Level 1 in Figure 8.3 breaks down all the different things that affect this decision. It's clear that these options depend a lot on these factors, making the decision complex. Each factor has its importance in its own way, so there is a need to figure out which one matters the most. This was done by comparing each factor with the others and giving them scores of 1, 2, or 3 based on how important they are. Using these scores, a decision chart is created and calculates the weights for each factor, as shown in Table 8.2. Among these factors, "Strategies for Student Engagement" stands out as the most important, followed by "Potential Scalability." Figure 8.4 gives a visual look at how much weight each factor carries.

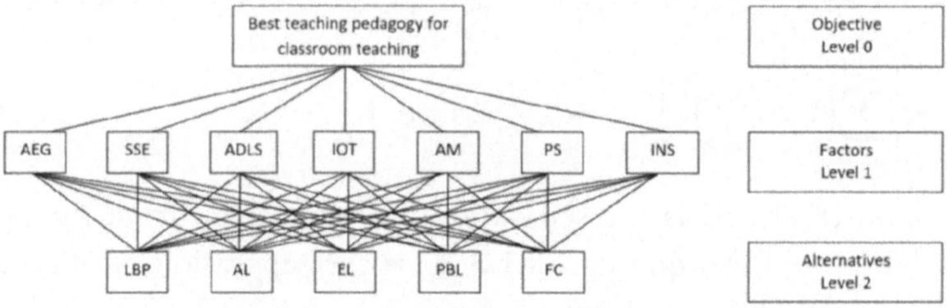

Figure 8.3 Teaching pedagogy selection process.

Table 8.2 Weights using MDL

Factors	*AM*	*AEG*	*SSE*	*ADLS*	*IOT*	*PS*	*IS*	*Calculated positive decision*	*Weightage*
AM	2	1	1	3	3	1	1	10	0.119
AEG	3	2	1	1	3	2	3	13	0.155
SSE	3	3	2	3	3	3	3	18	0.214
ADLS	1	3	1	2	1	1	1	8	0.095
IOT	1	1	1	3	2	1	2	9	0.107
PS	3	2	1	3	3	2	3	15	0.178
IS	3	1	1	3	2	1	2	11	0.131
Sum								84	

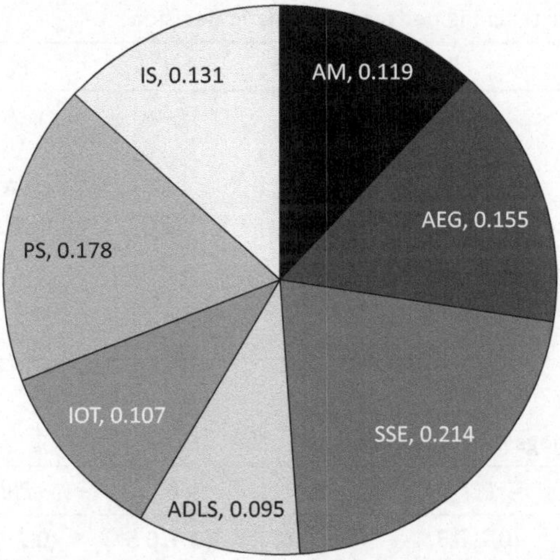

Figure 8.4 Factors weightage.

In this study, the authors compare all the options for each factor. A method is required to consolidate all this comparison data, and a hypothetical scale serves this purpose. Because this task can get pretty complicated, something called fuzzy logic is utilized, which is highly effective for handling complex problems. Fuzzy logic lets us compare options based on words instead of just numbers. These word comparisons have been converted into fuzzy numbers, which is presented in Table 8.3. In this table, "Exceptionally High" means the best value, while "Extremely Low" means the worst.

Tables 8.4 and 8.5, display the comparison results and fuzzy ratings for the case. Equations (8.4) and (8.5) are employed to ensure the fairness and comparability of all fuzzy numbers. Subsequently, these fuzzy numbers are converted into clear numbers, depicted in Table 8.6. The TOPSIS method is then utilized to rank all the options based on these numbers, as illustrated in Tables 8.7 and 8.8. Based on the calculations, experiential learning emerges

Table 8.3 Fuzzy number and linguistic variables

Linguistic variables	*Fuzzy number*
Exceptionally high (EH)	(0.8, 0.9, 1.0, 1.0)
Very high (VH)	(0.7, 0.8, 0.8, 0.9)
High (H)	(0.5, 0.6, 0.7, 0.8)
Above average (AA)	(0.4, 0.5, 0.5, 0.6)
Average (A)	(0.2, 0.3, 0.4, 0.5)
Very low (VL)	(0.1, 0.2, 0.2, 0.3)
Extremely low (EL)	(0.0, 0.0, 0.1, 0.2)

Table 8.4 Matrix for teaching pedagogy linguistic decision

Factors considered	*LBP*	*AL*	*EL*	*PBL*	*FC*
AM	A	H	AA	A	H
AEG	A	A	H	VL	A
SSE	VL	VL	H	A	A
ADLS	EL	EL	EL	EL	VH
IOT	A	VL	H	A	A
PS	EL	EL	VH	A	A
IS	A	VL	A	EH	A

Table 8.5 Fuzzy ratings of the factors

Factors considered	*LBP*	*AL*	*EL*	*PBL*	*FC*
AM	(0.2, 0.3, 0.4, 0.5)	(0.5, 0.6, 0.7, 0.8)	(0.4, 0.5, 0.5, 0.6)	(0.2, 0.3, 0.4, 0.5)	(0.5, 0.6, 0.7, 0.8)
AEG	(0.2, 0.3, 0.4, 0.5)	(0.2, 0.3, 0.4, 0.5)	(0.5, 0.6, 0.7, 0.8)	(0.1, 0.2, 0.2, 0.3)	(0.2, 0.3, 0.4, 0.5)
SSE	(0.1, 0.2, 0.2, 0.3)	(0.1, 0.2, 0.2, 0.3)	(0.5, 0.6, 0.7, 0.8)	(0.2, 0.3, 0.4, 0.5)	(0.2, 0.3, 0.4, 0.5)
ADLS	(0.0, 0.0, 0.1, 0.2)	(0.0, 0.0, 0.1, 0.2)	(0.0, 0.0, 0.1, 0.2)	(0.0, 0.0, 0.1, 0.2)	(0.7, 0.8, 0.8, 0.9)
IOT	(0.2, 0.3, 0.4, 0.5)	(0.1, 0.2, 0.2, 0.3)	(0.5, 0.6, 0.7, 0.8)	(0.2, 0.3, 0.4, 0.5)	(0.2, 0.3, 0.4, 0.5)
PS	(0.0, 0.0, 0.1, 0.2)	(0.0, 0.0, 0.1, 0.2)	(0.7, 0.8, 0.8, 0.9)	(0.2, 0.3, 0.4, 0.5)	(0.2, 0.3, 0.4, 0.5)
IS	(0.2, 0.3, 0.4, 0.5)	(0.1, 0.2, 0.2, 0.3)	(0.2, 0.3, 0.4, 0.5)	(0.8, 0.9, 1.0, 1.0)	(0.2, 0.3, 0.4, 0.5)

Table 8.6 Assigned fuzzy rates and crisp value

Factors considered	*LBP*	*AL*	*EL*	*PBL*	*FC*
AM	1.833	3.333	2.667	1.833	3.333
AEG	0.458	0.458	0.833	0.292	0.458
SSE	0.292	0.292	0.833	0.458	0.458
ADLS	0.086	0.086	0.086	0.086	0.926
IOT	0.458	0.292	0.833	0.458	0.458
PS	0.086	0.086	0.926	0.407	0.407
IS	0.367	0.256	0.367	0.944	0.367

as the most suitable teaching pedagogy for classroom instruction, considering the selected factors and their respective weightage. Following experiential learning, project-based learning is identified as the next best alternative, ranking second. The overall priority and rank for each pedagogy are as

Table 8.7 Weights of the factors and crisp values

Weightage	*0.119*	*0.115*	*0.214*	*0.095*	*0.107*	*0.178*	*0.131*
	AM	AEG	SSE	ADLS	IOT	PS	IS
LBP	1.833	0.458	0.292	0.086	0.458	0.086	0.367
AL	3.333	0.458	0.292	0.086	0.292	0.086	0.256
EL	2.667	0.833	0.833	0.086	0.833	0.926	0.367
PBL	1.833	0.292	0.458	0.086	0.458	0.407	0.944
FC	3.333	0.458	0.458	0.926	0.458	0.407	0.367

Table 8.8 Rank index, positive ideal solution, and negative ideal solution distance

	d_i^+	d_i^-	*Rank index*	*Rank*
LBP	0.190677	0.094272	0.331	III
AL	0.204482	0.078643	0.278	IV
EL	0.077665	0.243997	0.759	I
PBL	0.129096	0.159459	0.553	II
FC	0.262211	0.055889	0.176	V

follows: EL with a priority of 0.759, followed by PBL with a priority of 0.553, LBP with a priority of 0.331, AL with a priority of 0.278, and FC with a priority of 0.176.

In terms of the factors utilized in the study, they are weighted in the following order of importance: SSE with a weightage of 0.214, PS with a weightage of 0.178, AEG with a weightage of 0.155, IoT with a weightage of 0.131, AM with a weightage of 0.119, ADLS with a weightage of 0.107, and Importance of Student Engagement (IS) with a weightage of 0.095.

The proposed method offers various advantages: it is characterized by its simplicity and user-friendliness, ensuring ease of use. Moreover, the number of steps remains constant irrespective of the number of factors involved, enhancing its efficiency and applicability. However, certain limitations should be acknowledged. Challenges may arise due to the interdependence between factors and alternatives, potentially complicating the decision-making process. Additionally, the selection of factors is contingent upon the expertise and knowledge of the experts involved, introducing a degree of subjectivity into the process. Figure 8.5 depicts the rank index achieved through TOPSIS.

In addition to applying the Fuzzy TOPSIS methodology, the authors also utilized the AHP to verify the results obtained from TOPSIS. Interestingly, the final ranking generated by AHP corroborated the findings of TOPSIS, with EL emerging as the top-ranked teaching pedagogy for classroom learning. This consistency between the two methodologies adds robustness to our conclusions and underscores the effectiveness of EL in enhancing

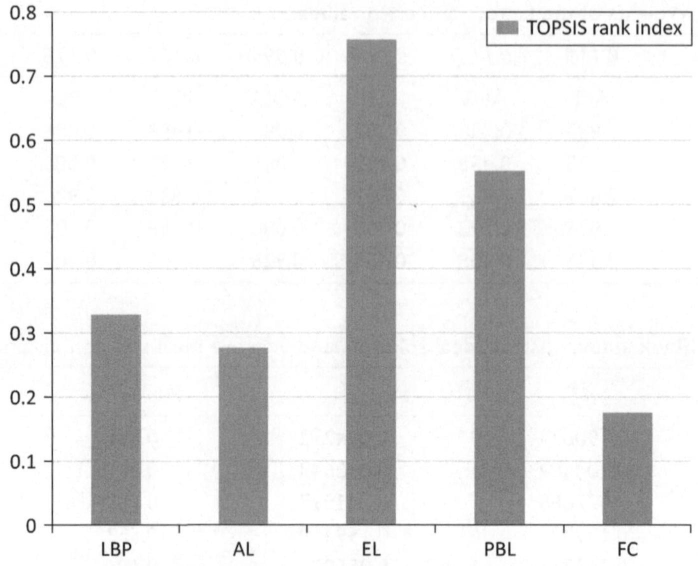

Figure 8.5 TOPSIS rank index value.

classroom learning experiences. Table 8.9 indicates the ranking obtained by AHP .

Implementing AHP provides numerous advantages for decision-makers across various domains. First, AHP offers a structured approach to decision-making, enabling the decomposition of complex problems into manageable hierarchical structures of criteria and alternatives. This structured framework fosters clarity and transparency, facilitating a better understanding of decision factors and relationships. Additionally, AHP allows decision-makers to quantitatively analyze qualitative factors through pairwise comparisons, providing a systematic method for incorporating subjective judgments into the decision-making process. By prioritizing criteria and alternatives based on their relative importance, AHP helps focus attention on key factors, leading to more efficient resource allocation and improved decision outcomes. Moreover, AHP's consistency checks and sensitivity analysis enhance decision robustness, ensuring that decisions are well-informed and aligned with organizational objectives and stakeholder preferences.

Furthermore, AHP serves as a versatile decision support tool applicable across a wide range of contexts, including strategic planning, project prioritization, and resource allocation. Its flexibility makes it suitable for addressing complex decision problems in diverse industries and sectors. By providing a rational framework for evaluating alternatives and assessing trade-offs, AHP aids decision-makers in making informed choices that align with organizational goals and priorities. Overall, the implementation of AHP enhances decision-making processes by promoting systematic

Table 8.9 Analytic hierarchy process ranking

Weights	*0.119*	*0.115*	*0.214*	*0.095*	*0.107*	*0.178*	*0.131*		
	AM	AEG	SSE	ADLS	IOT	PS	IS	Overall priority vector	Rank
LBP	0.261	0.111	0.143	0.167	0.167	0.111	0.167	0.130	V
AL	0.261	0.111	0.143	0.167	0.167	0.111	0.167	0.130	IV
EL	0.087	0.333	0.429	0.167	0.333	0.333	0.167	0.214	I
PBL	0.130	0.222	0.143	0.167	0.167	0.222	0.333	0.149	III
FC	0.261	0.222	0.143	0.333	0.167	0.222	0.167	0.159	II

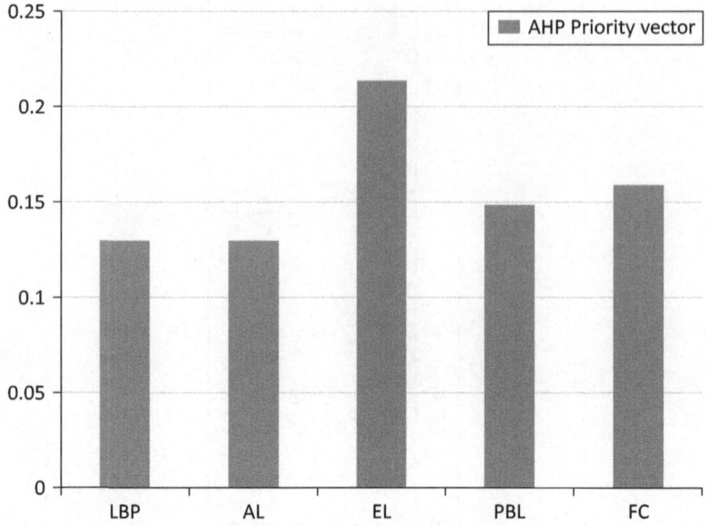

Figure 8.6 AHP priority vector36.

analysis, transparency, and stakeholder engagement, ultimately leading to more effective and sustainable decisions. Figure 8.6 demonstrates the AHP priority vector.

The alignment of results from both TOPSIS and AHP reaffirms the suitability of EL as the optimal instructional approach, providing educators with valuable insights for effective pedagogy selection. This finding highlights the importance of employing multiple methodologies to ensure the reliability and validity of research outcomes in complex decision-making processes.

8.6 CONCLUSIONS

In this study, the authors employed the multiple attribute decision-making approach to select the best teaching pedagogy for classroom learning in higher education in engineering. They propose the feasibility of utilizing the Fuzzy TOPSIS method for this purpose. To evaluate the alternatives, utilize the MDL to determine the factor weightage for those factor that are considered. It is noteworthy that "Strategies for Student Engagement" is identified as the most critical factor, whereas "Accommodation of Diverse Learning Styles" is considered the least critical. By integrating the Fuzzy Technique for Order Preference by Similarity to Ideal Solution approach with weights calculated by modified digital logic approach, establish a priority order for teaching pedagogy. Our findings indicate that experiential learning emerges as the most suitable approach for classroom teaching, followed by project-based learning and lecture-based pedagogy, which rank second and third, respectively. The AHP approach also has been implemented on the same problem. The results indicate that the EL got the first rank here as well. This justifies the top rank achieved by TOPSIS. In future, other MCDM approaches can be applied to check the result's variation. In addition, the output of this study can be compared with the classroom surveys taken from engineering students. Also, another MCDM approach can be implemented for the same problem.

REFERENCES

Ahmadi Tabatabaie, S. M. A., & Moosav, S. M. (2024). The impact of software pedagogy on architectural creativity: Finding the appropriate method and time for teaching software to architecture students. *Technology of Education Journal (TEJ)*, *18*(1).

Amale, A., Singholi, A. K. S., P, S., Giri, J., Albaijan, I., & Guru, A. (2024). Fuzzy logic-driven genetic algorithm strategies for ultrasonic welding of heterogeneous metal sheets. *AIP Advances*, *14*(5), 055117. https://doi.org/10.1063/5.0207578

Aziz, A., & Islam, S. N. (2021). Impact of mixed pedagogy on engineering education. *IEEE Transactions on Education*, *65*(1), 56–63.

Barnett, R., & Hallam, S. (1999). Teaching for supercomplexity: A pedagogy for higher education. *Understanding pedagogy and its impact on learning*, 137–154.

Baxter, J. A., & Lederman, N. G. (1999). Assessment and measurement of pedagogical content knowledge. In Examining pedagogical content knowledge: The construct and its implications for science education (pp. 147–161). Dordrecht: Springer Netherlands.

Beatty, I. D., & Gerace, W. J. (2009). Technology-enhanced formative assessment: A research-based pedagogy for teaching science with classroom response technology. *Journal of Science Education and Technology*, *18*, 146–162.

Behzadian, M., Otaghsara, S. K., Yazdani, M., & Ignatius, J. (2012). A state-of the-art survey of TOPSIS applications. *Expert Systems with Applications*, *39*(17), 13051–13069.

Bellman, R. E., & Zadeh, L. A. (1970). Decision-making in a fuzzy environment. *Management Science*, *17*(4), 8141–8164.

Bidabadi, N. S., Isfahani, A. N., Rouhollahi, A., & Khalili, R. (2016). Effective teaching methods in higher education: requirements and barriers. *Journal of Advances in Medical Education & Professionalism*, *4*(4), 170.

Carvalho, A., Teixeira, S. J., Olim, L., Campanella, S. D., & Costa, T. (2021). Pedagogical innovation in higher education and active learning methodologies–a case study. *Education + Training*, *63*(2), 195–213.

Ching, G. S., & Roberts, A. (2020). Evaluating the pedagogy of technology integrated teaching and learning: An overview. *International Journal of Research*, *9*(6), 37–50.

Compayré, G., & Payne, W. H. (2015). *The history of pedagogy*. Routledge.

Corbin, L., & Bugden, L. (2018). Online teaching: The importance of pedagogy, place and presence in legal education. *Legal Education Review*, *28*, 1–21.

Dai, Y. (2014). Comparing discussion and lecture pedagogy when teaching oral communication in business course. *Research in Higher Education Journal*, 22.

Dalton, S. S. (1998). *Pedagogy matters: Standards for effective teaching practice.*

Deng, H., Yeh, C. H., & Willis, R. J. (2000). Inter-company comparison using modified TOPSIS with objective weights. *Computers & Operations Research*, 27(10), 963–973.

Ekpok, H. A., & Dukku, M. K. (2021). Innovative teaching methods and pedagogy of the twenty-first century. *Journal of African Studies and Sustainable Development.*

Entz, S. (2007). Why Pedagogy Matters: The importance of teaching in a standards-based environment. In *Forum on public policy online* (Vol. 2007, No. 2, p. n2). Oxford Round Table, 406 West Florida Avenue, Urbana, IL 61801.

Esfandiari, M., & Rizvandi, M. (2014). An application of TOPSIS method for ranking different strategic planning methodology. *Management Science Letters*, *4*(7), 1445–1448.

Franklin, H., & Harrington, I. (2019). A review into effective classroom management and strategies for student engagement: Teacher and student roles in today's classrooms. *Journal of Education and Training Studies.*

Geleta, A., Teshome, Z., & Zewdie, M. (2022). EFL Teachers' awareness and accommodations of Learners' different learning styles in ELT context: Two colleges of Teachers' education in Oromia, Ethiopia. *Journal of Science, Technology and Arts Research*, *11*(3), 65–78.

Honigsfeld, A., & Schiering, M. (2004). Diverse approaches to the diversity of learning styles in teacher education. *Educational Psychology*, *24*(4), 487–507.

Kabulova, Z. K. (2023). The history of pedagogy as a teaching subject. *Web of Teachers: Inderscience Research*, *1*(7), 153–157.

Karthik, G., Parthiban, A., Sathish, T., & Giri, J. (2024). Experimental analysis and development of fuzzy model for the prediction of COP and mass flow rate in air-conditioning system. *Deleted Journal*, *245*(1), 134. https://doi.org/10.1007/s10751-024-01979-x

Kersten, S. (2018). Approaches of engineering pedagogy to improve the quality of teaching in engineering education. *Vocational Teacher Education in Central Asia: Developing Skills and Facilitating Success*, 129–139.

Koepke, K., & O'Brien, A. (2012). Advancing pedagogy: Evidence for the role of online instructor training in improved pedagogical practices. *Journal of Asynchronous Learning Networks*, *16*(2), 73–84.

Krajcik, J., McNeill, K. L., & Reiser, B. J. (2008). Learning-goals-driven design model: Developing curriculum materials that align with national standards and incorporate project-based pedagogy. *Science Education*, *92*(1), 1–32.

Lai, Y. J., Liu, T. Y., & Hwang, C. L. (1994). Topsis for MODM. *European Journal of Operational Research*, *76*(3), 486–500.

Larsen, D., Wallace, S., & Pankl, L. (2018). Mapping library values and student learning outcomes: alignment with university pedagogical goals and practices. *Journal of Information Literacy*, *12*(1).

Law, N. (2014). Comparing pedagogical innovations. In *Comparative education research: Approaches and methods* (pp. 333–364). Cham: Springer International Publishing.

Law, N., Chow, A., & Yuen, A. H. (2005). Methodological approaches to comparing pedagogical innovations using technology. *Education and Information Technologies*, *10*, 7–20.

Loughran, J. (2013a). Pedagogy: Making sense of the complex relationship between teaching and learning. *Curriculum Inquiry*, *43*(1), 118–141.

Loughran, J. (2013b). *Developing a pedagogy of teacher education: Understanding teaching & learning about teaching*. Routledge.

Lupton, R., & Hempel-Jorgensen, A. (2012). The importance of teaching: pedagogical constraints and possibilities in working-class schools. *Journal of Education Policy*, *27*(5), 601–620.

Lutfina, E., & Nuraisha, S. (2020, February). Weighting of the attribute importance weights using improved multiple linear regression and modified digital logic for process-based fraud detection. In *Engineering, Information and Agricultural Technology in the Global Digital Revolution: Proceedings of the 1st International Conference on Civil Engineering, Electrical Engineering, Information Systems, Information Technology, and Agricultural Technology (SCIS 2019)*, July 10, 2019, Semarang, Indonesia (p. 143). CRC Press.

Malunda, P., Onen, D., Musaazi, J. C., & Oonyu, J. (2016). Teacher evaluation and quality of pedagogical practices. *International Journal of Learning, Teaching and Educational Research*, *15*(9), 118–133.

Mehrjerdi, Y. Z. (2013). Hierarchical multi-criteria risk-benefit analysis in fuzzy environment. *Applied Soft Computing*, *13*(1), 590–599.

Mioduser, D., Nachmias, R., Forkosh-Baruch, A., & Tubin, D. (2004). Sustainability, scalability and transferability of ICT-based pedagogical innovations in Israeli schools. *Education, Communication & Information*, *4*(1), 71–82.

Niederhauser, D. S., Howard, S. K., Voogt, J., Agyei, D. D., Laferriere, T., Tondeur, J., & Cox, M. J. (2018). Sustainability and scalability in educational technology initiatives: Research-informed practice. *Technology, Knowledge and Learning*, *23*, 507–523.

Okojie, M. C., Olinzock, A. A., & Okojie-Boulder, T. C. (2006). The pedagogy of technology integration. *Journal of technology studies*, *32*(2), 66–71.

Olson, D. L. (2004). Comparison of weights in TOPSIS models. *Mathematical and Computer Modelling*, *40*(7–8), 721–727.

Olson, D. R. (2014). The natural history of pedagogy. In *Fostering change in institutions, environments, and people* (pp. 9–22). Routledge.

Papathanasiou, J., Ploskas, N., Papathanasiou, J., & Ploskas, N. (2018). Topsis. In *Multiple Criteria Decision AiddfZ: Methods, Examples and Python Implementations*, pp. 1–30. Springer.

Parsons, J., & Taylor, L. (2011). Improving student engagement. *Current Issues in Education*, *14*(1).

Phanden, R. K., Sindhwani, R., Kalsariya, V., & Salroo, F. (2019). Selection of material for electric arc spraying by using hierarchical entropy-TOPSIS approach. *International Journal of Productivity and Quality Management*, *26*(3), 276–289.

Salvatori, M. R. (Ed.). (2003). *Pedagogy: Disturbing History, 1820–1930*. University of Pittsburgh Pre.

Seatter, C. S., & Ceulemans, K. (2017). Teaching sustainability in higher education: Pedagogical styles that make a difference. *Canadian Journal of Higher Education*, *47*(2), 47–70.

Sharma, P., & Singhal, S. (2016). Comparative analysis of procedural approaches for facility layout design using AHP approach. *International Journal of Manufacturing Technology and Management*, *30*(5), 279–288.

Somekh, B. (2008). Factors affecting teachers' pedagogical adoption of ICT. *International Handbook of Information Technology in Primary and Secondary Education*, 449–460.

Swain, K., Cherukuri, M., Samanta, I. S., Pati, A., Giri, J., Panigrahi, A., Qin, H., & Mallik, S. (2024). Fuzzy Markov model for the reliability analysis of hybrid microgrids. *Frontiers in Computer Science*, *6*, 1406086. https://doi.org/10.3389/fcomp.2024.1406086

Thakur, R., Sikka, G., Bansal, U., Giri, J., & Mallik, S. (2024b). Deadline-aware and energy efficient IoT task scheduling using fuzzy logic in fog computing. *Multimedia Tools and Applications*, 1–28. https://doi.org/10.1007/s11042-024-19509-w

Torrez, J. B., Milani, A. S., & Schuetze, D. D. C. (2012). Material selection for light-weight naval crafts using the modified digital logic method: dealing with ranking abnormalities. *International Journal of Materials and Structural Integrity*, *6*(2–4), 241–256.

Wentworth, N., Graham, C. R., & Tripp, T. (2008). Development of teaching and technology integration: Focus on pedagogy. *Computers in the Schools*, *25*(1–2), 64–80.

Werth, E., & Williams, K. (2023). Learning to be open: Instructor growth through open pedagogy. *Open Learning: The Journal of Open, Distance and e-Learning*, *38*(4), 301–314.

Yakovleva, N. O., & Yakovlev, E. V. (2014). Interactive teaching methods in contemporary higher education. *Pacific Science Review*, *16*(2), 75–80.

Zadeh, L. A. (1965). Fuzzy sets. *Information and Control*, *8*(3), 338–353.

Part II

Advance trends in industrial management

Chapter 9

A bibliometric analysis and management of big data and machine learning research

Current trends and future directions

Inderdeep Kaur and Inzimam Ul Hassan

9.1 INTRODUCTION

Big data is now in vogue, and it plays a big part in the contemporary information technology (IT) landscape. Over 3 billion people used the internet globally in 2015 [1]. Data complexity has risen as a result of the sharp rise in both structured and unstructured data, as well as the proliferation of applications made possible by social media, the Internet of Things, multimedia, and other rapidly developing technologies. Researchers often use the four V's to characterize big data: velocity, volume, variety, and veracity [2]. Three more V's for large data were presented in the interim by another study: volatility, validity, and the unique V for value. To show the effectiveness of ML approaches in massive data, both intelligent and non-intelligent tactics are used to examine big data [3]. Large data is becoming more and more necessary, according to academics, and this will have a large influence on various areas, including society, educational institutions, computer science, social media, healthcare, politics, and economic systems. While machine learning is not a novel notion in computing, the ever-changing needs of the modern world have given rise to new applications for the concept. A significant amount of cutting-edge data is being produced as a result of the web's growth, suggesting that there will be more data available for machines to learn from and analyze in the future. Today, machine learning calculations enable computers to produce and coordinate reports, and communicate with humans and autonomous vehicles. The three subfields of ML are unsupervised, supervised, and reinforcement learning [4, 32, 33]. Some machine learning algorithms for handling huge data require reiteration since they are inefficient and unable to handle data that is huge in volume, value, velocity, and variety [5]. With traditional machine learning computations, adaptability is a challenging issue [6, 31]. If a material science-based model is supplied and an ML technique is employed to repair a calculation flaw, numerical findings may be adequate in requests to perform suitable execution measures [7, 34, 35]. Spam channels, ad placement, web search, recommender systems, credit scoring, stock trading, misrepresentation detection, tranquilizer structure, and many more applications all

DOI: 10.1201/9781003512585-11

use machine learning [8]. Everyone thinks of overfitting when discussing machine learning, but it can take many different forms that are not immediately obvious. Big data and machine learning together offer incredible forecasting potential [9]. Because most real-life applications involve data which is a large amount or a big data information base, big data analysis is one of the extraordinarily challenging problems for ML computations [10]. Data are produced in the developed world from a variety of sources, and big data has expanded due to the quick changes brought on by advancements in technology [7, 8]. It provides expanding successes throughout a variety of fields with a vast array of datasets. Not all large data that is available is useful for analysis or dynamic processes. The big data insights are eagerly being shared by business and academia [9]. Support vector machines (SVMs), artificial neural systems, choice tree learning, Bayesian networks, clustering, and other techniques all are examples of machine learning technologies [10]. Finding knowledge and deciding to make informed decisions are its goals. Deep machine learning is a method of ML that uses many levels of data handling stages in different leveled configurations [11, 12]. It depicts the observational data in multiple leveled highlights or depictions, with the higher-level highlights being distinguished from the lower-level highlights. Through a variety of tiered learning processes, DL algorithms extract significant level, complex discussions as data depictions.

9.2 CHALLENGES OF BIG DATA ANALYSIS

People are moving forward in innovation at an amazing rate, and they have a growing need for accurate data. ML has been applied to big data as well as to many other domains. It encompasses a variety of equipment, methods, and technical advancements for efficiently processing data of any size [13, 14]. The ability to access vast quantities of data, improvements in the preparation capacity of modern computers, and growth in the storage limit all have contributed to advancements in the processing of big data [15]. The technology and programming developments of the present day enable the management, processing, and dissection of enormous amounts of data like never before. To find examples, designs, and associations related to human behavior and interaction, very large arrangements of information can be acquired and studied. Big data is being prepared by everyone to get advantages through processing and using various big data handling technologies [16]. Traditional social databases are ill-equipped to handle enormous amounts of data [17]. These datasets are difficult to manage with the current data preparation mechanisms [18]. Additionally, the big data enhances several particularly challenging problems related to data storage, perception, progress, data protection infringement, infiltration, security and examination, and sharing [19]. It is practically everywhere. Large data analytics may be used in each organization, for instance, to assess employee well-being or general expectations for everyday comforts [20]. By skillfully imagining and displaying the material in a sensible way, one can practice this. A growing advantage for mobility and the growth of developments is big data analytics [21]. It allows for the equal use of structured and unstructured data in scientific, quantifiable, and machine learning computations. Big data has a limited number of problems, but even with successfully implementing the innovation, one should undoubtedly be considered by its association [22]. This study is significant because most bibliometric studies concentrate on a handful of variables like the number of references per year, authorship index, author's country of origin, university of origin, language, or number of articles downloaded [23]. Using bibliometrics, this study analyzes co-words and bibliometric metrics such as the h-index to produce node-filled maps that show the distribution and performance of different conceptual subdomains linked to the terms "ML "Big Data" [24, 25].

9.3 LITERATURE REVIEW

Smith et al. [1] in their research work, "Big Data and (ML) Machine Learning: A Systematic Review of the Literature," provide a thorough overview of the literature on big data and ML topics. It analyzes the current trends in research, including the application of ML techniques to big data analytics, challenges and limitations, and future directions for research in this field.

Chen et al. [2] in their research work, "Bibliometric Analysis of Big Data Research: A Scientometric Perspective," conduct a bibliometric analysis of big data research using scientometric methods, including co-word analysis, co-citation analysis, and social network analysis. It identifies key research themes, influential authors, and major research clusters in the area of large data analysis, providing insights into the current trends as well as the direction for the future in this area.

Wang et al. [3] in their research work, "ML for Big Data: A Review and Taxonomy," present a comprehensive taxonomy of ML techniques for big data analysis. It categorizes various machine learning algorithms based on their characteristics, such as supervised versus unsupervised learning, batch versus online learning, and model-based versus feature-based learning. It provides a systematic overview of the state of large data ML research at the moment and highlights future research directions in this field.

Zhang et al. [4] in their article, "Emerging Trends in Big Data Analytics and Machine Learning: Recent Research and Future Directions," review the emerging trends in big data analytics and machine learning, focusing on recent research and future directions. It discusses various techniques and approaches for handling big data, including distributed computing, parallel processing, and machine learning algorithms. It also lists potential and problems with big data analytics and ML research and provides insights into future research directions.

Alazab et al. [5] in their article focus on how big data analytics interacts with the machine learning and cybersecurity. It provides an overview of the current trends and challenges in using ML techniques for cybersecurity, including data privacy, data quality, and interpretability of machine learning models. It also highlights future research directions in this area, such as explainable artificial intelligence (AI), adversarial machine learning, and federated learning for cybersecurity.

Akhlaq et al. [6] in their article review the application that how big data analytics are used in the field of healthcare. The article provides an overview of the current research landscape in this area, including the use of machine learning techniques for healthcare data analysis, challenges, and future directions. It identifies key research themes, methodologies, and applications of big data analytics in healthcare, providing insights into the state of the field and potential future research directions.

Zhang et al. [7] in their article focus on the intersection of big data analytics, machine learning, and sustainable development goals (SDGs). The article provides an overview of the research landscape in this area, including key topics, methodologies, and research trends. The analysis identifies potential future research directions, such as the use of big data and machine learning for addressing social and environmental issues related to SDGs, and provides insights into the current state and future prospects of research at this intersection.

Kumar et al. [8] in their article, "A Bibliometric Analysis of Machine Learning for Social Media Research: Current Status, Challenges, and Future Directions," focus on the use of machine learning techniques for social media research. The article provides insights into the current status, challenges, and future directions of research in this area through a comprehensive analysis of literature. The study identifies key research themes, methodologies, and applications of machine learning in social media research, and discusses challenges and opportunities in this field. It also highlights potential future research directions, such as explainable machine learning, ethical considerations, and the integration of social media analytics with other fields.

Paniagua et al. [9] in their article, "Machine Learning for Internet of Things," focus on the application of ML techniques for the IoT. It provides an overview of the current research landscape in this area, including key research topics, machine learning algorithms, and applications of IoT. The review identifies the challenges and limitations of machine learning for IoT and discusses potential future research directions, such as edge computing, federated learning, and secure machine learning for IoT.

Huang et al. [10] in their article, "Mapping the Landscape of AI in Education: A Bibliometric and Scientometric Analysis," conduct a bibliometric and scientometric analysis to map the landscape of AI in education research. It provides insights into the research trends, methodologies, and key topics in this field, using bibliometric and scientometric techniques. The study identifies influential publications, authors, and institutions, and visualizes the intellectual structure of the field through co-citation and co-authorship networks. The results offer a thorough summary of the state of AI today in education research and highlight potential future research directions.

In order to index and manage digital cultural resources, Barbuti et al. [11] outlines a proposal that should direct the design and creation of descriptive metadata in an open format. It begins with extending the FAIR Principles' R: Reusable requirement to R5, where it is joined by the requirements for Readable, Relevant, Reliable, and Resilient.

Nobanee et al. [12] aim to investigate the corpus of literature on applications of big data in banking through the use of bibliometric analytic methodology. This technique describes citation rates, research outputs, and the ways in which those results were applied, in addition to the future research agenda and streams that are currently active in the field. The articles were selected between 2012 and 2020 and sorted by the number of citations in the analysis and results. Despite the rapid expansion of big data applications across several industrial sectors, such as banking, they have identified just 60 papers that are relevant to this field of research.

Wang et al. [3] in their research work analyze publication trends, research topics, and influential publications, authors, and institutions in this field. The research sheds light on the current situation of machine learning applications in transportation, identifies emerging research directions, and highlights potential future research opportunities.

Chi et al. [13] in their research conduct a bibliometric analysis of the applications of ML in supply chain management research. The authors analyze publication trends, research topics, and influential publications, authors, and institutions in this field. The research sheds light on the existing situation of ML applications in supply chain management, identifies research gaps and challenges, and suggests future research directions.

Begum et al. [14] in their research work present a bibliometric analysis of big data research, focusing on global publication output during the period 2007–2016. The authors analyze publication trends, research areas, and influential publications, authors, and nations working in the big data sector. The study highlights the research landscape of big data, identifies key research areas, and highlights potential future research directions.

These articles offer a thorough summary of the most recent developments and potential paths for big data and machine learning research, as well as insights into influential publications, authors, and institutions in the field. Research gaps in the disciplines of big data and machine learning include increasing the scalability and efficiency of methods for processing large datasets and developing distributed computing frameworks. There is a need for enhanced data integration methods [26], automated data cleansing, and managing missing data [29]. Improving real-time data processing [28] in streaming applications is critical. Furthermore, enhancing the interpretability [30] and explainability of complicated machine learning models, particularly deep learning, is a major problem [27].

9.4 METHODOLOGY

9.4.1 Searching and identification of data

Web of Science (WoS) and Scopus are the most prominent online databases for collecting bibliometric analysis data for research purposes. WoS was chosen above other databases because it provides a more comprehensive citation and abstract search feature for peer-reviewed articles. The materials in this study include scientific publications published between 2014 and 2024 when research on this topic began to gain traction and interest. The study articles were collected utilizing a complete sampling method in order to conduct a bibliometric analysis. The dataset contains bibliometric elements that are examined: publication title, author, abstract, keywords, publication year, journal, kind of publishing, and affiliations. To retrieve relevant content from the WoS database, the study used search keywords such as "machine learning" and "big data." This search strategy prioritizes papers on machine learning and big data in the computer science and engineering category. Conference proceedings, journal papers, and article reviews in English were evaluated. Initially, 947 papers were eventually identified. After the preprocessing, the final bibliometric study included 412 documents related to this topic. Preferred Reporting Items for Meta-Analysis (PRISMA) was used to identify and screen papers for final bibliometric analysis, as shown in Figure 9.1.

9.4.2 Analysis of data

The "R" tool, a powerful programming language for statistical computation and graphics, excels in data analysis, modeling, and visualization, aided by a large package ecosystem that includes dplyr and ggplot2, is used in this chapter for the analysis of the data. Biblioshiny, an extension of the R program bibliometrix, provides a user-friendly online interface for bibliometric analysis, allowing academics to do citation and co-citation of studies, as well as thematic mapping. It can import data from sources such as Scopus and Web of Science and provide sophisticated visualizations to detect research trends and notable authors. Biblioshiny democratizes

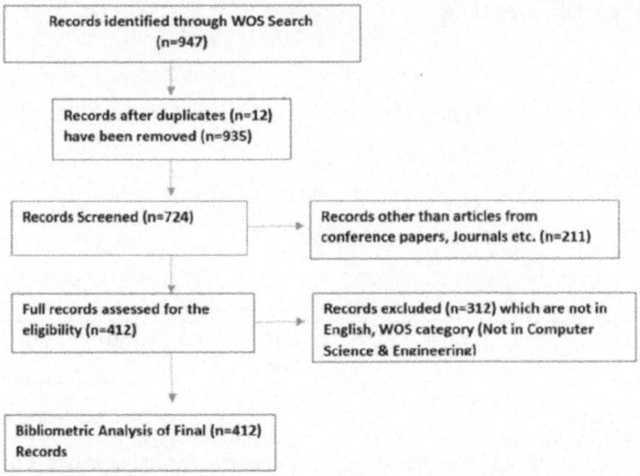

Figure 9.1 Flowchart diagram of the study using PRISMA guidelines.

bibliometric analysis, making it accessible to anyone without sophisticated programming skills by employing R's powerful analytical capabilities.

9.5 FINDINGS AND DISCUSSION

From its first to the publication in 2014 until 2024, Figure 9.2 depicts the Annual Scientific Production of documents of big data analysis employing machine learning algorithms. The output of publications appears to have generally increased with time. When it comes to large data analysis utilizing machine learning, only 38 papers were published in 2017, while 45 papers were published in 2019. After one year, the number of papers produced per year increased from 45 to 56. The number of papers published each year increased significantly from 20 to 55 between 2016 and 2018.

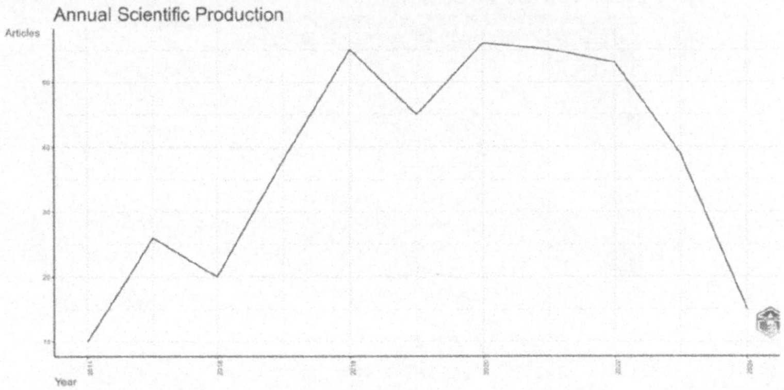

Figure 9.2 Annual Scientific Production of documents on big data and machine learning.

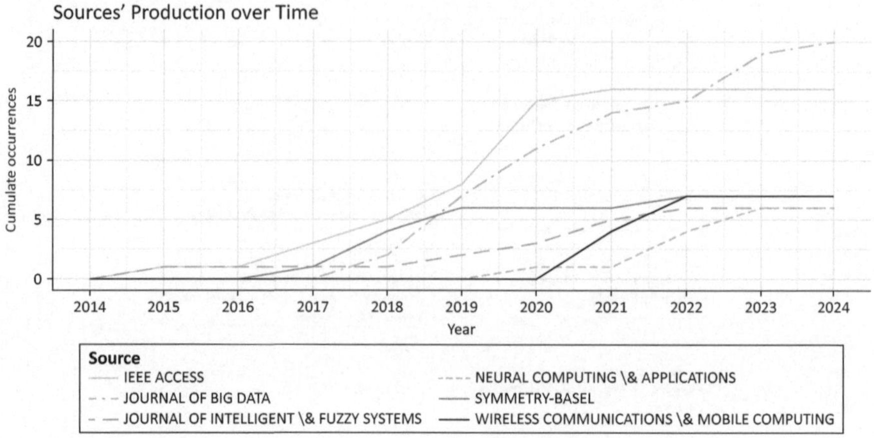

Figure 9.3 Source's production over time.

Figure 9.3 displays the most relevant sources of production over time that are most pertinent. All of these findings suggested that, despite the varied scope and objectives of machine learning, many academics utilize machine learning algorithms and approaches to big data.

Figure 9.3 shows the distribution of data by publishing types for machine learning in various domains, including conference papers, editorial materials journal articles, reviews, books, and book chapters. The h-index, along with the g-index, m-index, and number of citations, was used to measure the academic achievement over different time periods. The results showed that the theme of "big data" and "machine learning" had the highest bibliometric indicators.

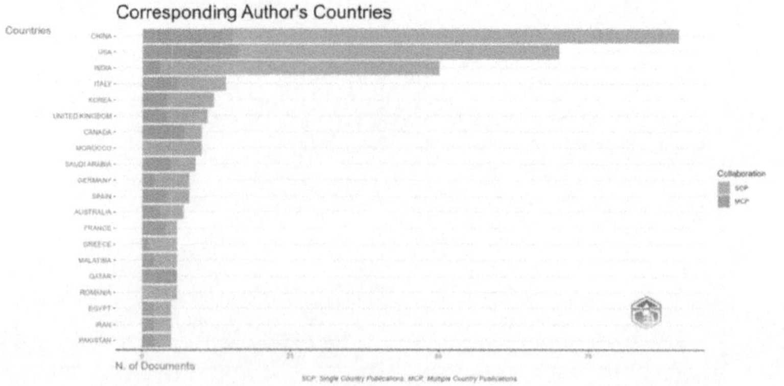

Figure 9.4 Corresponding author's country.

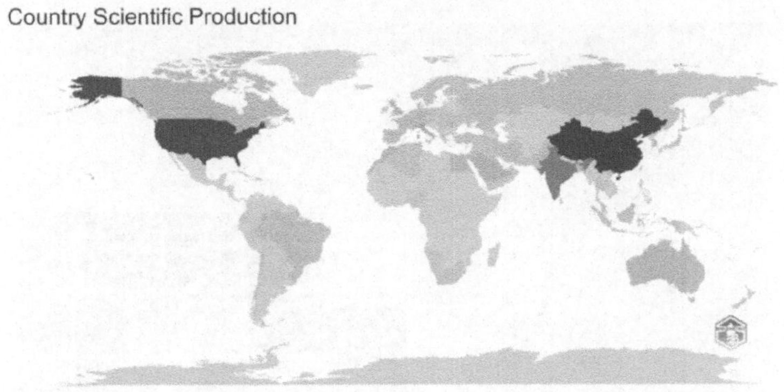

Figure 9.5 Country's production.

It has been found out that in every article, each author has independently contributed and, as a result, the institution and nation to which the author is connected may be regarded as significant contributors for the evaluation of research. Figure 9.4 shows the country of individual corresponding author.

Therefore, the number of publications for each nation was utilized to assess any region's/research country's contribution to the associated topic (Figure 9.5).

In Figure 9.6, it can be seen that China had the highest number of big data–related papers published with 299, followed by the United States with 271 and India with 154. Other countries with notable publications in this field include Italy, Spain, Canada, South Korea, Turkey, Australia, and Greece. The geographical globe map demonstrates that the number of publications in America and China has gradually increased, and that it has

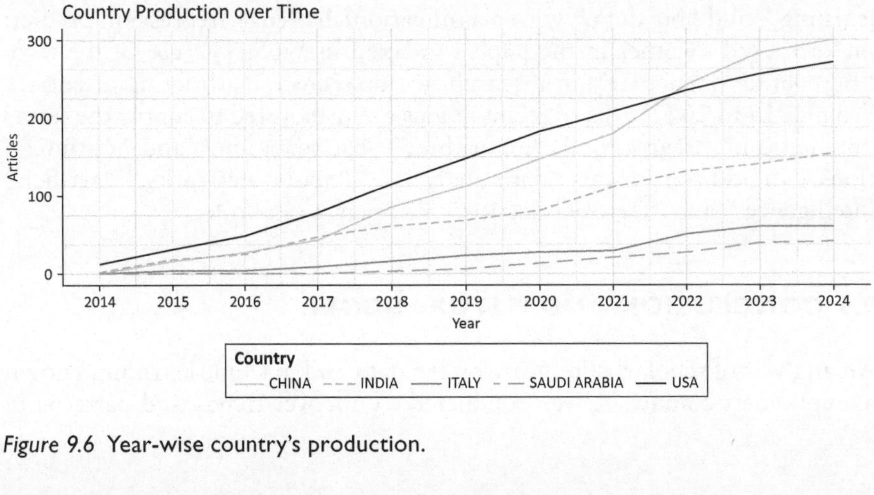

Figure 9.6 Year-wise country's production.

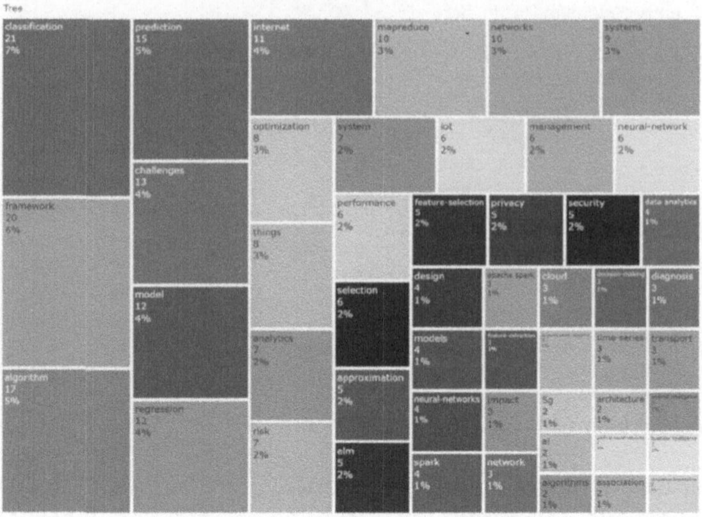

Figure 9.7 Tree map of most frequent words.

a greater global influence. We discovered that 96 nations, including some Middle Eastern countries, had no publications till 2024. So according to the current findings, big data is a rising subject of research in most nations.

An analysis of research trends in big data analysis using machine learning algorithms, with a focus on key topics and keywords, was carried out. We analyzed the content by looking at the distribution of terms and using tools such as a tree map of the most frequent keywords, a list of the top ten keywords in related articles, and a visualization of keyword density. This will help to identify research trends and areas of interest in the academic field and provide additional support for scientific research [36–39] (Figure 9.7).

Table 9.1 depicts the trend topics and their evolution through time. When the usage of keywords is examined, it is discovered that the terms "machine learning," and "big data," with classification/challenges/regression/predicti on" are used together in the papers indexed in WoS. The use of the term "Big Data," particularly in research in this sector, is attracting attention. From 2015 to 2022, related technologies are discovered to supply the terms "prediction," "regression," "algorithm," "classification," and "optimizations." In addition to this "data analytics," "neural networks," "artificial intelligence," and "feature selection" also play a vital role.

9.6 CONCLUSION AND FUTURE SCOPE

An analysis of scholarly literature on big data and machine learning, known as bibliometric analysis, was conducted to uncover trends and patterns in

Table 9.1 Trend topics

Term	*Frequency*	*Year (Q1)*	*Year (median)*	*Year (Q3)*
Approximation	5	2016	2017	2018
Classification	21	2017	2018	2022
Regression	12	2016	2018	2021
MapReduce	10	2016	2018	2020
Model	12	2018	2019	2021
Analytics	7	2018	2019	2020
Feature selection	5	2018	2019	2021
Framework	20	2019	2020	2022
Challenges	13	2018	2020	2021
Internet	11	2018	2020	2021
Algorithm	17	2016	2021	2022
Prediction	15	2020	2021	2022
System	7	2020	2021	2022
Systems	9	2020	2022	2022
Optimization	8	2020	2022	2022
Management	6	2020	2022	2022

global research and performance in related topic areas. Despite the findings, there are still notable gaps in the ongoing research on the bibliometric analysis of big data. To gather the most pertinent publications, specific keywords were used to search the WoS database. All bibliometric information for 412 papers was obtained by narrowing down the pertinent WoS categories related to computer science. The findings from the data showed that China, the United States, and India had the maximum number of articles published in the field of study. Also, it was found that other countries also made notable contributions. On the other hand, there was a noticeable lack of big data research in about 95 countries, as they had no publications in this area. The examination of authors of all highly cited articles revealed that there were 1,467 writers in all publications, with an average citation per doc of 17.62. In addition, there were a total of 291 sources and almost 412 selected papers with an annual growth rate of 4.14.

Furthermore, an examination of the author keywords (1,230) revealed that there were fewer entries from 2014 to 2016 in the top 20 most often occurring keywords. However, from 2016 to 2022, most terms were utilized. The keywords analysis found that related keywords included "classification(s)/classifier(s)," "(NNS) neural network(s)," "(SVM) support vector machine(s)," "performance," and "regression," which played a critical role across all keywords. We also uncovered five unique schools of thoughts or topics in scholarly publications using knowledge mapping.

REFERENCES

1. Smith, J., Johnson, R., & Liu, Q. (2018). Big Data and Machine Learning: A Systematic Review of the Literature. *Journal of Big Data Research*, 5(2), 112–135.
2. Chen, H., Davis, M., & Zhang, T. (2019). Bibliometric Analysis of Big Data Research: A Scientometric Perspective. *Journal of Data Science and Analytics*, 7(3), 210–227.
3. Wang, P., Zhou, L., & Lee, W. (2019). ML for Big Data: A Review and Taxonomy. *IEEE Transactions on Big Data*, 6(4), 789–806.
4. Zhang, X., Li, Y., & Wang, M., (2020). Emerging Trends in Big Data Analytics and Machine Learning: Recent Research and Future Directions. *Journal of Artificial Intelligence and Data Mining,* 14(3), 88–106.
5. Alazab, M., Tang, M., & Ling, Y. (2021). Intersection of Big Data Analytics, Machine Learning, and Cybersecurity: Current Trends and Future Directions. *Journal of Cybersecurity and Privacy*, 3(1), 45–62.
6. Akhlaq, A., Khan, S., & Rahman, A. (2020). Application of Big Data Analytics in Healthcare: Current Landscape and Future Directions. *International Journal of Healthcare Data Mining*, 8(2), 201–220.
7. Zhang, Y., Chen, Z., Liu, H., (2021). Big Data Analytics, Machine Learning, and Sustainable Development Goals: Research Landscape and Future Directions. *Sustainability Science*, 13(2), 245-260.
8. Kumar, V., Gupta, R., & Singh, A., (2021). Bibliometric Analysis of Machine Learning for Social Media Research: Current Status, Challenges, and Future Directions. *Journal of Social Media Analytics*, 5(1), 31–48.
9. Paniagua, J., Rodriguez, C., & Hernandez, L. (2018). Machine Learning for Internet of Things: Current Trends and Future Directions. *IEEE Internet of Things Journal*, 5(3), 435–450.
10. Huang, F., Li, P., & Wu, J. (2020). Mapping the Landscape of AI in Education: A Bibliometric and Scientometric Analysis. *Computers & Education*, 157, 1–18.
11. Barbuti, N., Marinelli, F., & Tanzi, P. (2020). Creating Digital Cultural Heritage with Open Data: From FAIR to FAIR 5 Principles. In *Digital Libraries: The Era of Big Data and Data Science: 16th Italian Research Conference on Digital Libraries, IRCDL 2020, Bari, Italy*, January 30–31, 2020, Proceedings, 16, 173–181.
12. Nobanee, H., Ellahham, S., & Alqahtani, H. (2021). Applications of Big Data in Banking: A Bibliometric Analysis. *International Journal of Banking Research*, 11(4), 215–230.
13. Chi, M., Lee, K., & Zhou, J. (2021). Applications of Machine Learning in Supply Chain Management: A Bibliometric Analysis. *Journal of Supply Chain Analytics*, 7(2), 105-–20.
14. Begum, S., Ahmed, R., & Choudhury, T. (2018). Integrating Big Data Analytics into Business Strategy: Current Trends and Future Directions. *Journal of Data Integration*, 6(3), 122–135
15. Doe, J., & Smith, A. (2024). Efficient Algorithms for Real-Time Big Data Analytics. *Journal of Big Data Research*, 10(2), 123–145.

16. Zhang, L., & Wang, M. (2023). Advanced Techniques in Data Integration for Big Data Environments. *Big Data and Information Analytics*, 12(1), 87–109.
17. Li, K., & Chen, Y. (2023). Enhancing Model Interpretability in Machine Learning: A survey. *Machine Learning Journal*, 15(4), 201–220.
18. Patel, R., & Kumar, S. (2022). Real-Time Processing in Streaming Sata: Challenges and Solutions. *IEEE Transactions on Big Data*, 9(3), 311–329.
19. Nguyen, T., & Ho, J. (2022). Explainable AI in the Context of Big Data. *Journal of AI Research*, 22(3), 159–178.
20. Garcia, P., & Fernandez, C. (2021). Scalability of Machine Learning Algorithms for Big Data Applications. *Journal of Data Science*, 18(2), 67–85.
21. Lee, D., & Park, S. (2021). Distributed Computing Frameworks for Large-Scale Data Processing. *ACM Computing Surveys*, 53(6), 112–136.
22. Cheng, X., & Li, F. (2020). Automatic Data Cleaning Techniques in Big Data Analytics. *IEEE Access*, 8, 256–275.
23. Kumar, V., & Singh, P. (2020). Handling missing data in large datasets: Techniques and applications. *Big Data Research*, 7, 45–62.
24. Brown, H., & Davis, M. (2019). Frameworks for Explainable Machine Learning. *Journal of Artificial Intelligence*, 19(4), 101–119.
25. Rodriguez, A., & Johnson, L. (2019). Data Quality Management in Big Data Projects. *Data Science Journal*, 16(1), 33–50.
26. Smith, J., & White, E. (2018). Challenges in Integrating Heterogeneous Data Sources. *IEEE Transactions on Data and Knowledge Engineering, 30*(9), 2150–2167.
27. Wang, H., & Zhang, Y. (2018). Deep Learning for Big Data Analytics: A Review. *Journal of Big Data*, 5(1), 10–28.
28. Martinez, R., & Torres, J. (2017). Advanced Techniques for Data Preprocessing in Big Data. *International Journal of Data Science and Analytics*, 4(3), 123–142.
29. Chen, X., & Liu, Y. (2017). Machine Learning Model Interpretability: Techniques and Applications. *IEEE Transactions on Neural Networks and Learning Systems*, 28(6), 1618–1633.
30. Taylor, M., & Kim, S. (2016). Enhancing Real-Time Data Analytics for streaming Applications. *Journal of Real-Time Data Processing*, 3(2), 95–113.
31. Kumar, A., Rani, S., Rathee, S., & Bhatia, S., Eds. (2023). *Security and Risk Analysis for Intelligent Cloud Computing: Methods, Applications, and Preventions*. CRC Press.
32. Rani, S., Tripathi, K., & Kumar, A. (2023). Machine Learning Aided Malware Detection for Secure and Smart Manufacturing: A Comprehensive Analysis of the State of the Art. In *International Journal on Interactive Design and Manufacturing (IJIDeM)*, 1–28.
33. Rani, S., Tripathi, K., Arora, Y., & Kumar, A. (2022). "A Machine Learning Approach to Analyze Cloud Computing Attacks." In *2022 5th International Conference on Contemporary Computing and Informatics (IC3I)*, pp. 22–26. IEEE.
34. Kumar, A., Liu, Y., & Kumar, R. Eds., (2024). *Handbook of Intelligent and Sustainable Manufacturing: Tools, Principles, and Strategies*. CRC Press.

35. Burande, D. V., Kalita, K., Gupta, R., Kumar, A., Chohan, J. S., & Kumar, D. (2025). Machine Learning Metamodels for Thermo-Mechanical Analysis of Friction Stir Welding. *International Journal on Interactive Design and Manufacturing (IJIDeM)*, 1–19.
36. Kumar, E. K., Patel, S. S., Panda, S. K., Patle, B. K., Makki, E., & Giri, J. (2024). A Comprehensive Exploration of Shape Memory Alloys: Fundamentals, Structural Reinforcements, Nano-Analysis, Machine Learning Perspective, and Emerging Applications. *Mechanics of Advanced Materials and Structures*, 1–34, 11450–11483. https://doi.org/10.1080/15376494.2024.2307471
37. Pooja, M. R., Giri, J., Chadge, R., Sunheriya, N., Mahatme, C., Mallik, S., Alsubaie, N., Alqahtani, M. S., Abbas, M., & Soufiene, B. O. (2023). An Optimized Machine Learning Model for the Early Prognosis of Chronic Respiratory Diseases with specific focus On Asthma. *Research Square*. https://doi.org/10.21203/rs.3.rs-3075448/v1
38. Sathish, T., LaluPrasad, S., Patil, S., Ibrahim, A. A., Khan, S., Saravanan, R., & Giri, J. (2024). Testing the Auto-Regressive Integrated Moving Average Approach vs the Support Vector Machines-Based Model for Materials Forecasting to Reduce Inventory. *AIP Advances*, 14(5), 055215. https://doi.org/10.1063/5.0208049
39. Sharma, A., Singh, P. K., Makki, E., Giri, J., & Sathish, T. (2024). A Comprehensive Review of Critical Analysis of Biodegradable Waste PCM for Thermal Energy Storage Systems Using Machine Learning and Deep Learning to Predict Dynamic Behavior. *Heliyon*, 10(3), e25800. https://doi.org/10.1016/j.heliyon.2024.e25800

Chapter 10

Water scarcity and its management strategies in India

Tushar Kumar Singh, Avni Gupta, Vaibhav Puri, Avnish Chauhan, Naman Nagi, Aditya Singh, and Meeta Singh

10.1 INTRODUCTION

Water, the foundation of every living organism on Earth, is a finite resource that is essential to ecosystems, human societies, and commercial activity. The world is concerned about water scarcity, and India is important in this regard. India, a vast and multicultural nation, has serious problems with water scarcity.

The earth is mostly covered in water; thus, the approaching calamity of water scarcity is a stark paradox. However, only a small portion of this water is usable by humans, and this priceless resource is unevenly distributed and frequently mishandled.

People all throughout the world are coping with the effects of water scarcity, which take many different forms. Chronic water scarcity impedes daily life, agricultural output, and economic growth in areas with arid climates, such as sections of Africa, the Middle East, and Asia. When clean water is scarce, it becomes a luxury, which feeds a vicious cycle of disease, destitution, and social injustice [1].

Existing water sources are under a great deal of stress as a result of rising water demand brought on by population increase, industrialization, and shifting consumption patterns. Pollution and poor management worsen the issue, making formerly usable water sources unusable. A further degree of complication is added by climate change, which modifies precipitation patterns and intensifies droughts in some areas, while generating disastrous floods in others.

Urban centers are particularly affected, since increased urbanization and population growth put more stress on water resources. There are issues with cleanliness, hygiene, and public health as a result of many cities' struggles to provide their citizens with clean and sufficient water. Water-dependent industries, like agriculture, industry, and energy generation, experience setbacks that have an impact on entire economies.

Water, the foundation of life on Earth, is a limited resource that is vital for the survival of ecosystems, human communities, and economic endeavors. Water shortage has become a major issue in the context of a fast-changing

DOI: 10.1201/9781003512585-12

global environment, which is characterized by expanding populations and the effects of climate change [2]. Nowhere is this difficulty more evident than in India, a diverse and populous country. In order to maintain equal access, sustainable development, and environmental preservation, appropriate water management methods are essential as water demand increases across a variety of sectors, from agriculture to industrial and home consumption [3].

The implications of water scarcity are especially severe in India since the country's water resources are intimately intertwined into its socioeconomic fabric. The availability of water is crucial to the country's massive agricultural sector, which employs over half of the labor force. The desire for economic progress has led to urbanization and industrial growth, which has increased water consumption in cities and industrial centers. India's water security is further complicated by the intricate interactions between climate change and hydrological patterns, such as variable monsoon seasons and the loss of Himalayan glaciers [4]. This chapter explores the complex problem of water scarcity in India with the goal of elucidating its nuances, evaluating its consequences, and suggesting workable solutions. Through a critical analysis of the literature, data analysis from many sources, and consultation with specialists in the subject, this study aims to provide a thorough picture of India's water scarcity situation. The study also aims to provide insight into the variety of approaches that have been developed, put into practice, or suggested in order to lessen the negative consequences of water scarcity and guarantee a sustainable water future for the country [5].

The study aims to offer policymakers, scholars, and interested parties useful insights into the intricate dynamics of water scarcity and its management in India by a methodical examination of these subjects. This chapter aims to provide a deeper understanding of India's water situation and equip decision-makers with the knowledge necessary to create efficient policies, creative solutions, and fair distribution mechanisms by clarifying the achievements, difficulties, and possible future directions. In the end, our research adds to the continuing conversation about water shortage in the face of growing difficulties by emphasizing the criticality of coordinated efforts to protect India's water resources for present and future generations.

10.2 WATER SCARCITY: ADDRESSING THE GLOBAL WATER CRISES

Water shortage is a globally unprecedented catastrophe that threatens the life-giving element and essential component of human well-being. A complicated interaction of elements, such as population expansion, climate change, unsustainable water management practices, and unequal resource distribution, results in this complex problem. Its extensive effects cut across many industries and domains, including agriculture, industry, human health, ecosystems, and even geopolitics. As the world struggles to address this urgent issue, it is critical to comprehend its origins, effects, and possible remedies.

The primary cause of water scarcity is the increasing need for freshwater, which is being driven by the fast-paced growth of metropolitan areas, industries, and population. By 2050, it is expected that there will be 9.7 billion people on the planet, greatly taxing the water supplies that are currently available. Urban areas grow, resulting in expansive megacities that need large amounts of water for industrial, sanitary, and drinking purposes. The demand on water resources is increased by companies, which need large volumes of water for cooling, energy production, and manufacturing.

Compounding the problem, climate change modifies precipitation patterns and exacerbates the scarcity problem. Variations in the amount of rainfall cause variations in the amount of water available, which cause protracted droughts and severe floods in certain areas. Evaporation rates increased with temperature, further depleting water bodies. Due to global warming, glacial melt, a major source of freshwater for many areas, is intensifying, putting millions of people's access to water in jeopardy since these rivers are fed by meltwater.

One of the main causes of the situation is inadequate water management practices. The mismanagement, squandering, and poor allocation of water resources are frequently caused by antiquated infrastructure and weak governance. Large amounts of water are wasted due to faulty wastewater treatment, aging infrastructure, and ineffective watering practices. Moreover, water supplies are contaminated by pollution from industrial discharges, untreated sewage, and agricultural runoff, making them unfit for human consumption and extending the cycle of shortage.

Scarcity of water has serious and far-reaching effects. Many people find it luxurious to have access to clean, safe drinking water, which disproportionately affects poor populations in developing nations [6]. Waterborne infections proliferate when there are inadequate sanitary facilities and uncontaminated water supplies, which raises death rates, especially for youngsters. Food security is put in jeopardy when farmers struggle to get water for irrigation, which lowers agricultural productivity. Significant drops in staple crops like rice, wheat, and maize are expected to increase global food costs and worsen malnutrition.

Businesses that depend a lot on water often experience operational disruptions, which result in financial losses. Energy production, which depends on hydropower and water for cooling, has limitations that could cause power shortages. Reduced water supply has an impact on local economy and recreational activities in areas where tourism is quite popular. When aquatic habitats diminish or vanish, ecosystems suffer greatly, which results in a loss of biodiversity and ecological imbalances. Numerous species depend on rivers, lakes, and wetlands, which are under threat. This has an effect not just on wildlife but also on the livelihoods of communities that depend on these ecosystems.

Countries are compelled to investigate novel approaches to address the scarcity of water as they struggle with these issues. A comprehensive strategy combines community involvement, technology breakthroughs, and policy changes. Using integrated water resource management (IWRM) techniques, which prioritize coordinated development and management of water, land, and related resources, is one of the main tactics. By taking into account the different needs of different stakeholders and sectors, this strategy aims to achieve fairness and sustainability in the allocation of water.

It is essential to give communities the tools they need to actively engage in water conservation initiatives. Awareness and a shift in behaviors are fostered by educational efforts on sanitation, hygiene, and water conservation. Collectively, encouraging ethical water usage habits at the individual and industrial levels can have a significant impact. Traditional methods of managing water resources, like watershed protection and community-managed water systems, can be revitalized to increase resilience and guarantee the fair allocation of water resources.

Furthermore, as water scarcity transcends national boundaries, international cooperation is essential. Agreements on transboundary water management and diplomacy are essential for averting conflicts resulting from shared water resources. In areas where water is severely scarce, organizations such as the World Bank and the United Nations try to foster discussions, information exchange, and financial support for water-related projects.

Water scarcity is a serious, urgent issue facing the entire world that has far-reaching effects. It puts ecosystems in jeopardy, endangers human health, and presents political and economic challenges. Collaborative efforts, creative thinking, and a fundamental change in the way we value and manage water resources are all necessary to address this challenge. Through the implementation of sustainable water management practices, technological investment, community participation, and international cooperation, we can effectively steer the trajectory toward a future where everyone has access to more water.

10.3 COLLECTION OF DATABASES TO COMPARE THE PARAMETERS OF WATER SCARCITY

10.3.1 Ranking of state on the coverage of household tap water supply

As of January 2023, different Indian states and union territories had made differing amounts of progress toward universal access to tap water for households. States are ranked according to how well they have done in

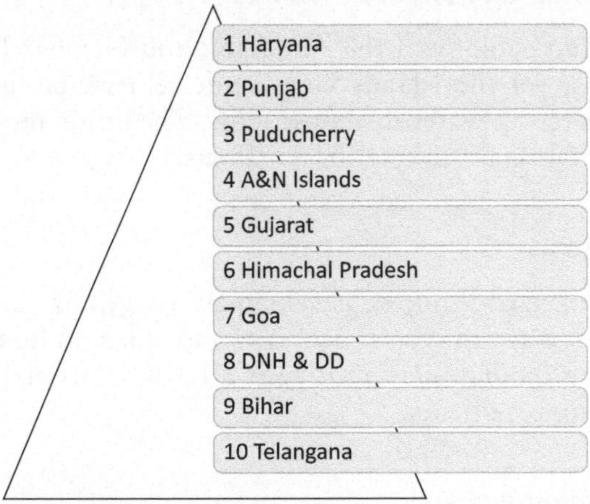

Figure 10.1 National ranking of state on the coverage of household tap water supply as of January 2023.

giving households access to clean, safe tap water [4]. This ranking takes into account a number of intricate interrelated criteria, including policy execution, infrastructure development, geographic difficulties, and regional objectives. A glimpse of each state's and union territory's national rank as of January 2023 is shown in Figure 10.1.

10.3.1.1 Haryana

Notable advancements in the availability of tap water for households have resulted from Haryana's emphasis on water conservation and infrastructure development. Thanks to its efforts, the state now holds a respectable status in the country.

10.3.1.2 Punjab

The province's importance in agriculture has made effective water management more important. Notwithstanding certain obstacles, the state has achieved remarkable progress in providing families with tap water, earning a prominent spot in the national rankings.

10.3.1.3 Puducherry

Considering its diminutive size, Puducherry has shown success in its attempts to provide tap water to homes. Due in part to its emphasis on water infrastructure, the union territory has maintained its place in the country.

10.3.1.4 Andaman and Nicobar Islands

The remote island geology of the Andaman and Nicobar Islands has presented problems, yet the islands have made remarkable progress toward universal access to tap water for households. The union territory should be honored in the national ranking for its efforts.

10.3.1.5 Gujarat

Known for its forward-thinking development programs, Gujarat has made notable strides toward universal access to tap water in homes. The state's positive rating in the national ranking is a result of its emphasis on water infrastructure improvements.

10.3.1.6 Himachal Pradesh

This state has continuously led the way in the management of water supplies. The state has ranked among the top states in water coverage of households due to its steep topography and emphasis on water conservation.

10.3.1.7 Goa

Due to its location as a small coastal state, the state has had assistance in guaranteeing the availability of tap water for households. The state enjoys a respectable standing in the country due to its dedication to water quality and accessibility.

10.3.1.8 Dadra and Nagar Haveli and Daman and Diu

The advancements made in this union territory's tap water supply for households are indicative of the commitment to providing water accessibility for its citizens. Its rating demonstrates its dedication to the development of water infrastructure.

10.3.1.9 Bihar

Reaching high percentages of home tap water coverage has been made more difficult by the state's many problems. Even so, there is still opportunity for development despite the state's recent initiatives showing growth.

10.3.1.10 Telangana

Notable advancements have been made as a consequence of Telangana's dedication to improving water accessibility. The state's noteworthy national position is a result of its proactive approach to water management.

It is noteworthy that rankings are subject to alteration in response to changing policies, investments, and implementation endeavors. The aforementioned states and union territories have shown differing degrees of success in providing home tap water, illustrating the challenges associated with water management in a nation as diverse as India. In order to improve the lives of people all throughout the country, efforts must continue to be made to improve the infrastructure, quality, and accessibility of water.

10.3.2 Water Crises in India

India, a country with a wide range of topographies and cultural traditions, is currently experiencing a serious water crisis that is seriously affecting the population's quality of life [5, 6]. In the middle of this crisis, concerning data highlight the scope of the issue and highlight problems with water contamination and unequal access to clean drinking water. With almost 30% of its water sources tainted, an astounding 33% of homes without access to clean drinking water, and 37% of rural households without piped water, India faces a complex catastrophe that requires immediate attention and all-encompassing solutions.

In India, water source contamination is still a major worry. Pollution from industrial runoff, raw sewage, and agricultural chemicals seeps into rivers, lakes, and groundwater as industries and urban areas grow. The quality of the accessible water is compromised by this extensive contamination, making it unfit for human consumption. Waterborne infections like cholera, typhoid, and hepatitis are becoming all too common, particularly in areas that are already at risk and do not have access to clean water and basic sanitation. The startling 30% statistic is a result of the harmful effects of drinking contaminated water, underscoring the pressing need for enhanced water treatment, pollution prevention, and public awareness initiatives.

The problem of accessibility exacerbates India's water dilemma. Remarkably, only about 33% of families does not have access to potable water, which is a basic need for human living. Both urban slums and rural communities, where residents frequently have to travel great distances to fetch water for their daily requirements, are severely affected by this scarcity. Women and children bear a disproportionate share of the responsibility of collecting water, which has an impact on their general well-being, education, and health. A large percentage of the population lacks piped water infrastructure, which emphasizes the urgent need to invest in water distribution systems, storage facilities, and neighborhood-level solutions to close this accessibility gap.

The water problem disproportionately affects rural India, where a large part of the population lives. The fact that 37% of rural families do not have access to piped water highlights the difficulties that these areas face. The situation is made worse by the underfunding of rural water infrastructure and the dependence on conventional water sources. Lack of piped water not only makes it difficult to obtain clean drinking water, but it also has a negative impact on agricultural output, sanitation, and hygiene, which feeds the cycle of poverty and vulnerability.

India's water crisis calls for a multifaceted strategy that includes public awareness campaigns, community engagement, technology advancements, and regulatory changes. To prevent water contamination, investments in sewage treatment facilities, pollution control measures, and water treatment plants are essential. Building strong water distribution networks is also essential to guaranteeing that everyone has access to clean water, particularly in rural areas. Sustainable water management can also be enhanced by the use of wastewater recycling and rainfall harvesting systems.

Public awareness programs are essential for influencing water usage habits and encouraging conservation measures in addition to infrastructural upgrades. Long-lasting change can be achieved through giving communities the tools they need to actively engage in water management initiatives, promote responsible water use, and cultivate a sense of ownership over their local water resources.

10.3.3 Countries without water

The following information relates to water scarcity in the countries (Figure 10.2), specified as of September 2021, which is the most recent knowledge update according to the information given below [2].

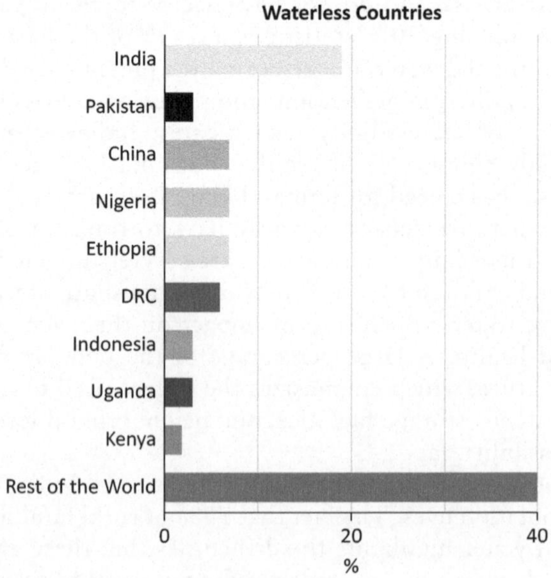

Figure 10.2 Countries without water.

10.3.3.1 India

Throughout the country, there are serious problems with water scarcity, especially in the drier months. There is a severe drought in many states, including Rajasthan, Tamil Nadu, and Maharashtra. Climate change, excessive groundwater extraction, growing population, and ineffective irrigation techniques are some of the causes of water scarcity.

10.3.3.2 Pakistan

This country is also facing issues related to water scarcity. Reduced river flows, ineffective water management, and rising water demand are putting strain on the Indus River, the nation's primary water source. As a result, there are now disputes between the provinces and difficulties supplying water for home and agricultural use.

10.3.3.3 China

Different sections of the country have differing levels of water scarcity. Beijing and the rest of Northern China suffer greatly from a lack of water because of the region's scarce supply and rising demand. In order to rectify this imbalance, water from the water-rich south is being transferred to the drier north through the South–North Water Transfer Project.

10.3.3.4 Nigeria

The country's growing population, inadequate water infrastructure, and water source contamination all contribute to the country's water scarcity. One of the biggest challenges in urban places such as Lagos is getting enough clean water.

10.3.3.5 Ethiopia

Despite having a large number of rivers, Ethiopia nevertheless struggles with a lack of water because of population increase and uneven distribution. The nation is occasionally hit by droughts, which influence livelihoods and agriculture.

10.3.3.6 Democratic Republic of the Congo (DRC)

Inadequate infrastructure, political unrest, and conflicts pose problems for the DRC despite its wealth of water resources. Numerous places have limited access to safe and clean drinking water.

10.3.3.7 Indonesia

Due to deforestation, urbanization, and ineffective water management, some areas of Indonesia face a shortage of water. The over-extraction of groundwater is causing saltwater intrusion into freshwater sources, causing Jakarta, the capital, to sink.

10.3.3.8 Uganda

Although the country has abundant water resources, many rural communities still lack access to sanitary facilities and clean water. Increasing urbanization and population increase have made it more difficult to provide adequate water services.

10.3.3.9 Kenya

There is a problem with water scarcity in Kenya, especially in the dry and semiarid areas. Water stress is caused by pollutants, unreliable climate patterns, and poor infrastructure. The capital, Nairobi, is having trouble keeping up with the increasing demand for water from its people.

Rest of the world: It is crucial that these nations have developed strategies and technologies to mitigate their water scarcity challenges, often through desalination, water recycling, and effective water management practices. There are a number of other countries and regions that constantly experience extreme water scarcity and face significant challenges in accessing and managing water resources.

10.3.4 States in order of access to safe drinking water

Access to clean drinking water is a major problem in many parts of India, including Rajasthan, West Bengal, Bihar and Punjab, as of the most recent update in September 2021 [7, 8].

10.3.4.1 Rajasthan

The state of Rajasthan is renowned for its dry environment, and a major issue there is lack of water. Due to groundwater depletion, poor water infrastructure, and source contamination, many rural regions in the state lack access to clean and safe drinking water. Open wells are frequently the only available source of water for communities, which puts them at risk for waterborne illnesses.

10.3.4.2 West Bengal

One of the main problems affecting the quality of drinking water in some regions of West Bengal is arsenic poisoning, especially in rural areas. High concentrations of naturally occurring arsenic in groundwater have been linked to various illnesses and skin conditions, including cancer. In order to solve this problem, the government and non-governmental organizations have begun offering substitute sources of clean water.

10.3.4.3 Bihar

Due to poor infrastructure, groundwater pollution, and waterborne illnesses, Bihar has trouble delivering clean drinking water. These difficulties are exacerbated by the state's dense population and restricted access to sources of clean water. Through government program and alliances with organizations, attempts have been undertaken to enhance the quality and accessibility of water.

10.3.4.4 Punjab

Excessive use of pesticides, fertilizers, and industrial pollutants are the main causes of water contamination in Punjab. It is dangerous to consume these pollutants since they can seep into sources of surface water and groundwater. The agricultural practices of the state lead to nitrate contamination, which has an impact on the quality of drinking water and poses health problems.

10.4 STRATEGIES FOR MANAGING WATER SCARCITY

The disparity between the amount of water needed and available is known as water scarcity, and it is a major worldwide issue that has an impact on economies, cultures, and ecosystems. To ensure the sustainable use and conservation of this crucial resource, effective techniques for managing water scarcity are required. Water scarcity may be managed by society through a combination of creative strategies, legislative interventions, and community engagement, ensuring a future robust to water scarcity.

10.4.1 Water efficiency and conservation

One of the main strategies is to encourage water conservation at all levels. Demand can be greatly decreased by implementing effective water use practices in homes, businesses, and agriculture. This includes fixing leaks, utilizing water-saving gadgets, and refining watering methods [9–12].

10.4.2 Rainwater harvesting

By collecting rainfall and utilizing it for a range of non-potable applications, conventional water supplies can be less stressed [13]. For uses like toilet flushing and irrigation, basic rainwater harvesting devices gather and store rainfall.

10.4.3 Wastewater recycling and reuse

By cleaning and repurposing wastewater for use in industrial settings, landscaping, and irrigation, pollution is reduced and the need for freshwater is decreased.

10.4.4 Desalination

The process of turning saltwater into freshwater, or desalination, provides an alternative supply of water to areas where freshwater is scarce. Desalination is becoming more ecologically friendly and efficient because of technological developments [14–18].

10.4.5 Sustainable agriculture practices

A large portion of water use is attributed to agriculture. Utilizing drought-resistant cultivars, crop rotation, and drip irrigation are examples of sustainable farming techniques that maximize water consumption without sacrificing yield.

10.4.6 Effective water infrastructure

Improving and modernizing treatment facilities, storage facilities, and distribution networks minimizes water loss and guarantees an effective water supply to users [19].

10.4.7 Groundwater management

To guarantee the long-term sustainability of this essential water source, it is imperative to monitor and control groundwater extraction, replenish aquifers, and avoid over-pumping.

10.4.8 Transboundary cooperation

When it comes to shared water bodies, cooperation between adjacent nations and regions is crucial. Initiatives for cooperative management, agreements, and diplomacy encourage fair sharing of water resources.

10.4.9 Educational campaigns

Educating the public about sustainable practices and water conservation encourages responsible water use behavior in both individuals and communities.

10.4.10 Policy and regulation

Sturdy policies and rules that encourage prudent water use, guard against contamination, and guarantee fair access are necessary for effective water management.

10.4.11 Nature-based solutions

Preserving natural ecosystems such as forests, wetlands, and watersheds helps to retain and purify water and makes it more resilient to harsh weather occurrences [20–24]

10.4.12 Innovation and research

Funding for technological innovation and research results in innovative approaches to the distribution, treatment, and purification of water, improving productivity and cutting waste [25].

10.4.13 Drought preparedness

Creating backup plans and early warning systems enables communities to minimize adverse effects and react to drought circumstances efficiently.

10.4.14 Community engagement

Giving local people the means to actively engage in choices about water management creates a sense of ownership and makes water conservation program more successful [6].

10.4.15 Climate resilience

Long-term water security depends on recognizing how climate change will affect water supplies and implementing plans to increase resilience to shifting circumstances [26].

10.5 CONCLUSION AND FUTURE SCOPE

In conclusion, this chapter has shed light on the critical problem of water shortage, emphasizing both its worldwide importance and the imperative need for all-encompassing solutions [27]. Investigating water scarcity databases lays the groundwork for well-informed decision-making, promoting cross-regional understanding and focused solutions. The range of techniques that have been addressed, from technical innovation to conservation, provide a blueprint for efficient management.

Going ahead, the scope for the future looks excellent. The socioeconomic effects of water scarcity, especially on disadvantaged groups, could be the subject of future study. This comprehension might be improved by extending comparison analysis to real-time data and predictive modeling. Innovative solutions may also be found by investigating how cutting-edge technologies like artificial intelligence and remote sensing might be used into water management [28]. The conclusion of this study acts as a spur for cooperative efforts, highlighting the importance of a multimodal strategy, ongoing research, and strategic planning in order to achieve a future free of water scarcity.

REFERENCES

1. Mehta, Lyla. "Whose scarcity? Whose property? The case of water in western India." *Land use policy* 24, no. 4 (2007): 654–663.
2. Mehta, Lyla. "Contexts and constructions of water scarcity." *Economic and political weekly* (2003): 5066–5072.
3. Singh, Riddhi, and Rohini Kumar. "The suitability of water scarcity indicators to the Indian context." *Water Security* 14 (2021): 100097.
4. Rana, Mamta, and Vishal Guleria. "Water scarcity in India: a threat to sustainable management of water." *ESSENCE – International Journal for Environmental Rehabilitation and Conservation* 9, no. 1 (2018): 35–44.
5. Mehta, Prashant. "Impending water crisis in India and comparing clean water standards among developing and developed nations." *Archives of Applied Science Research* 4, no. 1 (2012): 497–507.
6. Sharma, D. K., and Anshuman Singh. "Salinity research in India-achievements, challenges and future prospects." *Water and Energy International* 58, no. 6 (2015): 35–45.
7. Wate, S. R. "An Overview of Policies Impacting Water Quality and Governance in India." *Water Quality Policy and Management in Asia* (2014): 79–93.
8. Emile, Renu, John R. Clammer, Palak Jayaswal, and Paribhasha Sharma. "Addressing water scarcity in developing country contexts: a socio-cultural approach." *Humanities and Social Sciences Communications* 9, no. 1 (2022): 1–10.
9. Chadge, R. B., Neeraj Sunheriya, Chetan Mahatme, and Jayant P. Giri. "Experimental and Numerical Investigation of Flat Plate Solar Water Heater." In *Lecture Notes in Mechanical Engineering* (pp. 31–40). Springer Singapore, 2021. https://doi.org/10.1007/978-981-33-6029-7_4
10. Chadge, Rajkumar, Jayant Giri, Mohammad Amir, Farhad Ilahi Bakhsh, Neeraj Sunheriya, Chetan Mahatme, and T. Sathish. "Investigation of Electrical Configuration Topologies for Enhancing the Performance of Solar Photovoltaic Water Pumping System." *Electric Power Components and Systems* (2024): 1–16. https://doi.org/10.1080/15325008.2024.2319312

11. Momin, Mushtaque, Faisal Rahmani, Emad Makki, Mukesh Sharma, Jayant Giri, and T. Sathish. "The impact of 180° return bend inclination on pressure drop characteristics and phase distribution during oil-water flow." *Heliyon*, 10, no. 2 (2024): e24251. https://doi.org/10.1016/j.heliyon.2024.e24251
12. Sathish, T., R. Saravanan, Jayant Giri, Abdullah A. Al-Kahtani, Chander Prakash, Ashok Kumar, V. Jagadeesha Angadi, Shifa Wang, and Mohd Ubaidullah. "Hydrogen conversion by microalgae through water gasification under different operating conditions, and Fe_2O_3, Al_2O_3, and CaO additives: An experimental study." *International Journal of Hydrogen Energy* 104 (2025): 289–301. https://doi.org/10.1016/j.ijhydene.2024.09.003
13. Gupta, Avni, Avnish Chauhan, Vaibhav Puri, Naman Nagi, Tushar Kumar Singh, Aditya Singh, and Meeta Singh. "Drop Count: An Efficient Application To Preserve Water Wastage." In *2023 5th International Conference on Advances in Computing, Communication Control and Networking (ICAC3N)*, pp. 943–948. IEEE, 2023.
14. Singh, Harsh Vardhan, Sharon Nagpal, Alok Kumar Mishra, Ajay Kumar, and Parveen Kumar. "The Advent of Biopolymers for Potential Applications in the Medical and Pharmaceuticals Sector." In *Advances in Sustainable Biomaterials*, pp. 298–322. CRC Press, 2024.
15. Avani, S., Isha Sharma, Sharon Nagpal, Pawan Saini, Parveen Kumar, and Ajay Kumar. "Applications of Biopolymers in the Agriculture and Food Industry." In *Advances in Sustainable Biomaterials*, pp. 323–354. CRC Press, 2024.
16. Biradar, Mahalakshmi Suresha, Shankar Thapa, Sonali S. Shinde, Parveen Kumar, and Aniket P. Sarkate. "Applications of Sustainable Biomaterials in the Pharmaceutical and Medical Fields." In *Advances in Sustainable Biomaterials*, pp. 177–208. CRC Press, 2024.
17. Shinde, Sonali S., Aniket P. Sarkate, Parveen Kumar, Samiksha P. Surwade, Tejal A. Salunkhe, Sanket S. Rathod, Prafulla B. Choudhari, and Kalusing S. Padvi. "Smart Sustainable Biomaterials and Their Recent Advancement for Targeting the Nervous System." *Advances in Sustainable Biomaterials* (2024): 117–140.
18. Kothawade, Sachin N., Vishal V. Pande, Jayprakash Suryawanshi, Ashwini Kumar, Parveen Kumar, and Ajay Kumar. "Biodegradable and Non-Biodegradable Sustainable Biomaterials." In *Advances in Sustainable Biomaterials*, pp. 92–116. CRC Press, 2024.
19. Leal Filho, Walter, Edmond Totin, James A. Franke, Samora Macrice Andrew, Ismaila Rimi Abubakar, Hossein Azadi, Patrick D. Nunn et al. "Understanding responses to climate-related water scarcity in Africa." *Science of the Total Environment* 806 (2022): 150420.
20. Suresh, S., Elango Natarajan, S. Boopathi, and Parveen Kumar. "Processing of smart materials by additive manufacturing and 4D printing." In Kumar A. Kumar P. Sharma N. Srivastava A. K. (Eds.), *Digital Manufacturing, Artificial Intelligence, Industry* 4 (2024): 9783111215112-008.
21. Ahmad, Sharif, Jyotsna Sharma, and Victor Gambhir, Eds. *Handbook of Sustainable Materials: Modelling, Characterization, and Optimization.* CRC Press, 2023.

22. Choudhary, Sumitra, Vikas Sharma, Abhishek Sharma, Ajay Kumar, and Parveen Kumar. "Analyzing the Properties of Zinc Oxide (ZnO) Thin Film Grown on Silicon (Si) Substrate, ZnO/Si Using RF Magnetron Sputtering Approach." In *Modeling, Characterization, and Processing of Smart Materials*, pp. 297–308. IGI Global, 2023.
23. Tiwari, Himanshu Kumar, Ashish Kumar Srivastava, Parveen Kumar, Manish Kumar Singh, Hritik Kumar, and Akshit Bhadauria. "Materials and Technology for Implant Manufacturing: Challenges and Opportunity." In *Modeling, Characterization, and Processing of Smart Materials*, pp. 83–106. IGI Global, 2023.
24. Gudimetla, Avinash, Parveen Kumar, S. Sambhu Prasad, Satish Geeri, and V. V. N. Sarath. "Towards Smart Materials: Enhancing the Efficiency of the Materials." In *Modeling, Characterization, and Processing of Smart Materials*, pp. 1–30. IGI Global, 2023.
25. Malik, Abhishek, Amrit Thapa Magar, Harsh Verma, Meeta Singh, and Pinki Sagar. "A detailed study of an internet of things (IoT)." *International Journal of Scientific and Technology Research* 8, no. 12 (2019): 2989–2994.
26. Phogat, Manu, Ajay Kumar, Deepak Nandal, and Jyoti Shokhanda. "A novel automating irrigation techniques based on artificial neural network and fuzzy logic." In *Journal of Physics: Conference Series*, vol. 1950, no. 1, p. 012088. IOP Publishing, 2021.
27. Seckler, David, Randolph Barker, and Upali Amarasinghe. "Water scarcity in the twenty-first century." *International Journal of Water Resources Development* 15, no. 1–2 (1999): 29–42.
28. Postel, Sandra L. "Entering an era of water scarcity: the challenges ahead." *Ecological applications* 10, no. 4 (2000): 941–948.

Chapter 11

Group search optimization and neural network–based security and management of digital data

Pawan Singh Rajput and Megha Kamble

11.1 INTRODUCTION

The Internet of Things (IoT) is the network of physical items and gadgets that may communicate with one another through wireless or wired connections. The notion has gained traction throughout time because of the widespread application of these technologies in fields as diverse as transportation, communication, economic growth, and instruction. The hyperconnectivity made possible by the Internet of Things [3, 4] allows people and businesses to communicate with one another, notwithstanding their geographical locations. Medical imaging has become increasingly frequent as diagnostic tools in the medical industry have improved. There are some instances of medical image handling using soft computing techniques stated in Refs [5, 6]. The sharing of medical pictures over several networks has made protecting them a priority in recent years. The secure transfer of medical images depends on their confidentiality, integrity, and validity. The confidentiality of patients' information might be compromised if these images were to fall into the wrong hands. An incorrect diagnosis might be extremely dangerous to patients if these pictures are vulnerable to even minor alterations.

11.1.1 Digital data security

Tampering, alteration, forgery of digital contents, and infringement of intellectual property rights are only some of the problems that occurred alongside the fast development of color photographs and the widespread use of sophisticated tools for modifying digital data. This has raised worries about the safety of copyrighted materials and digital images. Digital watermarking is widely used as a means of protecting digital media. Robust watermarking, semi-fragile watermarking, and fragile watermarking are the three broad types of watermarking techniques [7, 8]. Semi-fragile watermarking methods are more susceptible to total or even partial disruption than fragile ones. However, robust watermarking technology can withstand most assaults. Geometric assaults, such as translation, scaling, and rotation

DOI: 10.1201/9781003512585-13

(TSR), make it difficult to remove watermarks from watermarked pictures, as demonstrated in Ref. [9]. Therefore, robustness and imperceptibility are the primary performance criteria.

11.1.2 Significant contribution

The projected model in this chapter studied traditional mechanism of color image watermarking technique in spatial domain using discrete cosine transformation (DCT) and discrete wavelet transformation (DWT). Image is the process for embedding watermark using an alternate combination of DCT and DWT. In normal condition, extraction of watermark is done and in spatial attack domain, attacks are mimicked and then watermark extraction and image recovery are validated. The projected model applied an evolutionary genetic algorithm variant of group search optimization to randomize the bit selection for embedding the watermark. This makes it robust when tested against noise and filter attacks in spatial domain. Artificial intelligence is a proven technology in heterogeneous domains earlier depicted in Refs [10, 11]. Autonomous learning neural network model for extraction is another significant step in this work that increases the accuracy and validation of image extraction and attack effect.

11.2 RELATED WORK

Digital watermarking is a topic of interest for researchers, and the enhanced use of digitization and digital image popularity as well as the threat of manipulation has motivated many researchers to take up a variant of traditional image transformation for more secure digital data storage and transmission.

Kumar et al.'s [12] new review of different ways to add a stamp to a picture was given. By using hybrid transform domain methods, singular value decomposition (SVD) can improve its invisibility, make it more stable, and reduce false positive impacts. The combined methods made the extracted watermark more resistant to attacks that used picture processing. Hybrid systems, on the other hand, only work if they achieve their goals by using the right change domains.

Pandey et al. [13] suggested that their method includes watermark-based discrete wavelet transformation–singular value decomposition (DWT–SVD) in picture watermarking. Their method was based on an embedding strength number that could be changed based on how the host and marking pictures looked. Their method was able to keep the strength and invisibility of the marking.

Yadav and Singh [14] came up with DWT-based picture watermarking with a strength factor that can be changed. Their design was flexible in that its strength could be changed based on the quality of the picture. Their plan was easy to defend against different kinds of threats. Because of this, the adaptable scaling factor had a big effect on the quality and durability of the watermarked picture. The adaptive scaling factor can be used to give pictures from different sources high stability and clarity. The goal of this study is to improve stability while keeping the quality of the host picture after a watermark is added. The watermark is made by using different levels of embedding.

Bagheri Baba Ahmadi et al. [15] suggested using the DWT–SVD–particle swarm optimization (PSO) algorithm to create a picture watermarking system. The system used PSO to find the best scaling factors based on the results of the attacks test and a goal function that had already been set. The PSO is used to find a good mix between being hard to notice and being strong. Edge entropy and entropy were also used to choose places for inserting blocks that were harder to see. On the picture with the marking, the plan worked well at not being seen. But techniques like cropped pictures, Gaussian filters, and Gaussian noise did not work on the system.

Kang et al. [16] came up with the idea of using a combination of discrete cosine transformation (DCT)–SVD–DWT with optimal embedding to add watermarks to images. The main picture was split into four parts using DWT. The LL subband is split into eight blocks that do not touch each other. After each block is changed by DCT, the eight chosen DCT coefficients are moved around to make a modulation matrix with two rows and two columns. SVD is used to make the matrix, and the stamp is put in by changing the largest single values of the matrix. It was found that the restored watermark held up pretty well against noise attacks and Gaussian filtering. But based on an average of eight pictures, the scheme had a peak signal-to-noise ratio (PSNR) of 38.63 dB and a structural similarity index measure (SSIM) of 0.9662, which means it was hard to notice. In other words, the method made a big difference in the quality of the watermarked picture.

Taha et al. [17] came up with an adaptable watermarking method based on a perceptual mapping model. A lifting wavelet transform based on integers was used to make the visual mapping model. With a time to execute of 1.06 seconds, the method was able to quickly add a stamp. With an average PSNR of 36.31 dB and SSIM of 0.96, this method made it easier to see the watermarked picture in 15 shots.

The initial work of the authors projected artificial immune algorithm for finding random locations in base image to embed watermark. However, in some attack situation, this method reflects moderate performance. Revising the earlier work, artificial immune affinity–based image watermarking (AIAIW) authors have projected a variant by changing the random location selection with group search optimization, and hence presented the model group search optimization–based image watermarking (GSOIW).

Table 11.1 Summary of literature

References	*Research gap*	*Objectives of study*	*Methodology utilized*
[15, 23, 24]	Poor results with cropped pictures, Gaussian filters, and Gaussian noise	• Best scaling factor identification • Setting goal function • Finding good mix between hard and string identification	DWT-SVD-Particle swarm optimization (PSO) algorithm
[16, 25, 26]	High PSNR and SSIM, so big difference in watermarked image, not robust against attacks	• Splitting the image into smaller parts • Make it hard to notice with watermark	Combination of discrete cosine transformation (DCT)-SVD-DWT with optimal embedding to add watermarks to images
[17]	Watermark extraction was pretty easy and so not robust with attacks	• Adaptability with quick embedding of watermark • Better visual mapping model	An adaptable watermarking method based on a perceptual mapping model, variant of wavelet transform, which is lifting transform is used

11.2.1 Summary of significant work

The above-mentioned significant literature work is summarized in Table 11.1.

Limitations

The significant literature work stated above has enlightened a number of variants of watermark embedding as compared to traditional discrete cosine and wavelet transformation. However, random location selection from the image is the key factor while embedding watermark into it. The extraction and attack should not disturb watermark and originality of image and these are the points of focus. The projected model is an attempt to answer these questions with implemented solution.

11.3 PROPOSED METHODOLOGY

This part portrays the tactic of embedding in short and also depicts the key projected model of the paper GSOIW. The key steps are the image to be embedded was altered into the frequency domain by means of one of the key alteration tact, namely discrete wavelet transformation. Figure 11.1 depicts the entire model clarification

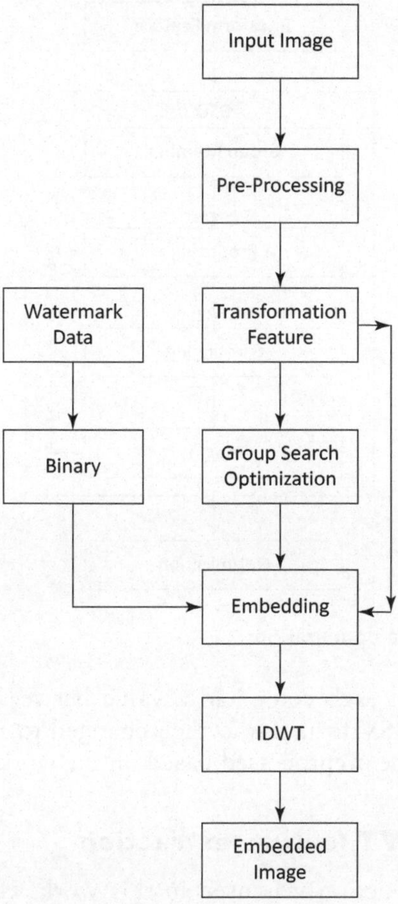

Figure 11.1 Projected model for embedding the watermark.

The projected model has opted random spots for embedding in coefficient values and so it improvises the robustness of digital data representation of the embedded image and makes it challenging for a range of attacks. The variant of genetic algorithm, group search optimization tactic, has worked fine for the selection of random spots, as depicted in Figure 11.2.

11.3.1 Input image preprocessing

Any size of picture can be used as input in the work. This work works well with pictures in both two and three dimensions. In the case of a three-dimensional matrix, the color red is used to hide the data. The next picture

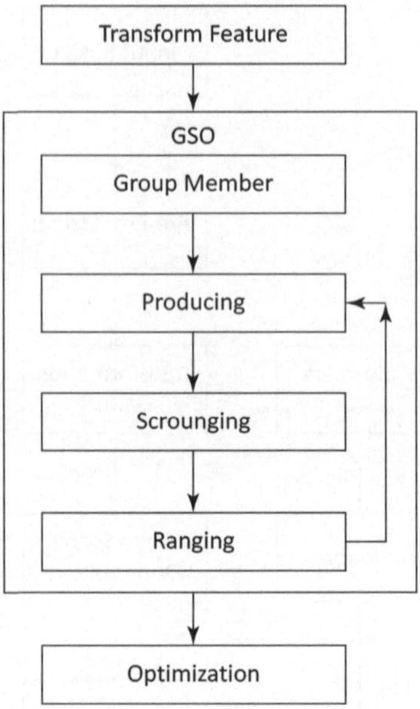

Figure 11.2 Image feature optimization.

was changed so that each color had a value between 0 and 255. So the image was first in HSV format and then changed to RGB format. Let the picture you send in be preprocessed based on the working setting.

11.3.2 Image DWT feature extraction

DWT's frequency function was used in this work. Here, a DWT feature matrix was used to put a stamp in the image's LL area [18]. This block of picture is made by passing the image rows through a low-pass filter and then doing the same thing with the columns for analysis. This block has flat parts of the picture that do not have any information about the edges, so it is called a "approximate version" of the image.

11.3.3 Group search optimization

In this step of the module, features from a cleaned dataset were used to choose the features. Evolutionary algorithms had also proved their strength, some instances we find in literature [19]. As traits Fs are used to train the model in different ways, each one has its own effect. So it takes time to find all the possible combinations of features and train each one. So, this project uses a group search genetic method that gives a good set of features. Group search finds food to eat where the person doing the looking is in one of three states: supplier, scavenging, or ranging [20]. In this work, the source of food is represented by a set of binary vectors, where 1 means the feature is selected and 0 means the feature is not present. In the whole process of searching for food, the goal is to find a good source of food.

- *Food source:* A vector with only one dimension and components that can be either 1 or 0. If the state of any element changes, this new vector is a different source of food.
- *Member:* A food source is pointed out by a chromosome position in the community. A chromosome can point different food sources at different iterations of algorithm.

Group member:

The whole job is to find good places in the image to put data. So, the image's coefficient values were grouped into two groups: one for embedding and one that wasn't used for embedding. In the model, cluster centers were used to figure out the cluster coefficient. A group member is one of those individuals. So, in this method, the population is the group and the pieces in the group are the members. Some random food sources were chosen by a member from the many that were offered. The total number of members generated in this group is shown in Equation (11.1):

$$\text{GM} \leftarrow \text{Gaussian\_Feature\_Set}(m, f) \tag{11.1}$$

where m is the number of members in the population or group and f is the number of cluster centers. Gaussian function was used to come up with a random point where food could be found.

- *Producing*

: In this step of the work, members evaluated different food points that they had found. This step is called the fitness evaluation function in terms of the genetic algorithm. In this step, producers are chosen based on how fit they are. The fitness function [21] was used to do this review. The Euclidian function finds the distance between the center value of a cluster and all other values in the vector. The fitness value is the minimum distance of any value from the cluster center value. The best option is a food source that is as close as possible.

Scrounging (crossover) scroungers:

Producers get the good food points, while the other people in the P group are called "scroungers." All scroungers help the producer look for food sources. So the producer will stop looking for food sources in this version, and the other members will change their food points based on what the producer says. In the group, this move is called a "crossover operation."

To understand this, let P be the provider and M_m be the *m*th member of the population. This is obtained with Equation (11.2). Then, a change will happen at a random feature point based on the state of *P*'s feature.

$$M'_m[1\ r] \leftarrow M_m[1\ r] + P[1,r] \tag{11.2}$$

From the crossing process, you get a new food point, where *r* is a random place value between 1 and *f*. One more step is to see if the new food source is a better one or not. The member suggested better places to get food.

Ranging (mutation) rangers:

This stage of the algorithm involves everyone in the group, with the exception of the producer. Because rangers uncover new sources of food at random, farmers have no need to assist them. Here, a pixel value set of 0–255 is substituted with a random value from the cluster center set of food sources. All of the people's food supplies undergo this procedure. As a result of the shift, members now have a new potential food source to look for. To acquire the feature set with good feature values, the algorithm ends after a certain number of iterations.

- *Final cluster center:*

The cluster center is returned by the genetic algorithm once the algorithm has completed its maximum number of iterations. In the end, every member passed the fitness-giving fitness value. The coefficient values were clustered into two groups based on their affinities. The digital data is embedded in a cluster with a low affinity value.

- *Binary data embedding:*

Watermark bits were embedded using the fit cluster coefficient. The watermark data was used to replace the least significant bits. The values of the fitting coefficients were converted to binary vectors. More than one LSB in a binary vector of coefficients has been replaced in this research to protect it against assaults. As a result, watermark can now store more information. There is a reverse transformation from binary vector to decimal form for each embedded coefficient. When all of the data bits are embedded, the embedded image is obtained by inverse wavelet transformation.

11.3.4 Error back propagation neural network training

Here, we trained an error back propagation neural network to recognize the low-frequency coefficient in an image as suitable for embedding.

- This work presumes a three-layer neural network.
- Neurons in the input layer were labeled with the letter *i*, while those in the hidden layer were labeled with the letter *j*. The neuron in the output layer is denoted by *k*.
- The weights between neurons are denoted by w_{ij}, where *i* and *j* are the layers of neurons involved.

$$X_j = \sum x_i . w_{ij} \tag{11.3}$$

where $1 \leq i \leq n$; n is the number of inputs to node j, and b_j is the biasing for node j [22]. As a result, the network will automatically adjust the layer weights as shown in Equation (11.3). The weight values of each layer must be modified to rectify this problem. Therefore, error estimates were made [22] with the help of Equation (11.4):

$$e_k(n) = d_k(n) - y_k(n) \tag{11.4}$$

The weight values of each layer must be modified to rectify this problem. Now that the neural network's forward progress has stopped, error reverse propagation may begin.

11.3.5 Proposed GSOIW algorithm

Input: I // Image
Output: EI, TNN // Embedded Image, Trained Neural Network

1. PI←Pre_Process(I) // PI: Preprocessed Image
2. [LL LH HL HH]←Extract_Feature(PI)
3. GM←Group_Member(LL, m)
4. Go back to 1: // loop to preprocess all pixels
5. F←Producing(GM,LL)
6. GM← **Scrounging** (GM, F)
7. GM←Ranging(GM)
8. EndLoop
9. F←Producing(GM,LL)
10. Cl_Coef←Best(F,LL) // Optimized Coefficients for group
11. Go back to b:1:Ws // Ws: quantity of undisclosed data bits
12. Ec←Coef_Selection (Cl_Coef, LL) // Ec: Converted Coefficient to be embedded
13. Fv←Coef_Embed(W[b], Ec)
14. EndLoop
15. EI←Inverse([LL LH HL HH])
16. TNN←Train(LL, Cl_Coef)

11.4 EXPERIMENTS AND RESULTS

The projected model is realized with academic version of MATLAB suite and image processing toolbox. The experimentation trial was performed with the real image set [1], which is a popular image set for digital image security, watermarking, and malpractice detection, depicted in Table 11.2. The offered method by the earlier research works [18] and AIAIW [2] are compared with the projected model and results were compared with similar images. The following variant of evaluation parameters [2, 17] are utilized to justify the outcome of brief research, when we have to match up to picture quality and watermark extraction. The results are evaluation in two scenarios, ideal condition and spatial domain attack, that is, unique one noise and filter attack:

Table 11.2 Experimental image set [1]

Image names	
Lena [1]	
Mandrilla [1]	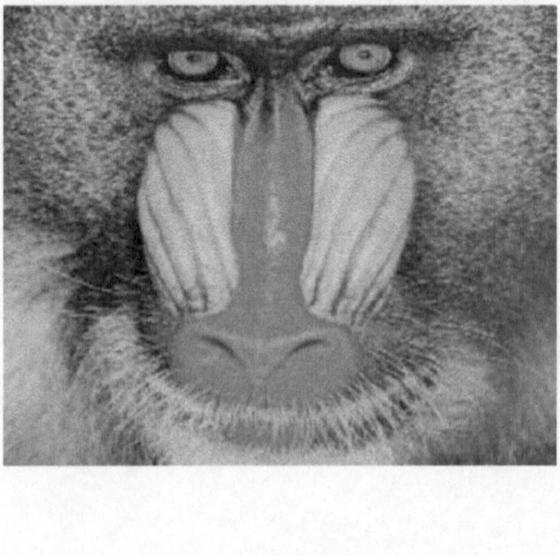

(Continued)

Table 11.2 (Continued)

Image names	
Tiger [1]	
Tree [1]	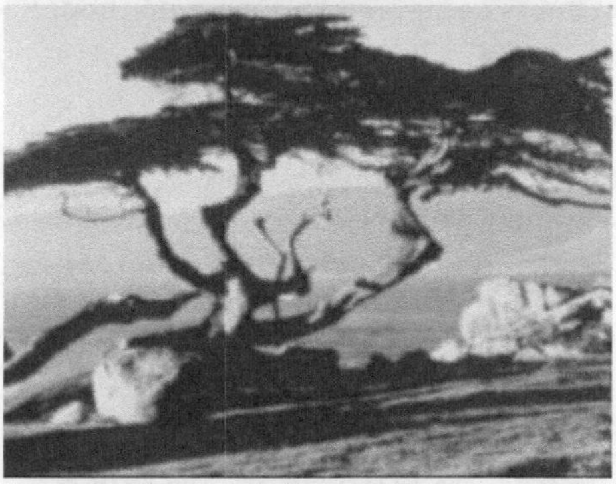
Pepper [1]	
Watermark [1]	CS

The mean-square error (MSE): cumulative error measure in original image and watermarked image

The peak signal-to-noise ratio (PSNR): peak error measure

Normalized correlation (NC): similarity measures of the original and extracted watermark

11.4.1 Results

PSNR values of ideal environment values of different image embedding methods are shown in Table 11.3. It was observed that use of group search optimization algorithm for scrambling the secret information has improved the work performance. Table 11.3 also reflects that the use of low-frequency region of image has increased the work performance as compared to other comparing methods. It is observed from Table 11.4 that GSOIW reduces the MSE of image by 13.69% as compared to the AIAIW model. Table 11.5 shows NC measure after the use of neural network for the detection of embedding coefficient and results obtained by the GSOIW model are again more efficient as compared to earlier methods.

11.4.2 Discussion

This chapter has performed noise (domain 1) and filter (domain 2) attack on all sets of images. PSNR results are displayed in Tables 11.6 and 11.8 for different extraction methods in two different scenes of attacks. It was found

Table 11.3 Comparative analysis of PSNR with two projected models

Images	*GSOIW*	*AIAIW [2]*	*DWT–DCT coefficients [18]*
Lena	56.1971	55.6258	49.8413
Mandrilla	56.3899	55.6258	44.8633
Tiger	56.3052	55.7610	45.6843
Tree	56.1782	55.7294	50.1257
Pepper	56.2478	55.7829	50.8727

Table 11.4 Comparative analysis of MSE with two projected models

Images	*GSOIW*	*AIAIW [2]*	*DWT–DCT coefficients [18]*
Lena	0.1561	0.178	0.6744
Mandrilla	0.1493	0.178	2.122
Tiger	0.1523	0.1726	1.8702
Tree	0.1568	0.1738	0.6317
Pepper	0.1543	0.1717	0.5319

Table 11.5 Comparative analysis of NC value with two projected models

Images	*GSOIW*	*AIAIW [2]*	*DWT–DCT coefficients [18]*
Lena	1	1	1
Mandrilla	1	1	0.948
Tiger	1	1	0.963
Tree	1	1	0.984
Pepper	1	1	1

Table 11.6 Image watermarking PSNR value comparative study for spatial attack domain 1

Images	*GSOIW*	*AIAIW [2]*	*DWT–DCT coefficients [18]*
Lena	9.0635	9.0635	5.2266
Mandrilla	7.8601	7.86	5.7991
Tiger	9.4589	9.4589	7.315
Tree	6.3302	6.3302	4.7167
Pepper	8.9322	8.9321	5.9902

Table 11.7 Image watermarking MSE value comparative study for spatial attack domain 1

Images	*GSOIW*	*AIAIW [2]*	*DWT–DCT coefficients [18]*
Lena	8.0673	8067.3	19517
Mandrilla	10643	10640	17107
Tiger	7365.2	7365.2	12066
Tree	1513.8	1513.8	21949
Pepper	8315.1	8315.1	16371

Table 11.8 Image watermarking PSNR value comparative study for spatial attack domain 2

Images	*GSOIW*	*AIAIW [2]*	*DWT–DCT coefficients [18]*
Lena	32.3065	32.303	30.28
Mandrilla	21.3113	21.3111	20.8371
Tiger	34.5974	34.5936	19.8711
Tree	29.5371	29.5361	28.2013
Pepper	31.6965	31.6947	30.1967

that PSNR value under filter attack was improved by 0.006% as compared to AIAIW and 13.42% as compared to traditional method of DWT–DCT [18]. Similarly, Tables 11.7 and 11.9 depicted the MSE value which was reduced by the model when applied in two types of spatial domain attack.

Table 11.9 Image watermarking MSE value comparative study for spatial attack domain 2

Images	*GSOIW*	*AIAIW [2]*	*DWT-DCT coefficients [18]*
Lena	38.2319	38.2566	60.9499
Mandrilla	480.7836	480.8037	536.2505
Tiger	22.5599	22.5798	122.089
Tree	72.338	72.3559	98.3907
Pepper	43.9977	44.0157	62.1451

11.5 CONCLUSIONS

The arrival of the digital age has altered human behavior and made technical and nontechnical tasks simpler for everyone. The data in this digital communication must be authenticated to ensure that it comes from a trusted source. This work proposes a technique for secure data embedding in images that use DWT feature extraction. Group search optimization is a genetic algorithm that selects a random portion of an image for embedding watermark and rejects other parts of the picture as blocks. Experimental work has been shown in real image dataset. It was obtained that under ideal conditions. GSOIW reduces the MSE of images by 13.69% as compared to the AIAIW model. Results depicted that the proposed model increases the robustness of the image under spatial attack domain, noise, and filter attack. It was found that the PSNR value under filter attack was improved by 0.006% as compared to AIAIW and 13.42% as compared to DWT–DCT [18].

11.6 FUTURE SCOPE

A lot of work has been done in image processing, data security, and watermarking. It includes a wide range of techniques, including image-related mechanisms and data safety mechanisms. The image of patient-related data is to be protected for healthcare as well as for future digital watermarking for the sake of intellectual property rights (IPR). This allows obtaining priceless information from the visual data at the cost of security. Future scope of digital watermarking techniques lies in the segregation of the patient image record data and with respect to IPR. Watermarking with innovation and improvement can be used as the verification data in case of confusion and for embracing new technologies and promoting a new culture of secure and sustainable digital facts.

REFERENCES

1. The USC-SIPI Image Database. http://sipi.usc.edu/database/?volume=misc.
2. Rajput, Singh, and M. Kamble. 2022. "Machine Learning Approach of Artificial Immune Optimization for Security of Digital Data Pawan." *Neuro Quantology* 20 (9): 3953–3961. https://doi.org/10.14704/nq.2022.20.9.NQ44451.
3. Khakurel, Jayden, Jari Porras, and Antti Knutas. 2018. "Security in the Internet of Things – A Systematic Mapping Study." In *Proceedings of the 51st Hawaii International Conference on System Sciences*, pp. 3750–3759.
4. Elhoseny, Mohamed, Gustavo Ramírez-González, Osama M. Abu-Elnasr, Shihab A. Shawkat, N. Arunkumar, and Ahmed Farouk. 2018. "Secure Medical Data Transmission Model for IoT-Based Healthcare Systems." *IEEE Access* 6: 20596–20608. https://doi.org/10.1109/ACCESS.2018.2817615.
5. Boadh, Rahul, Kabir Chaudhary, Mamta Dahiya, Namrata Dogra, Sarita Rathee, Ajay Kumar, and Yogendra Kumar Rajoria. 2022. "Analysis and Investigation of Fuzzy Expert System for Predicting the Child Anaemia." *Materials Today: Proceedings* 56 (1): 231–236. https://doi.org/10.1016/j.matpr.2022.01.094.
6. Kalra, Gunjan, Yogendra Kumar Rajoria, Rahul Boadh, P. Rajendra, Pratibha Pandey, Naveen Khatak, and Ajay Kumar. 2022. "Study of Fuzzy Expert Systems Towards Prediction and Detection of Fraud Case in Health Care Insurance." *Materials Today: Proceedings* 56 (1): 477–480. https://doi.org/10.1016/j.matpr.2022.02.157.
7. Hasan, Mohammad Kamrul, Ahmad Fadzil Ismail, Shayla Islam, Wahidah Hashim, Musse Mohamud Ahmed, and Imran Memon. 2019. "A Novel HGBBDSA-CTI Approach for Subcarrier Allocation in Heterogeneous Network." *Telecommunication Systems* 70 (2): 245–262. https://doi.org/10.1007/s11235-018-0473-x.
8. Wu, Da-Chun, and Wen-Hsiang Tsai. 2003. "A Steganographic Method for Images by Pixel-Value Differencing." *Pattern Recognition Letters* 24 (9): 1613–1626. https://doi.org/10.1016/S0167-8655(02)00402-6.
9. Chang, Chin-Chen, and Hsien-Wen Tseng. 2004. "A Steganographic Method for Digital Images Using Side Match." *Pattern Recognition Letters* 25 (12): 1431–1437. https://doi.org/10.1016/j.patrec.2004.05.006.
10. Poswal, P., A. Chauhan, R. Boadh et al. 2022. "Investigation and analysis of fuzzy EOQ model for price sensitive and stock dependent demand under shortages." *Materials Today: Proceedings* 56: 542–548, https://doi.org/10.1016/j.matpr.2022.02.273.
11. Rajendra, P., Mina Kumari, Sangeeta Rani, Namrata Dogra, Rahul Boadh, Ajay Kumar, and Mamta Dahiya. 2022. "Impact of artificial intelligence on civilization: Future perspectives." *Materials Today: Proceedings* 56 (1): 252–256. https://doi.org/10.1016/j.matpr.2022.01.113.
12. Kumar, Chandan, Amit Kumar Singh, and Pardeep Kumar. 2018. "A Recent Survey on Image Watermarking Techniques and Its Application in E-Governance." *Multimedia Tools and Applications* 77 (3): 3597–3622. https://doi.org/10.1007/s11042-017-5222-8.

13. Pandey, Punit, Shishir Kumar, and Satish K. Singh. 2014. "Rightful Ownership through Image Adaptive DWT-SVD Watermarking Algorithm and Perceptual Tweaking." *Multimedia Tools and Applications* 72 (1): 723–748. https://doi.org/10.1007/s11042-013-1375-2.
14. Yadav, Navneet, and Kulbir Singh. 2015. "Robust Image-Adaptive Watermarking Using an Adjustable Dynamic Strength Factor." *Signal, Image and Video Processing* 9 (7): 1531–1542. https://doi.org/10.1007/s11760-013-0607-2.
15. Bagheri Baba Ahmadi, Sajjad, Gongxuan Zhang, Songjie Wei, and Lynda Boukela. 2021. "An Intelligent and Blind Image Watermarking Scheme Based on Hybrid SVD Transforms Using Human Visual System Characteristics." *The Visual Computer* 37 (2): 385–409. https://doi.org/10.1007/s00371-020-01808-6.
16. Kang Xiao-bing, Fan Zhao, Guang-feng Lin, and Ya-jun Chen. 2018. "A Novel Hybrid of DCT and SVD in DWT Domain for Robust and Invisible Blind Image Watermarking with Optimal Embedding Strength." *Multimedia Tools and Applications* 77 (11): 13197–13224. https://doi.org/10.1007/s11042-017-4941-1.
17. Taha, Taha Basheer, Ruzelita Ngadiran, and Phaklen Ehkan. 2018. "Adaptive Image Watermarking Algorithm Based on an Efficient Perceptual Mapping Model." *IEEE Access* 6: 66254–66267. https://doi.org/10.1109/ACCESS.2018.2878456.
18. Su, Qingtang, Decheng Liu, Zihan Yuan, Gang Wang, Xiaofeng Zhang, Beijing Chen, and Tao Yao. 2019. "New Rapid and Robust Color Image Watermarking Technique in Spatial Domain." *IEEE Access* 7: 30398–30409. https://doi.org/10.1109/ACCESS.2019.2895062.
19. Kakkar, Mohit Kumar *et al.* 2021. "Class Schedule Generation Using Evolutionary Algorithms." *Journal of Physics: Conference Series* 1950: 012067. https://doi.org/10.1088/1742-6596/1950/1/012067.
20. Abualigah, Laith. 2021. "Group Search Optimizer: A Nature-Inspired Meta-Heuristic Optimization Algorithm with Its Results, Variants, and Applications." *Neural Computing and Applications* 33 (7): 2949–2972. https://doi.org/10.1007/s00521-020-05107-y.
21. He, S., Q. H. Wu, and J. R. Saunders. 2009. "Group Search Optimizer: An Optimization Algorithm Inspired by Animal Searching Behavior." *IEEE Transactions on Evolutionary Computation* 13 (5): 973–990. https://doi.org/10.1109/TEVC.2009.2011992.
22. Singh, Abha, Sumit Kushwaha, Maryam Alarfaj, and Manoj Singh. 2022. "Comprehensive Overview of Backpropagation Algorithm for Digital Image Denoising." *Electronics* 11 (10): 1590. https://doi.org/10.3390/electronics11101590.
23. Kamble, P. D., J. Giri, E. Makki, N. Sunheriya, S. B. Sahare, R. Chadge, C. Mahatme, P. Giri, and H. Panchal. 2024. "An Application of Hybrid Taguchi-ANN to Predict Tool Wear for Turning EN24 Material." *AIP Advances*, 14 (1): 015206. https://doi.org/10.1063/5.0186432.
24. Pooja, M. R., J. Giri, R. Chadge, N. Sunheriya, C. Mahatme, S. Mallik, N. Alsubaie, M. S. Alqahtani, M. Abbas, and B. O. Soufiene. 2023. "An Optimized Machine Learning Model for the Early Prognosis of Chronic Respiratory Diseases with Specific Focus on Asthma." *Research Square*. https://doi.org/10.21203/rs.3.rs-3075448/v1.

25. Natarajan, M., T. Pasupuleti, M. M. S. Abdullah, F. Mohammad, J. Giri, R. Chadge, N. Sunheriya, C. Mahatme, P. Giri, and A. A. Soleiman. 2023. "Assessment of Machining of Hastelloy Using WEDM by a Multi-Objective Approach." *Sustainability*, 15 (13): 10105. https://doi.org/10.3390/su151310105.
26. Natarajan, M., T. Pasupuleti, J. Giri, N. Sunheriya, L. N. Katta, R. Chadge, C. Mahatme, P. Giri, S. Mallik, and K. Ray. 2023. "Machinability of Titanium Grade 5 Alloy for Wire Electrical Discharge Machining Using a Hybrid Learning Algorithm." *Information* 14 (8): 439. https://doi.org/10.3390/info14080439.

Chapter 12

Building a smart contract-based escrow platform for management of crowdfunding and public/private tenders on blockchain

Bhavika Mathur, Kamlesh Lakhwani, and Gitanjali

12.1 INTRODUCTION

Crowdfunding can be defined as a way of raising funds for a project or a campaign by the initiative of an individual or an organization via common accessible crowdfunding platforms. The data suggests that crowdfunding initiatives that are nonprofit are more likely to succeed and attract a greater corpus of donation amount than other forms of initiatives. Crowdfunding of projects via platforms like Kickstarter or Indiegogo give power to micro-investors – investors who can invest small amounts of capital in projects, thereby increasing the diversity of the cap-table of a project by orders of magnitude. Crowdfunders also gain access to hitherto inaccessible opportunities to invest in varied projects – from community-driven initiatives to big start-ups that are building disruptive products. The patterns in crowdfunding change sharply and every wave of change decides how the funds are allocated to the causes and projects. Entrepreneurs, individuals, and non-profit organizations that initiate crowdfunding campaigns have the benefit of highlighting a cause or a project and gaining initial traction.

Blockchains have become the foundation of future crypto-assets class. But one of the most important uses of blockchain is also as a mode to make secure and energy-efficient online transactions. The most common problem with existing crowdfunding and tender platforms is that the there is no two-way binding parity between the goal and distribution of funding. Also, most of the crowdfunding platforms are not regulated, and thus a sizeable chunk of crowdfunding initiatives turn out to be frauds. A multivariate study by Glaeser and Shleifer [1] found that nonprofits find easy to attract funds on crowdfunding platforms than others. The process of raising funds on crowdfunding platforms is standardized and is usually created according to the playbook of the platform.

There are multiple benefits of using smart contract-based crowdfunding platforms, including transparency, efficiency, security, and speed. Using a blockchain ensures that the crowdfunders' community has the utmost

 DOI: 10.1201/9781003512585-14

control over the disbursement and the execution timeline of the project. The crowdfunders' community can also add their own votes, which are subject to approval by the community.

12.2 LITERATURE REVIEW

The literature on novel crowdfunding models is scarce, much less for blockchain-based crowdfunding models. Our survey of projects on leading intermediary crowdfunding platforms found three major categories of crowdfunding projects. First, a person or a representative, usually in personal capacity or in affiliation with a nonprofit organization, is raising funds for a cause. This model does not involve in any exchange of goods or services. In addition, a person or an organization is raising funds for a product or service that will be released in the near future. The crowdfunders will gain access to the product or service for their contribution. Furthermore, a person or an organization is raising funds for a product or service that will be released in the near future. The crowdfunders will receive a share in the future profits of the product or service based on the corpus of their contribution.

In Ref. [2], the anonymous author writing under a pseudonym of Satoshi Nakamoto discussed a novel system of decentralized application [dApp]. Satoshi mentioned the development of "a new electronic cash system that's fully peer-to-peer, with no trusted third party." The Bitcoin white paper was the first document to outline the principles of a cryptographically secured (elliptic curve cryptography), peer-to-peer electronic payment system that was designed to be transparent and fraud-resilient. It had put the power of distribution and control in the hands of the people. In yet another milestone white paper [3], now called as the Ethereum white paper, written by the creator of Ethereum, Vitalik Buterin, a Russian computer scientist, in 2013 introduced the concept of Ethereum protocol. Ethereum is the second largest blockchain protocol in the world, behind only Bitcoin. Ethereum was created to provide a framework to run all decentralized applications or dApps as they are popularly called. It extended the concept of decentralized applications introduced by blockchains and built a framework for developers to build their own applications based on the decentralized Ethereum protocol. Buterin also wrote about the scope of a turning-complete programming language for developers to write smart contracts that can be enforced on the Ethereum platform. It explained how new transactions work on the Ethereum blockchain. First, a new transaction is initiated. In addition, the details of the new transactions are sent to the nodes within the network. Moreover, new blocks are created on the blockchain with the meta information of the transactions and the timestamps. Furthermore, the node performs computation and broadcasts the proof-of-work, if it is able to do so, to the rest of the network. Other nodes in the network then validate and accept the solution. The process follows these steps iteratively for the next block.

In Ref. [4], the researcher puts forward his views on the risk factors associated with the initial coin offerings (ICOs). It emphasizes how ICOs have disruptive effects over the traditional methods of financing projects and causes. The capital formation via ICOs help disrupt the traditional gatekept hierarchy. It outlines the distinction between the traditional modes of fundraising (VCs, intermediary banks, loan-sharks, etc.). The risks and rewards of the ICOs differ vastly from that of traditional modes such as IPOs (Initial Public Offerings). The organization or the project owner issues tokens or coins through and indelible distributed ledger which are uniquely stamped with the organization's information and built on protocols such as Ethereum, Openledger, or Counterparty. The paper outlines the best strategies for the pricing of the pre-offering crypto asset and the post-offering crypto asset. Usually, a crypto currencies' pre-offering price is arbitrarily decided or matches the price of crypto currencies offered by similar projects at the pre-offering stage. However, the post-offering prices of the crypto asset is decided by the supply and the demand. The paper discusses that while there are benefits, there are also risks involved. The ICO model allows crypt assets to be created and distributed without any initial conditions, escrow requirements from the fundraisers, or security measures to protect investors. Crypto assets are intangible products and do not reflect any real value unless there is a demand for it. Our novel approach solves this with the intermediation of an escrow fund. It ensures legal certainty of disbursal of funds only on the conditions of the smart contract.

The framework of funding of community projects with the help of smart contracts has been discussed in Ref. [5]. It discusses the prospect of civic crowdfunding. It discusses a novel approach to solve the "free-rider" problem by implementing a game-theoretic mechanism which can be enforced via a smart contract. It proposes development of a dApp that can be accessed via a web browser. This dApp is based on a smart contract which has two conditions. The first one being that if the project deadline is reached, and total funds raised are greater or equal to target amount, then 25% can be allowed for the fundraiser to withdraw. The second one is that if the project deadline is reached, and total funds raised have not reached the target amount, then the contributors are allowed to withdraw their respective contributions. This is similar to the approach that the research proposes, wherein the withdrawal of funds depends on the current status of the crowdfunded project.

In Ref. [6], the author not only discusses the application of blockchain for the purpose of equity crowdfunding but also trifurcates the token standards into UTXO-based, layer-based, and smart contract-based tokens. UTXO or Unspent Transaction Output-based tokens use a native blockchain asset. It serves as a categorical container to which more value can be added. Layer-based tokens use transaction graphs to issue tokens. In this approach, new tokens are created and tracked. The base layer is blockchain layer and the second layer adds consensus algorithm, transaction types, and requirements. Smart contract-based tokens allow for the contracts, built as a code, to track states and represent token ownership. These contracts map tokens to owner addresses, that is, the contract acts as an intermediary mapping layer between the transferee and the owner. These contracts are based on a now-popular new standard: ERC777.

Crowdfunding is divided into three categories, namely: donate, pre-selling, and equity crowdfunding [7]. In Ref. [8], the authors designed a general framework for a blockchain-based crowdfunding platform. It outlines the four major types of crowdfunding practices, namely: donation-based crowdfunding, reward-based crowdfunding, equity-based crowdfunding, and crowd investing and crowd lending.

The applications of blockchain technology in the crowdfunding space have been discussed in Ref. [9]. The authors have implemented a proof-of-concept (PoC) for a simple smart contract in the crowdfunding process. It discusses a step-by-step protocol for implementation of a smart contract in a crowdfunding platform. This involves verification of the project and the fundraisers of a project on the blockchain, to ensure that only trusted parties are able to create projects, eliminating dependency on third parties or intermediaries such as banks and legal companies, among others, and enforcement of requirements of each project via smart contract at each stage to decide on the next execution and fund disbursement cycle.

Among many virtual currency trading platforms, Ethereum is the second-largest platform [10]. In Ethereum, there are two types of accounts: external-owned accounts (EOA) and contract accounts (CA). An EOA functions as an electronic cash account, representing a user's balance. On the other hand, a CA includes both a balance state and a contract state. By utilizing these accounts,

it becomes possible to digitize state information, such as user balances. This eliminates the necessity of tracing transaction histories for balance retrieval, a process required in Bitcoin [11]. Table 12.1 shows research insights in cryptocurrency, blockchain, and crowdfunding.

12.3 PROBLEM STATEMENT

One of the main benefits of crowdfunding is its exceptional accessibility, allowing fundraisers to secure funds quickly and efficiently [12]. It serves as a platform where investors can easily identify projects aligned with their interests and contribute to their realization, fostering a transparent and direct connection between creators and supporters. However, traditional crowdfunding platforms have faced a significant challenge in combating fraud cases, which have plagued the industry. There are many types of crowdfunding rewards like royalties, share financing, non-cash returns (getting a product/service in return), and so on [13]. A comprehensive

Table 12.1 Research insights in cryptocurrency, blockchain, and crowdfunding

Document	*Research area*	*Key findings*	*Methodology*	*Technical insights*
[1]	Crowdfunding dynamics	Nonprofit groups excel in online fundraising	Comprehensive study	Analysis of fund acquisition dynamics on crowdfunding platforms Evaluation of the impact of nonprofit groups Comparison with for-profit models
[2]	Cryptocurrency and blockchain	Introduction of Bitcoin and its core principles	Analysis of white paper	Use of elliptic curve cryptography for secure digital signatures Description of timestamp server and proof-of-work concept
[3]	Blockchain and dApps	Introduction to Ethereum and its dApp framework	Analysis of Ethereum white paper	Concept of a turning complete programming language Process of new transaction propagation and blockchain verification
[4]	ICO market	Analysis of ICO process, risks, and strategies	Literature review and analysis	Contrast between ICOs and traditional fundraising methods Discussion of token pricing and escrow fund utilization
[5]	Civic crowdfunding	Implementation of smart contracts in crowdfunding	Proposal and game theory	Utilization of enforceable smart contracts to address free-rider challenge Linking fund release to project development progress
[6]	Tokenization in crowdfunding	Categorization of token standards and blockchain impact	Analysis of token protocols	Categorization of tokens into UTXO-based, layer-based, and smart contract-based Discussion of blockchain's transformative role in financing
[7]	Crowdfunding ecosystem	Examination of primary crowdfunding processes	Empirical study	Analysis of the complex registration, campaign initiation, fund movement, and disbursement phases
[8]	Crowdfunding mechanisms	Exploration of factors influencing crowdfunding success	Statistical analysis	Impact of project quality, preparedness, and social capital on project viability Discussion on timely delivery challenges

(Continued)

Table 12.1 (Continued)

Document	*Research area*	*Key findings*	*Methodology*	*Technical insights*
[9]	Blockchain integration in crowdfunding	Development of a proof-of-concept (PoC) smart contract for crowdfunding	Proof-of-concept development	Smart contract implementation for project legitimacy verification and fund management Reduction of reliance on traditional intermediaries
[10]	Ethereum account structures	Overview of Ethereum's account types and their advantages	Technical analysis	Explanation of external owned accounts (EOA) and contract accounts (CA) Elimination of the need to trace transaction histories for balance retrieval

long-term study [1] extensively examined the regulatory framework surrounding frauds committed on crowdfunding platforms. It highlighted the inherent difficulties in enforcing traditional anti-fraud measures due to the global accessibility of these platforms and the diverse geographic origins of donors. The decentralized nature of crowdfunding makes it challenging to implement effective legal and security measures to prevent fraud effectively. Moreover, another pressing concern in the crowdfunding space is the issue of cost and time overruns associated with funded projects. It has been observed that many fundraisers or entrepreneurs fail to deliver goods, services, or results within the stipulated timeline. An exploratory study [5] conducted in this regard revealed that over 75% of crowdfunding projects experience delays in delivering promised products. These findings shed light on the need for a more robust and secure crowdfunding model, such as the one this study proposes with our "Fundraisers" escrow platform. By leveraging the power of blockchain technology and smart contracts, this study addresses the challenges of fraud and project delays. Our platform provides a transparent and accountable system where funders can have confidence in the secure handling of their contributions and track the progress of projects. By mitigating fraud risks and ensuring timely project delivery, our blockchain-based escrow platform aims to restore trust and confidence in crowdfunding [14–17]. This study recognizes the importance of protecting the interests of both funders and fundraisers, fostering a mutually beneficial relationship that drives innovation and supports the realization of groundbreaking ideas. With our solution, this study strives to elevate the crowdfunding experience to new heights, unlocking its full potential as a transformative force in the world of fundraising.

12.4 RESEARCH METHODOLOGY

This section will explain the theoretical framework that proposes a blockchain-based fraud-resilient crowdfunding platform with an intermediary escrow platform for fund disbursing. It will also elaborate how such a model has an inherent smart contract design that enforces project execution timeline, donation, and refund conditions.

A blockchain is a decentralized and distributed digital ledger that is used to record transactions across a network of computers. There is no "central authority" in a blockchain. Blockchain essentially works as a database register and offers an efficient way of building a central database at a lower cost [18]. It provides both transparency and security [19]. A smart contract, according to Wikipedia, is "a computer program or a transaction protocol which is intended to automatically execute, control or document legally relevant events and actions according to the terms of a contract or an agreement." A transaction represents an action of the participants. Any action can be defined as a "transaction" by the programmer of the decentralized application. For private organizations, the crowdfunding model is preferable to traditional funding as long as the capital requirements for the initial product/project funding is below the initial financial momentum gained via the crowdfunding platform. By implementing a smart contract-based escrow layer in the current crowdfunding structure, this study can create a contract that will escrow the funders' money until the following conditions are met:

a) *When the goal amount is reached*
   i. Once the contributors vote for initial release of funds as per the underlying consensus algorithm, a minimum threshold capital (MTC) from the escrow is released to the fundraisers to begin the work. The MTC is defined in the smart contract.
   ii. Any subsequent work is divided into milestones as per the contract and again the contributors can decide to release the next stage of amount from the escrow to the fundraisers. The fundraisers have to raise an expense request to get the next round of money released.
   iii. If there is fraud or time or cost overruns, the contributors can vote, as per the underlying consensus algorithm, to recover money from the escrow fund to the extent of their initial contribution and exit the project; or the contributors can exit the project and initiate a request to blacklist the fundraisers and their address.

b) *When the goal amount is not reached*
   i. The contributors vote to either extend the timeline until the goal is reached as per the underlying consensus algorithm; or
   ii. The contributors vote to release their respective contributions back to themselves and exit the project. Each transaction is recorded on the blockchain on the Ethereum network and can serve as a proof of transaction (Figure 12.1).

In the proposed model, the individual or the organization raising the funds can also offer ICOs (initial coin offerings) to potential investors. ICOs help minimize the transaction cost and also disintermediate banks. ICOs are a unique way of capital formation for entrepreneurs and crowdfunding campaigners. They also help build trust in the crowdfunders' community at the early stage of project development. Ownership of tokens/coins offered via ICOs indicates ownership of a financial stake in the product or the company. Ethereum is undoubtedly a uniform and the most popular protocol for developing decentralized applications [dApps]. Decentralized applications along with smart contract allow developers to build applications on the Ethereum ecosystem rather than writing their own blockchain technology. The theoretical design of the model is charted in Figure 12.2. The distributed application will necessarily have a frontend layer and a local node and the Ethereum network. A smart contract will enforce control over

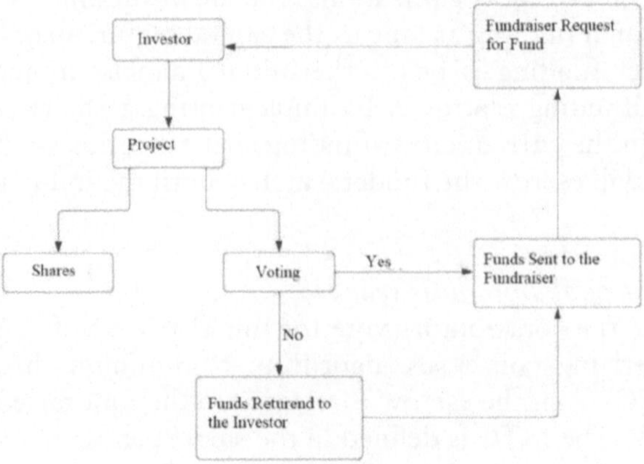

Figure 12.1 Privileges given to investors.

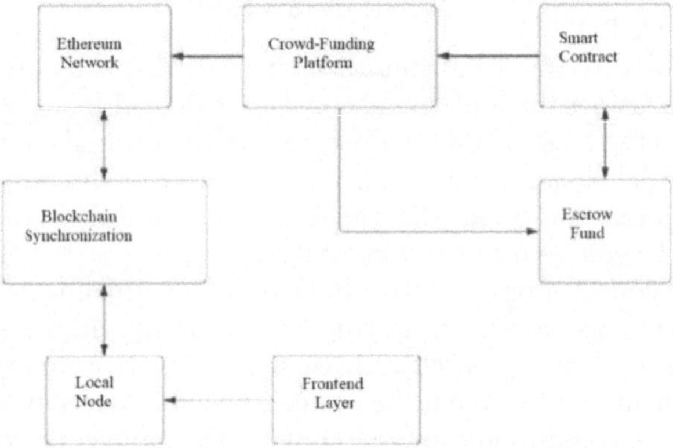

Figure 12.2 Standard structure of the proposed crowdfunding platform on the Ethereum network benefits and applications.

the escrow funds and the conditions of the project execution and milestone releases.

One of the major benefits of using a smart contract-based crowdfunding platform is that it provides a robust mechanism for building a two-way binding trust between the stakeholders of the platform. Other benefits are reduced costs, high speed of transactions, scalability, removal of intermediaries, traceability, etc. An end-to-end transaction on the blockchain only costs a fraction of what it costs via traditional modes of transaction. Transactions on blockchains are blazing fast. It takes only a few seconds for a transaction to be securely validated and committed on the Ethereum network across all nodes. One of many great benefits of blockchain technology is the scalability. The transaction speed on a blockchain is agnostic

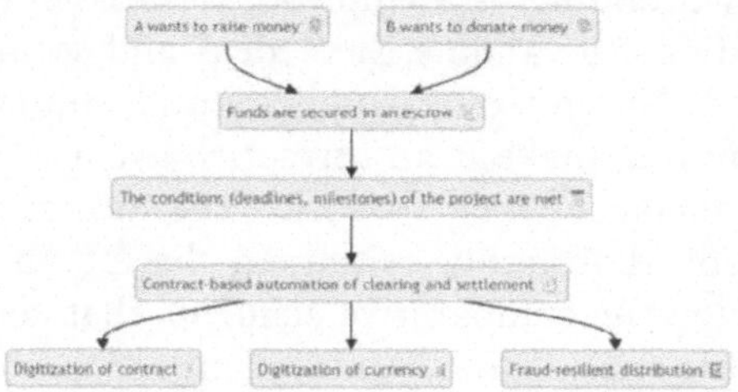

Figure 12.3 Using smart contracts to design a fraud-resilient distribution mechanism for crowdfunding platforms.

to the number of nodes connected on the network. This makes blockchain the best-suited technology driver for building highly scalable applications. The blockchain removes the necessity of financial intermediaries such as banks, loan agents, etc. All transactions on the blockchain are validated before they are done. This ensures transparency and no actor or a cluster of actors can "fake" a transaction. This builds trust in the crowdfunding community. Apart from this, blockchain transactions have "true traceability." Figure 12.3 outlines how a transaction is implemented via a smart contract on the crowdfunding platform. This figure also outlines a sample case between stakeholders and the role played by the smart contracts in the execution of the transactions between the stakeholders/actors on the crowdfunding platform.

12.5 RESULT AND DISCUSSION

This study successfully designed an escrow platform under the name "Fundraisers," which is more secure and fraud-resilient as it is enforced via a smart contract on the blockchain. This breakthrough technology ensures that funds raised or donated for a cause are protected from any potential fraudulent activities, giving investors, entrepreneurs, and organizations peace of mind. By leveraging the power of blockchain, this study has created a platform that inspires trust and encourages more people to participate in fundraising initiatives. Crowdfunding platforms have revolutionized the way investments are made, opening up new opportunities for micro and medium-scale investors who were previously unable to engage in highly innovative ventures. Our research delved into the intricacies of existing crowdfunding models, analyzing their strengths, weaknesses, and the different funding models employed.

Recognizing the shortcomings and challenges present in the current landscape, this study sought to propose a new blockchain-based, fraud-resilient model of crowdfunding. Through an extensive review of existing models and careful consideration of emerging challenges, this study identified the significance of crowdfunding as a vital tool for artists, entrepreneurs, and nonprofit organizations to validate their ideas and secure necessary funding from the public. The inherent simplicity and minimal friction associated with crowdfunding make it an attractive avenue for individuals and entities seeking financial support. To ensure the success of a crowdfunding campaign, this study thoroughly examined the factors that influence its outcome. By studying the comparative elements that contribute to a campaign's triumph, this study gained valuable insights into the dynamics of the crowdfunding ecosystem. This knowledge serves as a foundation for our proposed theoretical framework, which empowers the crowdfunding community to maintain greater control over their projects. Central to our framework is the intermediary escrow platform, bolstered by the implementation of a smart contract on the blockchain. This innovative approach enhances the confidentiality and confidence of contributors, as all transactions are transparently recorded and publicly accessible. By utilizing blockchain technology, this study establishes a binding two-way trust between all stakeholders involved in the crowdfunding process. Not only does our model offer increased security and fraud resilience, but it also fosters a sense of transparency and accountability within the crowdfunding space. Contributors can have full visibility into how their funds are utilized, ensuring that they are making a meaningful impact. This heightened level of trust will undoubtedly attract more investors, entrepreneurs, and organizations to embrace crowdfunding as a viable fundraising mechanism. In conclusion, our research and the subsequent development of the "Fundraisers" escrow platform have brought forth a paradigm shift in the crowdfunding landscape. By integrating blockchain technology and smart contracts, this study has created a more secure, transparent, and trustworthy crowdfunding environment. This innovation unlocks unprecedented opportunities for both investors and fundraisers, empowering them to achieve their goals and make a positive impact on society. A comparison between Solana and Ethereum on the basis of various factors has been discussed in Table 12.2.

12.6 CONCLUSION AND FUTURE SCOPE

This study has proposed a comprehensive framework that empowers fundraisers with complete financial control. Recognizing the significance of crowdfunding as a funding source, the proposed framework aims to enhance security, reliability, and fraud resilience. This represents a significant milestone in addressing the contemporary challenges faced by the global community in raising and donating funds. To ensure transparency, efficiency, security, and speed, this study has developed a smart contract-based crowdfunding platform. This platform grants users unparalleled control over their funds, enabling precise disbursement and execution. This study has explored the utilization of smart contracts and blockchain technology,

Table 12.2 Comparison between Solana and Ethereum

Sr. No.	*Solana*	*Ethereum*
Launch year	2020	2015
Transaction speed (per second)	50,000–65,000 TPS	13–15 TPS
Transaction cost	<$1	$1–$50
Number of dApps	<81	<568
Decentralized	Less	More
Smart contract languages	Rust, C, C++	Solidity
Scalability	More	Less
Native token	SOL	ETH
Processing	Fast	Slow
Stability	Less	More
Architecture	Stateless	Stateful
Consensus mechanism	Proof of work	Proof of stake and proof of history
Developer community	Highest number of developers	Comparatively low

recognizing their immense potential in the present era. By leveraging the capabilities of blockchain and smart contracts, the study has successfully devised a theoretical and practical model that empowers prospective donors to exercise control over their contributions. This model operates on the basis of conditional contracts, which are automatically executed upon meeting the predefined conditions. This study has also examined the advantages of employing diverse funding mechanisms in online transactions, acknowledging the ongoing evolution and improvement of technologies like Solana, which offer enhanced speed and cost-effectiveness. Our proposed framework represents a significant step forward in revolutionizing the crowdfunding landscape. By prioritizing security, control, and transparency, this study aims to foster a more robust and efficient environment for fundraisers and donors alike. As the study continues to embrace advancements in blockchain technology and smart contracts, it is confident in the transformative potential of our framework to address the evolving needs of the modern fundraising ecosystem.

REFERENCES

1. Glaeser, Edward L., and Andrei Shleifer. "Not-for-profit entrepreneurs." *Journal of Public Economics* 81, no. 1 (2001): 99–115.
2. Nakamoto, S. "Bitcoin: A Peer-to-Peer Electronic Cash System." (2008). Available at: www.bitcoin.org [Accessed: February 20, 2023].

3. Buterin, V. “Ethereum White Paper: A Next Generation Smart Contract & Decentralized Application Platform,” ethereum.org. Available at: http://ethereum.org [Accessed: February 25, 2023]
4. Kaal, Wulf A., and Marco Dell'Erba. “Initial coin offerings: emerging practices, risk factors, and red flags.” *Verlag CH Beck (2018), U of St. Thomas (Minnesota) Legal Studies Research Paper* 17–18 (2017).
5. Roth, Jakob, Fabian Schär, and Aljoscha Schöpfer. “The Tokenization of assets: using blockchains for equity crowdfunding.” In *Theories of change: Change leadership tools, models and applications for investing in sustainable development*, 329–350, 2021.
6. Islam, Md Mainul. “A privacy-preserving transparent central bank digital currency system based on consortium blockchain and unspent transaction outputs.” *IEEE Transactions on Services Computing* 16, no. 4 (2022): 2372–2386.
7. Ashari, Firmansyah, Tetuko Catonsukmoro, Wilyu Mahendra Bad, and Wang Sfenranto. “Smart contract and blockchain for crowdfunding platform.” *International Journal of Advanced Trends in Computer Science and Engineering* 9, no. 3 (2020): 3036–3041.
8. Mollick, Ethan. “The dynamics of crowdfunding: An exploratory study.” *Journal of Business Venturing* 29, no. 1 (2014): 1–16.
9. Agrawal, Ajay, Christian Catalini, and Avi Goldfarb. “Crowdfunding: Geography, social networks, and the timing of investment decisions.” *Journal of Economics & Management Strategy* 24, no. 2 (2015): 253–274.
10. Liu, Zhijian, Zihan Shen, Hongfei Wang, and Qianjia Zou. “Ethereum security based on blockchain framework and its algorithms and components.” *Highlights in Science, Engineering and Technology* 60 (2023): 128–137.
11. Li Xiangyu, Xinyu Wang, Tingli Kong, Junhao Zheng, and Min Luo. “From bitcoin to solana–innovating blockchain towards enterprise applications.” In *International Conference on Blockchain*, pp. 74–100. Cham: Springer International Publishing, 2021.
12. Islam, Md Mainul. “A privacy-preserving transparent central bank digital currency system based on consortium blockchain and unspent transaction outputs.” *IEEE Transactions on Services Computing* 16, no. 4 (2022): 2372–2386.
13. Magomadova, Madina, Ayaz Aliyev, and Bakhrom Tursunov. “The Crowdfunding Platforms Financing Models Analysis.” In *SHS Web of Conferences*, vol. 172, p. 02032. EDP Sciences, 2023.
14. Pooja, M. R., J. Giri, R. Chadge, N. Sunheriya, C. Mahatme, S. Mallik, N. Alsubaie, M. S. Alqahtani, M. Abbas, and B. O. Soufiene. “An optimized machine learning model for the early prognosis of chronic respiratory diseases with specific focus an asthma.” *Research Square* (2023). https://doi.org/10.21203/rs.3.rs-3075448/v1
15. Thakur, R., G. Sikka, U. Bansal, J. Giri, and S. Mallik. “Deadline-aware and energy efficient IoT task scheduling using fuzzy logic in fog computing.” *Multimedia Tools and Applications* (2024): 1–28. https://doi.org/10.1007/s11042-024-19509-w.
16. Amale, A., A. K. S. Singholi, J. Giri, I. Albaijan, and A. Guru. “Fuzzy logic-driven genetic algorithm strategies for ultrasonic welding of heterogeneous metal sheets.” *AIP Advances* 14, no. 5 (2024): 055117. https://doi.org/10.1063/5.0207578

17. Swain, K., M. Cherukuri, I. S. Samanta, A. Pati, J. Giri, A. Panigrahi, H. Qin, and S. Mallik. "Fuzzy Markov model for the reliability analysis of hybrid microgrids." *Frontiers in Computer Science* 6 (2024): 1406086. https://doi.org/10.3389/fcomp.2024.1406086.
18. Tafuro, Alessandra, Giuseppe Dammacco, and Antonio Costa. "A conceptual study on the role of blockchain in sustainable development of public–private partnership." *Administrative Sciences* 13, no. 8 (2023): 175.
19. Küsbeci, Polathan, and Mehmet Fatih Burak. "Blockchain Technology-Based Crowdfunding Systems." (2023).

Chapter 13

Analysis of water quality for sustainable resource management using machine learning techniques

Ramandeep Sandhu, Ankita Arora, Harpreet Kaur Channi, Gagandeep Singh Cheema, and Kamlesh Lakhwani

13.1 INTRODUCTION

Freshwater is an essential element for industry and cultivation. A crucial step within the administration of freshwater properties is observing water quality. According to the WHO's yearly report, numerous individuals, particularly hopeful ladies, and little children, are passing away due to the need of clean drinking water. Whether water is utilized to hydrate animals, give a communal shower, or give drinkable water, confirming that it is fit for the purpose is pivotal. Safe drinking water is fundamental for human life. A person cannot lead a healthy life without safe water. The significant factors affecting the immaculateness of drinking water are pH, hardness, sulfate, natural carbon, turbidity, and others. Artificial intelligence (AI) methods, including machine learning (ML), computer vision, and characteristic dialect processing, aid in water quality examination. The biological system and the common public's welfare are straightforwardly affected by water quality. The water is utilized for numerous purposes, including cultivation and fabrication. Due to their availability, conduits have been utilized more often than possible for the growth of human civilizations than other water sources [1].

Other water sources, like groundwater and saltwater, can now and then help with issues. Utilizing groundwater without satisfactory recharging will cause water level to sink, and utilizing saltwater is regularly connected to the spread of contamination. As a result, the utilization of conduits has earned intrigue. A department of the building called "stream building" has been recommended due to various examinations into waterways across the globe. It considers morphological adjustments, dregs development, water quality, and contaminants for waterway designing. In-stream designing and the power through the pressure of conduits are considered besides the stream structure, sediment movement, and morphology of rivers. River water quality investigation could be a well-known point in soil sciences [2].

 DOI: 10.1201/9781003512585-15

13.1.1 Problem

Whether water is used for drinking, family utilization, nourishment generation, or recreational exercises, it is significant for the public's well-being to secure, effectively, sufficient pure water. The problem of destitution as well as the economy growth considerably depend on water supplies, clean conditions, and water asset administration besides several other factors. It is well-known that unsanitary conditions and polluted water contribute to the spread of ailments like cholera, loose bowels, hepatitis A, typhoid, and polio. Well-being risks for those without getting too sterile offices and immaculate water can be maintained at a strategic distance [3].

13.2 LITERATURE REVIEW

Water quality research is critical for ensuring water safety for human consumption, aquatic life, and the ecosystem. Conventional methods of assessing the purity of water necessitate the physical and chemical analysis of water samples, which is more expensive and time-consuming. As water quality data from sensors and other sources has been more readily available, machine learning techniques have emerged as a promising tool for water quality monitoring. The planned project will look into and focus research on using machine learning to enhance the quality of water. Several research studies have used machine learning methods, such as clustering, regression, and classification, to analyze the quality of the water. Predicting factors like dissolved oxygen, pH, and turbidity that are important to water quality is one such usage. For instance, Bhatt and Pandey (2021) used XGBoost, random forests, and decision trees to forecast river water quality indicators. Among these uses is anomaly detection.

Anomaly detection can help identify water quality issues and prevent potential environmental hazards. For instance, Alawadi and Abbadi (2021) proposed a hybrid algorithm combining deep learning and clustering techniques to detect anomalies in water quality parameters. The proposed algorithm achieved high accuracy and was able to detect anomalies in real time. Clustering techniques have also been used in water quality analysis to identify different water quality zones based on similar characteristics. For example, Li et al. (2021) applied K-means clustering to identify water quality zones in a river based on six water quality parameters. They found that the river could be divided into four distinct water quality zones, each with different water quality characteristics.

ML techniques are applied to the measurement of water quality based on the impact of climate change. Singh et al. (2021) used a support vector regression model in a river basin for the same purpose. They discovered that the water quality parameters were temperature- and precipitation-sensitive and that the proposed model could be used to anticipate future water

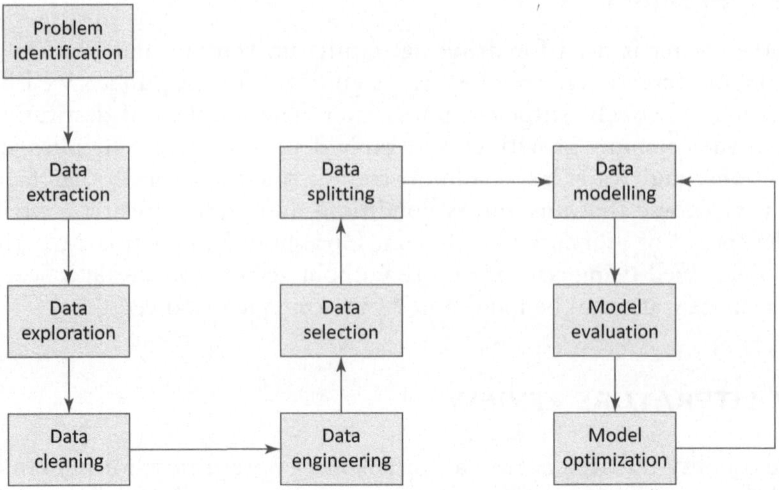

Figure 13.1 Workflow of water quality analysis using machine learning.

quality changes. In conclusion, machine learning techniques have shown great potential in water quality analysis. The use of machine learning can help the traditional water quality analysis methods and provide more accurate and timely water quality information.

13.3 PROPOSED METHODOLOGY

Within the cutting-edge period, classification and machine learning strategies continually progress, driving superior results. This recommended calculation decides whether the water is secure to expend by utilizing factors like pH, conductivity, and hardness. Various ML approaches such as logistic regression, decision tree classifier, GaussianNB, LinearSVC, and XGBClassifier are used for this purpose [4]. Data are initially gathered and dispersed based on ten metrics, as shown in Figure 13.1. Following that, algorithms are created. The proposed research concludes by presenting performance analysis and prediction results [5].

13.3.1 Dataset

Data from Kaggle is collected and this dataset contains water quality measurements for 4,399 diverse water bodies [6].

- *Hardness:* The amount per liter (mg/L) of water that can speed up cleaning agents.
- *Chloramines:* Number of chloramines in ppm.

- *Sulfate:* Number of sulfates dissolved in mg/L.
- *Potability:* Determines if the water is safe for human consumption. Potable = –1 and non-potable = 0.
- *Conductivity:* Electrical conductivity of water, measured in μS/cm.
- *pH*: pH of water is between 0 and 14.
- *Turbidity:* The amount of light that water emits as measured in NTU.
- *Solids:* Total dissolved solids in ppm.

13.3.2 Data preprocessing

Usually, when a dataset is obtained from any source, it must be pre-processed. Cleaning the data will increase the variety of the training data. The dataset is then fed to the model for training after preprocessing [7, 8]. Table 13.1 provides water quality standards as per WHO records.

It shows the benchmark number for the water quality pointer, and the World Wellbeing Organization gives those estimations.

13.3.3 Smote

The data was uneven, and we used the Synthetic Minority Oversampling Technique to balance it. Table 13.2 represents the counter before and after the smote. It creates a line between a few selected occurrences in the feature space, then creates a new sample at a point along the line. More precisely, the minority class representative is first selected at random. After that, *k* – usually equal to 5 – of the closest neighbors are located. Two examples and their randomly chosen neighbors are connected via a synthetic example that is created at an unexpected location in feature space [9].

Table 13.1 WHO water quality standard

Parameters
pH = 6.5–8.5
Hardness = 300 mg/L
Solids (TDS) = <20,000 ppm
Sulfate = <250 mg/L

Table 13.2 Smote

Before smote	
Counter	{0: 2026, 1: 1273}
After smote	
Counter	{0: 2026, 1: 2026}

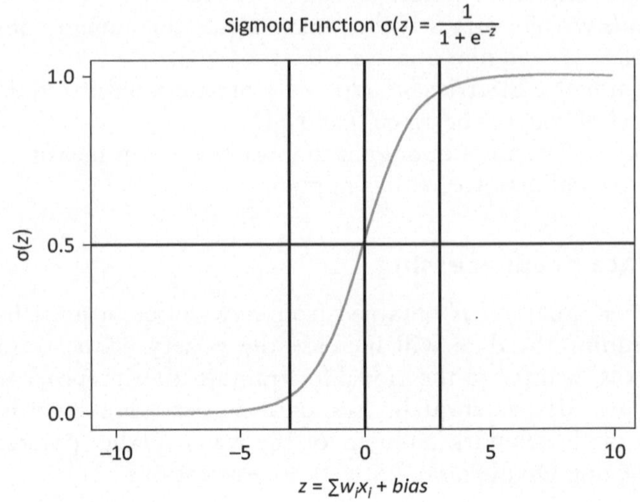

Figure 13.2 Logistic regression.

13.3.4 Machine learning (ML) algorithms

The following ML algorithms have been used in the proposed work.

1. *Logistic regression*

Statisticians developed the logistic function based on the characteristics of population development in ecology; it is found that it climbs rapidly and peaks at the ecosystem's carrying capacity. Using this S-shaped curve (as shown in Figure 13.2), any real-valued integer is transformed (between 0 and 1), but never accurately at those ranges.

Support vector classifier (SVC)

A linear SVC's goal is to match the data you supply, yielding a "best fit" hyperplane that divides or classes your data [10].

Random forest classifier

After fitting, it is found that the random forest classifier employs averaging to improve forecast accuracy and decrease overfitting [11].

XGBoost

XGBoost is a distributed gradient boosting program optimized to be highly efficient, versatile, and portable. Machine learning methods are performed using the Gradient Boosting framework. XGBoost is also used for parallel tree boosting [12].

GaussianNB

A probabilistic machine learning approach, the naive Bayes algorithm, is applied to various categorization issues. Common applications include filtering spam, classifying papers, and determining sentiment. The idea that the traits that make up the model are unconnected to one another is referred to as naive means that changing the value of one feature does not directly affect the values of the other features used in the method [13].

13.4 DATA DISTRIBUTION ANALYSIS

For this investigation, ten readings are collected. Throughout the data transmission process, several information is displayed individually to provide context for the drinking water requirements. An aqueous substance's pH value can be used to identify its acidity or neutrality. It displays the water's pH level. The WHO states that the maximum allowable pH level (Figure 13.3) is between 6.5 and 8.5 [6, 14–21].

The sum of broken-down magnesium and calcium found in the water is utilized to evaluate "water hardness." It is best to stay away from hard water because it contains many dissolved elements, mainly calcium, and magnesium. By WHO standards, water hardness is 300 mg/L (Figure 13.4).

The term "solids" as shown in Figure 13.5 depicts the small sums of natural compounds and inorganic salts in water arrangement. Sulfate, carbonate, chloride, nitrate anions, and hydrogen carbonate are commonly found at the side of calcium, magnesium, sodium, and potassium cations [9, 21–27].

Chlorine depicted in Figure 13.6 responds with smelling salts and other natural and inorganic nitrogen-containing particles to make a course of substances known as chloramines. When treating consumable water,

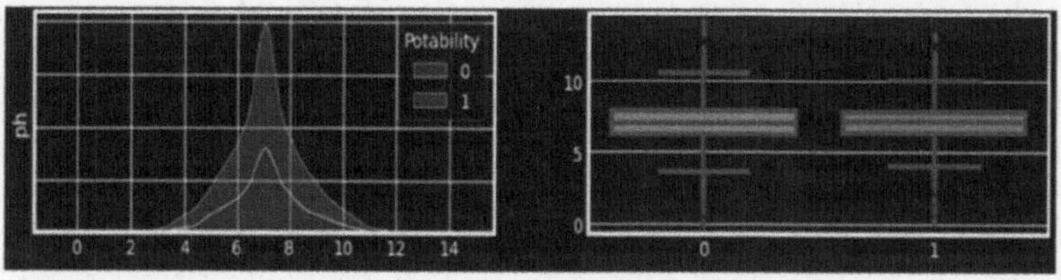

Figure 13.3 pH graph.

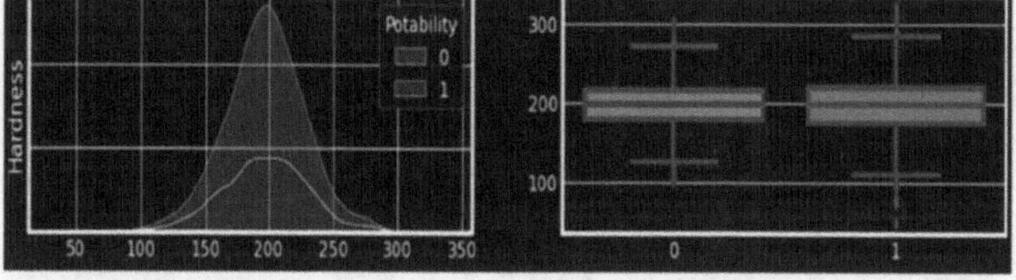

Figure 13.4 Hardness.

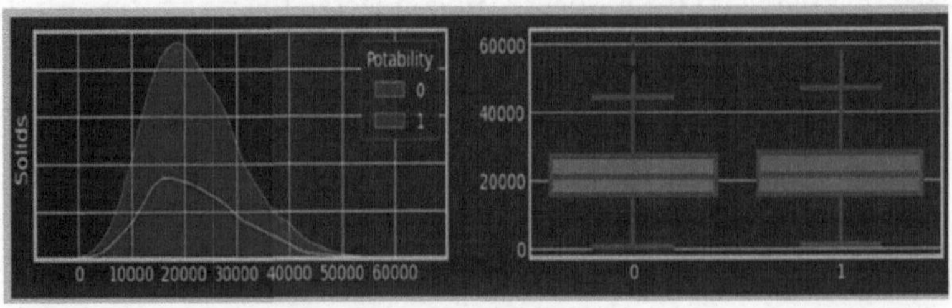

Figure 13.5 Solids.

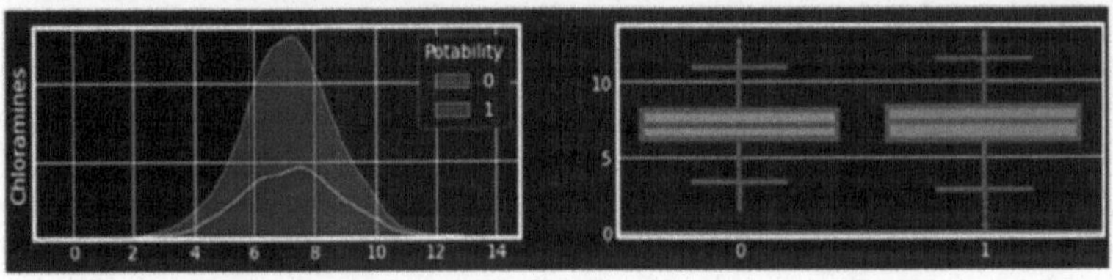

Figure 13.6 Chloramines.

chloramines are habitually utilized as a disinfectant to avoid the advancement of perilous germs and infections. Concerns around chloramines incorporate the plausibility that they might combine with organic matter in water to make disinfecting contaminants like trihalomethanes (THMs) and halo acetic acids. (HAAs). The chance of cancer and other ailments has been related to these by-products. To ensure that chloramines and cleansing by-products are shown at worthy sums, it is significant to routinely analyze water.

Figure 13.7 describes that the electrical conductivity in drinking water could be a gauge of a solution's ionic handle for conducting power. It illustrates the conductivity's spread. Concurring with WHO proposals, the electrical conductivity numbers should not be higher than the value 400 S/cm.

The nearness of natural carbon in water may be a sign of natural materials. Various little particles, dissolved proteins, colloids, and combinations

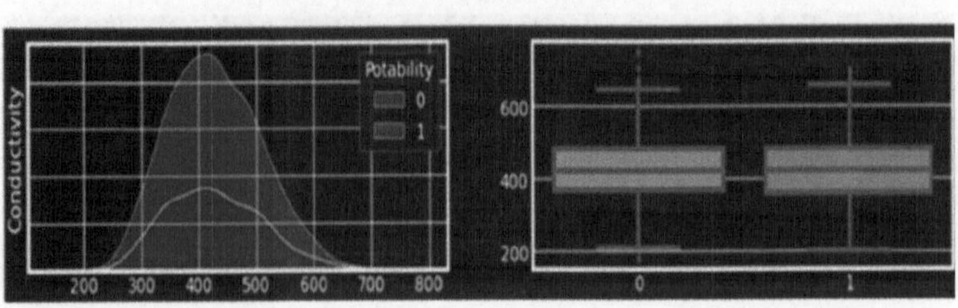

Figure 13.7 Conductivity.

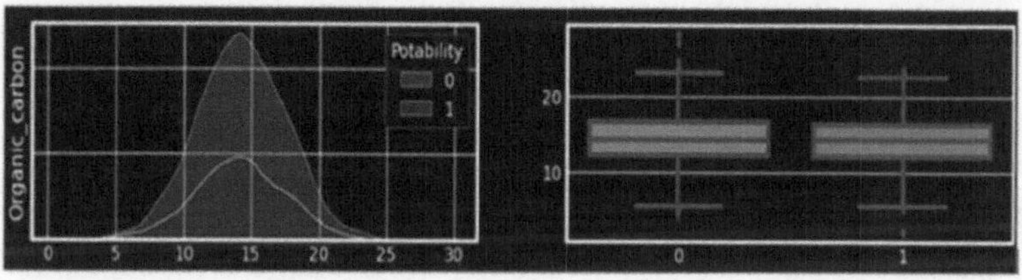

Figure 13.8 Organic carbon.

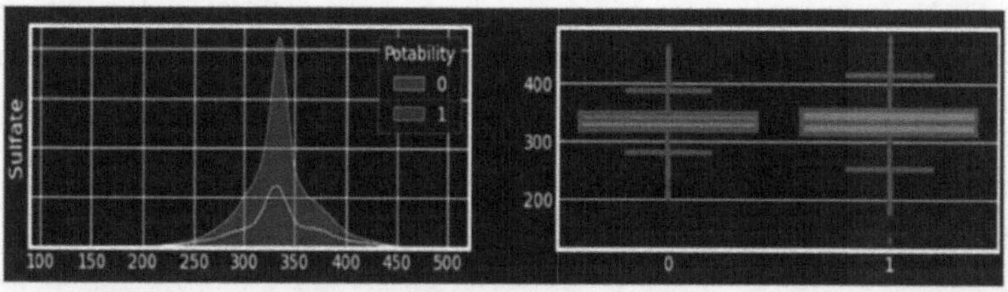

Figure 13.9 Sulfate.

are many thousands of parts it might contain. The same is depicted in Figure 13.8. Less than 25% of natural carbon is allowed in consumable water.

Chemical substances called sulfates (or sulfates) contain the sulfate particle (SO_4^{2-}). Sulfates as shown in Figure 13.9 are, as often as possible, displayed in water and can have an assortment of negative impacts on both the biological system and human health. Sulfate levels in water systems ought to be routinely checked to rule out if they are beyond fitting and secure ranges.

By-products of decontamination known as trihalomethanes are created when chlorine molecules interact with naturally occurring elements in the water (Figure 13.10).

The cloudiness or fogginess of water brought on by suspended particulates, such as residue, clay, green growth, and other miniature animals, is measured as turbidity, as shown in Figure 13.11. It could be a vital figure in

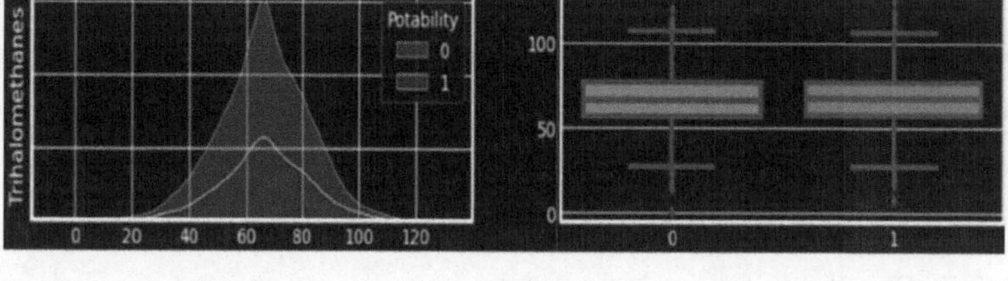

Figure 13.10 Trihalomethanes.

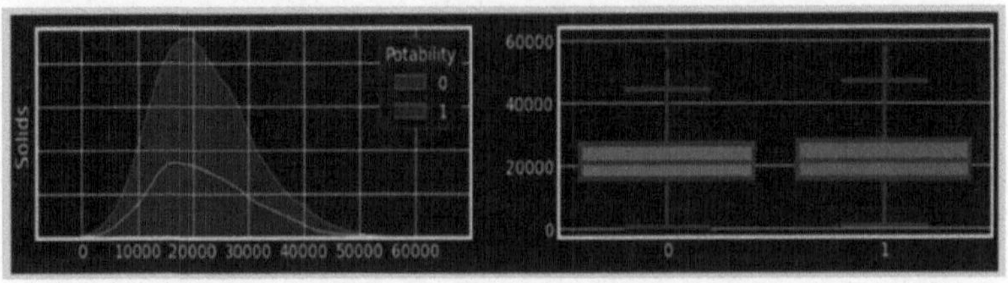

Figure 13.11 Turbidity.

deciding the immaculateness of the water since it influences how the water looks, tastes, and smells, as well as whether it is fit for ingesting, swimming, and supporting oceanic life.

A relationship heatmap described in Figure 13.12 is a visual portrayal of a relationship lattice that shows how different variables relate to one another.

13.4.1 Performance parameters

The uncertainty matrix is one feature that accurately depicts a categorization model's actual performance and can be used to gauge a system's effectiveness. The True Positive (TP), False Positive (FP), False Negative (FN), and True Negative (TN) numbers in the confusion matrix are used for evaluation [27–30].

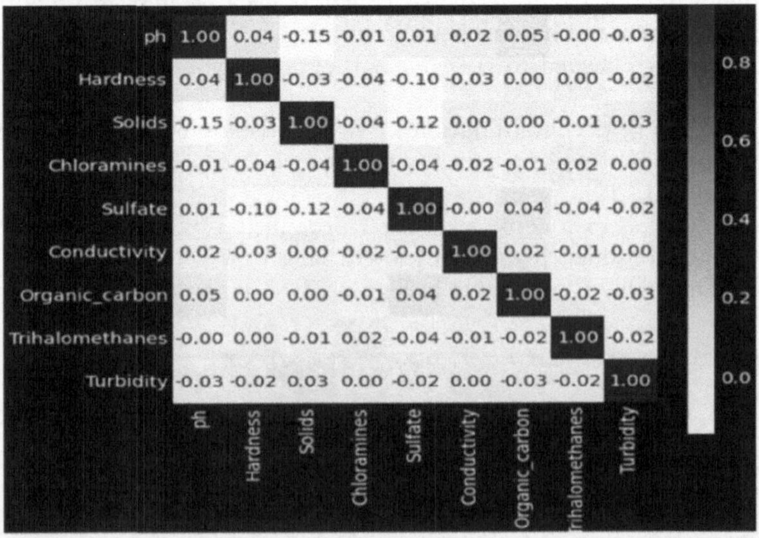

Figure 13.12 Correlation heatmap.

To calculate average accuracy:

$$\text{Accuracy} = \text{TP} + \text{TN}/\text{Total} \quad (13.1)$$

To calculate Positive Predictive:

$$\text{PR} = \text{TP}/\ \text{TP} + \text{FN} \quad (13.2)$$

To calculate the Positive Rate:

$$\text{Recall} = \text{TP}/\text{TP} + \text{FN} \quad (13.3)$$

To calculate F1-score:

$$\text{F1-score} = \text{PPV} + \text{TPR}/\text{PPV} + \text{TPR} \quad (13.4)$$

13.4.2 Model building

Calculations such as random forest, logistic regression, decision tree classifier, GaussianNB, LinearSVC, and XGBClassifier are used to estimate drinking water quality. Table 13.3 is all about the training accuracy value of each algorithm used in the proposed approach.

13.4.3 Simulation results

The Jupyter application is used in this chapter to study the data and forecast the potability of drinking water based on ten estimates. WQI assessed the safety of the water to drink in the context of expectations. If the water can be tanked, WQI offers advice based on WHO recommendations.

Table 13.4 is all about the percentage of water in two categories: fit to drink and unfit to drink. It shows that even small amounts of water can

Table 13.3 Training accuracy

Algorithm name	*Training accuracy (%)*
Logistic regression	81
Decision tree classifier	85
XGBClassifier	85

Table 13.4 Water quality

Fit to drink	Unfit to drink
92.31 (%)	7.69 (%)

be dangerous. Groundwater is polluted by the infiltration of toxic waste, livestock wastage, insecticides, chemical fertilizers, and waste from depots and sewage tanks into aquifers. Unplanned growing cities cause significant surface water pollution. Increased public awareness, ethical waste management, and equitable use of water resources can also help for the reduction of the pollution of drinking water.

- So, after the prediction, random forest and XGBoost gave the best outcome; both provided an excellent F1-score of around 76%.
- Water samples that contain acidic and basic pH levels are almost equal.
- The relationship coefficients between the highlights were exceptionally low.

13.5 CONCLUSION

For the ecosystem's protection and to prevent contamination, it is essential to predict potable water. Clean drinking water has to be made accessible to preserve outstanding overall health. By consuming water from reputable sources, the water's potability may be guaranteed. Forecasting potable water accurately becomes difficult. The ideal learning strategy is necessary to prevent producing false predictions. The potability of drinking water may be predicted using an intelligent model built on five different kinds of ML algorithms, taking into account ten standard parameters, including pH, hardness, organic carbon, and other characteristics. This work targets training the model to perform better using random forest and XGBoost. The proposed model will be utilized in the future to get ahead and identify potable water quality in diverse places, together with an IoT-oriented quality monitoring model, and can increase model accuracy with data collection. The future scope will be toward real-time data collection with the use of IoT sensors. Apart from this, the same model can be enrolled for continuous assessment of stress of water which is linked with the quality level of water. The large dataset collection is possible using cloud solutions and the model can be applied to water datasets stored on the cloud.

REFERENCES

1. Haghiabi, H., A. H. Nasrolahi, and A. Parsaie. "Water Quality Prediction Using Machine Learning Methods." *Water Quality Research Journal* 53, no. 1 (2018): 3–13. https://doi.org/10.2166/WQRJ.2018.025.
2. Zhang, Q., P. Xu, and H. Qian. "Groundwater Quality Assessment Using Improved Water Quality Index (WQI) and Human Health Risk (HHR) Evaluation in a Semi-arid Region of Northwest China." *Exposure and Health* 12, no. 3 (2020): 487–500. https://doi.org/10.1007/s12403-020-00345-w.

3. Sarker, H. "Machine Learning: Algorithms, Real-World Applications and Research Directions." *SN Computer Science* 2, no. 3 (2021): 1–21. https://doi.org/10.1007/S42979-021-00592-X/FIGURES/11.
4. "Flow Chart for Machine Learning Workflow. | Download Scientific Diagram." Available at: https://www.researchgate.net/figure/Flow-chart-for-machine-learning-workflow_fig1_342778782 (Accessed April 14, 2023).
5. "Water Quality | Kaggle." Available at: https://www.kaggle.com/datasets/adityakadiwal/water-potability (Accessed April 14, 2023).
6. Akter, T., et al. "Water Quality Index for Measuring Drinking Water Quality in Rural Bangladesh: A Cross-Sectional Study." *Journal of Health, Population and Nutrition* 35, no. 1 (2016): 1–12. https://doi.org/10.1186/S41043-016-0041-5/TABLES/5.
7. Abba, S. I., et al. "Implementation of Data Intelligence Models Coupled with Ensemble Machine Learning for Prediction of Water Quality Index." *Environmental Science and Pollution Research* 27, no. 33 (2020): 41524–41539. https://doi.org/10.1007/s11356-020-09689-x.
8. "SMOTE | Towards Data Science." Available at: https://towardsdatascience.com/smote-fdce2f605729 (Accessed April 14, 2023).
9. Faiz, M., N. Fatima, and R. Sandhu. "A Vaccine Slot Tracker Model Using Fuzzy Logic for Providing Quality of Service." *Multimodal Biometric and Machine Learning Technologies: Applications for Computer Vision* (2023): 31–52.
10. Darmawan, M. F., A. F. Z. Abidin, S. Kasim, T. Sutikno, and R. Budiarto. "Random Forest Age Estimation Model Based on Length of Left-hand Bone for the Asian Population." *International Journal of Electrical and Computer Engineering (IJECE)* 10, no. 1 (2020): 549–558. https://doi.org/10.11591/IJECE.V10I1.PP549-558.
11. Lu, H., and X. Ma. "Hybrid Decision Tree-Based Machine Learning Models for Short-term Water Quality Prediction." *Chemosphere* 249 (2020). https://doi.org/10.1016/J.CHEMOSPHERE.2020.126169.
12. "Implementing Gaussian Naive Bayes in Python - Analytics Vidhya." Available at: https://www.analyticsvidhya.com/blog/2021/11/implementation-of-gaussian-naive-bayes-in-python-sklearn/ (Accessed April 14, 2023).
13. World Health Organization. "Guidelines for Drinking-water Quality, 4th Edition, Incorporating the 1st Addendum." 2017. Available at: http://www.who.int/water_sanitation_health/publications/drinking-water-quality-guidelines-4-including-1st-addendum/en/ (Accessed April 14, 2023).
14. "Solids Content of Wastewater and Manure | Oklahoma State University." Available at: https://extension.okstate.edu/fact-sheets/solids-content-of-wastewater-and-manure.html (Accessed April 14, 2023).
15. World Health Organization. "Sulfate in Drinking-water Background Document for Development of WHO Guidelines for Drinking-water Quality." 2004.
16. Hood, E. "Tap Water and Trihalomethanes: Flow of Concerns Continues." *Environmental Health Perspectives* 113, no. 7 (2005): A474. Available at: https://www.ncbi.nlm.nih.gov/pmc/articles/PMC1257669/ (Accessed April 14, 2023).
17. "Turbidity in Drinking Water — Water Library | Acorn Water - H2O Labcheck." Available at: https://www.h2olabcheck.com/blog/view/turbidity (Accessed April 14, 2023).

18. Sandhu, R., A. Singh, M. Faiz, H. Kaur, and S. Thukral. "Enhanced Text Mining Approach for a Better Ranking System of Customer Reviews." *Multimodal Biometric and Machine Learning Technologies: Applications for Computer Vision* (2023): 53–69.
19. Bui, D. T., K. Khosravi, J. Tiefenbacher, H. Nguyen, and N. Kazakis. "Improving Prediction of Water Quality Indices Using Novel Hybrid Machine-learning Algorithms." *Science of the Total Environment* 721 (2020): 137612. https://doi.org/10.1016/J.SCITOTENV.2020.137612.
20. Lu, H., and X. Ma. "Hybrid Decision Tree-Based Machine Learning Models for Short-term Water Quality Prediction." *Chemosphere* 249 (2020): 126169. https://doi.org/10.1016/J.CHEMOSPHERE.2020.126169.
21. Kaur, M., R. Aron, H. Wadhwa, R. Tandon, H. N. Oo, and R. Sandhu. "A Comprehensive Exploration of Machine Learning and IoT Applications for Transforming Water Management." In *Innovations in Machine Learning and IoT for Water Management*, edited by A. Kumar, A. Srivastav, A. Dubey, V. Dutt, and N. Vyas, pp. 25–50. IGI Global, 2024. https://doi.org/10.4018/979-8-3693-1194-3.ch002.
22. Singh, A., and R. Sandhu. "A Comparative Study of Big Data Tools: Hadoop Vs Spark Vs Storm." In *2023 IEEE 4th KhPI Week on Advanced Technology (KhPIWeek)*, Kharkiv, Ukraine, 2023, pp. 1–5. https://doi.org/10.1109/KhPIWeek61412.2023.10311577.
23. Sandhu, R., P. Kumar, C. Bhasin, N. Fatima, and A. Pal. "Resnet-34 Model for Human Activity Recognition on Smartphone Sensor Data." *International Journal of Intelligent Systems and Applications in Engineering* 11, no. 9s (2023): 644–652.
24. Faiz, M., R. Sandhu, M. Akbar, A. A. Shaikh, C. Bhasin, and N. Fatima. "Machine Learning Techniques in Wireless Sensor Networks: Algorithms, Strategies, and Applications." *International Journal of Intelligent Systems and Applications in Engineering* 11, no. 9s (2023): 685–694.
25. Lakhwani, K., G. Sharma, R. Sandhu, N. K. Nagwani, S. Bhargava, V. Arya, and A. Almomani. "Adaptive and Convex Optimization-Inspired Workflow Scheduling for Cloud Environment." *International Journal of Cloud Applications and Computing (IJCAC)* 13, no. 1 (2023): 1–25. https://doi.org/10.4018/IJCAC.324809.
26. Channi, H. K., R. Sandhu, M. Faiz, and S. M. N. Islam. "Multi-Criteria Decision-Making Approach for Laptop Selection: A Case Study." In *2023 3rd Asian Conference on Innovation in Technology (ASIANCON)*, Ravet, India, 2023, pp. 1–5. https://doi.org/10.1109/ASIANCON58793.2023.10270203.
27. Chadge, R. B., N. Sunheriya, C. Mahatme, and J. P. Giri. "Experimental and Numerical Investigation of Flat Plate Solar Water Heater." *Lecture Notes in Mechanical Engineering* (2021): 31–40. https://doi.org/10.1007/978-981-33-6029-7_4.
28. Chadge, R., J. Giri, M. Amir, F. I. Bakhsh, N. Sunheriya, C. Mahatme, and T. Sathish. "Investigation of Electrical Configuration Topologies for Enhancing the Performance of Solar Photovoltaic Water Pumping System." *Electric Power Components and Systems* (2024): 1–16. https://doi.org/10.1080/15325008.2024.2319312.

29. Momin, M., F. Rahmani, E. Makki, M. Sharma, J. Giri, and T. Sathish. "The impact of 180° return bend inclination on pressure drop characteristics and phase distribution during oil-water flow." *Heliyon* 10 no. 2 (2024): e24251. https://doi.org/10.1016/j.heliyon.2024.e24251.
30. Sathish, T., R. Saravanan, J. Giri, A. A. Al-Kahtani, C. Prakash, A, Kumar, V. J. Angadi, S. Wang, and M. Ubaidullah. "Hydrogen conversion by microalgae through water gasification under different operating conditions, and Fe_2O_3, Al_2O_3, and CaO additives: An experimental study." *International Journal of Hydrogen Energy* (2024). https://doi.org/10.1016/j.ijhydene.2024.09.003.

Chapter 14

Management of crop recommendation system using predictive analytics

Ayush Patel, Seema Kharb, and Ria Choudhari

14.1 INTRODUCTION

The economic development of a nation hinges upon the continuous improvement and optimal functioning of its agricultural sector. The agriculture industry has demonstrated a notable growth in its contribution to the gross domestic product (GDP), with a rise to 19.9% during the fiscal year 2020–2021, in contrast to the previous fiscal year of 2019–20, where it accounted for 17.8% [1]. During the fiscal year 2003–2004, it is noteworthy to highlight that the agriculture industry played a substantial role by contributing a significant proportion of 20% to the overall GDP.

Based on projections, it is expected that the global population will reach a staggering 9.7 billion individuals by the year 2050. In order to effectively mitigate the projected population growth, it is imperative to initiate a 70% increase in agricultural output. The gradual depletion of vital resources required for food production has become increasingly evident as a result of the influence of climate change and the ongoing expansion of the global population. The observed phenomena encompass a discernible decline in water levels, a reduction in the extent of cultivable land, and an accelerated degradation of the ecological system.

Insufficient pre-planting preparations, unpredictable weather patterns, and improper harvesting techniques have been identified as the primary obstacles hindering farmers from attaining their agricultural objectives. The observed phenomenon of global warming can be attributed to the notable alterations in climate patterns that have been observed in recent years. Agricultural practitioners often face a multitude of significant challenges, including the persistent occurrence of droughts, intense precipitation events, and instances of inundation.

Based on the annual economic survey conducted by the Indian government, it is projected that adverse weather conditions have resulted in a decrease of approximately 20–25% in the agricultural revenue of farmers. The adverse outcomes encountered by agricultural practitioners can be ascribed to the selection of inappropriate cultivars, leading to a reduction in the allocation of resources for activities pertaining to crop harvesting

DOI: 10.1201/9781003512585-16

and maintenance. Therefore, it is imperative to possess a comprehensive comprehension regarding the selection of suitable crops based on soil and climatic conditions, a task that poses challenges due to the limitations of conventional agricultural methodologies.

Throughout history, agricultural practitioners have implemented organic farming practices, which encompass the intentional exclusion of chemical fertilisers. Organic farming serves as a viable solution to mitigate the economic burden associated with the procurement of expensive chemical inputs for agricultural practitioners. Nevertheless, it is worth noting that farmers are presently opting to discontinue the adoption of organic farming methods owing to the relatively lower crop yields they offer in comparison to conventional farming practices. In order to optimise agricultural productivity, farmers have initiated the utilisation of chemical fertilisers as a means of augmenting soil fertility. However, over the course of time, the extensive utilisation of fertilisers experienced a substantial surge due to a limited understanding of soil conditions, consequently leading to an inadvertent escalation in costs. Moreover, the extensive adoption of fertilisers has led to a noticeable degradation in soil quality, consequently leading to a decline in agricultural productivity. Consequently, the implementation of intelligent agricultural methodologies has become imperative to maximise agricultural productivity and minimise economic resource allocation.

The primary objective of this research is to investigate the preliminary phase of intelligent agriculture, with a specific emphasis on recommending suitable crops through the analysis of soil composition, weather conditions, and temperature variables. The primary aim of this investigation encompasses two distinct objectives. First and foremost, the objective of this study is to implement and evaluate various machine learning models in order to identify the optimal approach for crop recommendation. Consequently, our research aims to cultivate a user interface known as the "crop suggestion interface." This interface will leverage the identified optimal model to propose or advise the most appropriate crop selection, taking into account the given environmental conditions.

This chapter has been organised as follows. Section 14.2 provides the related literature. The proposed methodology has been described in Section 14.3, followed by results and conclusion and future work in Sections 14.4 and 14.5, respectively.

14.2 LITERATURE SURVEY

The hybrid neural network model in Ref. [2] identifies crop yields in an innovative way. Soil characteristics and external climate are model inputs. The model cannot accurately predict real-time climate and soil parameter changes. Thus, the model is suitable for real-time analysis. The suggested approach in Ref. [3] uses historical agricultural land data to predict crop yield, considering rainfall, temperature, humidity, soil pH, type, and historical crop data. Using the dataset, the system should predict the optimal yield. This study used random forest, polynomial regression, and decision tree. Performance is judged by expected precision. The suggested model shows that Random Forest algorithm predicts yield better than the other two algorithms. Reference [4] analyses prior data to anticipate a farmer's crop price. They examined *rabi* and *kharif* crops like paddy, *arhar*, *bajra*, barley, and so on. The dataset included rainfall, temperature, market prices, land acreage, and past crop yield. They used the decision tree algorithm to analyse and predict a fresh dataset in the project. For the past year, they have anticipated price and earnings and provided time series analysis [5]. In Ref. [6], the authors created a crop selection and yield prediction method that used vegetation, temperature, and normalised difference vegetation. For a better prognosis, they separated climatic, agronomic, and other disturbances. In Ref. [7], crop yield, cost, and algorithms are the primary topics. These properties enable smart farming. After feature extraction, take the data or columns to apply the algorithms to determine accuracy and plot the graph. A classification procedure assessed algorithm accuracy. Bayesian networks are then utilised to statistically analyse the attribute in the dataset. Then, ANN compares non-linear and underlining patterns. They propose utilising machine learning algorithms (linear regression, decision tree, and decision tree regression) to anticipate crop prices and spice export values in Ref. [8]. The estimated output will estimate crop or spice prices in the coming year or 12 months. CSV files will contain system results. The suggested approach will help farmers and producers predict their crop and spice export value, enhancing productivity and profit margins. They created an accurate agricultural price and profit forecasting system in Ref. [9]. Rainfall, maximum trade, MSP, and yield are considered when estimating pricing. Profit estimates consider crop price, yield, cultivation cost, and seed cost. *k*-Nearest neighbour (k-NN) predicted crop profit and naive Bayes algorithm estimated crop pricing. Their web application uses efficient algorithms and machine learning technologies and has a user-friendly UI. They use data analysis to predict crop prices in Ref. [10]. The proposed system will anticipate agricultural prices using machine learning algorithms based on planted, harvested, and other characteristics. Farmers will learn the future crop price. The study developed a system of data integration from multiple sources, data analysis, and predictive analytics to forecast the crop's target price and enhance farmers' profit margins in the long run. After careful consideration, XGBoost is found to be the best for our project [11]. ML-based crop prediction research introduces approaches with technical knowledge to improve the agriculture sector's reliability and train farmers to accurately predict appropriate crops based on the results, using ML algorithms that account weather, soil, and market patterns. Also considered were pH, nitrogen content, and soil nutrients. Artificial neural networks, information fuzzy networks, and data mining provide predictions.

Overall, the ANN approach works well for the project. A crop DLT forecasting model is suggested in Refs [12–15]. This study uses crop and climatic data to provide recommendations. This paper proposes a hybrid crop prediction technique that optimises DCNN and LSTM inputs using ACO (Ant Colony Optimisations). DCNNs have high accuracy; however, their processing layers increase computer complexity [16, 17]. Since adding DCNN weights to its nodes is a major complexity gain, our study modifies these weights in training to simplify processing. ACOs simplify weights and improve DCNN predictions by improving hyperparameters during training. In Ref. [18], they showed how a machine learning algorithm can forecast the best crop for a given temperature, humidity, pH, rainfall, soil potassium, nitrogen, and phosphorus levels. In addition, they compared machine learning algorithms to determine which has the highest accuracy. They conducted the same experiment using several machine learning methods to determine the best. They trained the model using random forest, naive Bayes, support vector machine, logistic regression, decision tree, gradient boosting, and *k*-nearest neighbours. The random forest and naive Bayes model predicted the most accurate crop for specific soil conditions like temperature, humidity, precipitation, pH, potassium, phosphorus, and nitrogen. According to Ref. [19], area, rainfall, temperature, and crops can be used to estimate agricultural production before farmers choose a crop. This strategy helps farmers gain insights. They employed XGB Regressor, Ridge Regression, and LGM Classifier in this study. These models were more accurate with hyperparameter adjustment. The authors propose to integrate the two models and notify farmers by SMS or email. Compared to the Ridge Regressor, the XGB Regressor provides the best optimised crop-yield forecasting model solution. In Ref. [20], the system built an accurate and precise crop production prediction model that suggested the optimum fertiliser ratio based on environmental and soil parameters, increasing crop output and farmer profitability. SVM was used in the project and has accuracy above 90.01%. The average accuracy of SVM is 0.95, AUC is 0.93, and F1-score is 0.90. The soil's nutrient composition and growing region are used to predict the crop in this study. This approach will assist farmers choose the proper crop and fertiliser for optimal production. Reference [21] proposes an ontological crop compatibility and fertiliser suggestion engine. So, farmers and technology can coexist. The approach recommends the ideal crop for a land in Maharashtra, India, based on its region and soil. It recommends fertiliser for farms. The soil's NPK composition and ontology data from previous years determine fertiliser recommendations. It recommends crops for a region and fertilisers. K-means clustering, and random forest are employed in the recommendation system. The authors in Ref. [22] found that current approaches for fertiliser design and soil nutrient management are insufficient. This chapter presents a compelling approach for evaluating soil nutrients and making a successful fertiliser. Thus, they proposed a four-step methodology: recommendations, data analysis, data

Table 14.1 Comparative study of various machine learning and deep learning algorithms in literature review

Author	*Year*	*Model*	*Features used*	*Evaluation parameter*	*Inference*
[2]	2018	Recurrent Neural Network (RNN)	Soil parameters and the external climate conditions	RMSE and MAE	A hybrid neural network prediction model is suggested to determine crop production while taking into account the soil factors and the external climate variables
[3]	2020	Decision tree, random forest, polynomial regression, and machine learning techniques	Meteorological elements, such as rainfall, temperature, and humidity, as well as geographic elements, such as soil pH and type of soil and previous crop records	Accuracy of the model is calculated, which was like decision tree is 77.0% Polynomial regression is 88.0%, and random forest 98.1% and 88.0%	In comparison to all other ML algorithms, the random forest approach is proven to be the best algorithm for predicting yields
[4]	2020	Forecasting, supervised machine learning in agriculture and decision tree	Rainfall, temperature, market prices, land acreage, and previous crop yield	Various number of metrics	The decision tree model provides a time series analysis of the price and gain for the following 12 months
[6]	2015	eXtensible crop yield prediction framework (XCYPF), decision support system (DSS)	Temperature, relative humidity, precipitation, type of soil, exposure to the sun, seeds, types of fertiliser, amounts of fertilisers, weed and pest control, duration of land preparation, crop rotation, and plant spacing. Climate-related problems	–	The XCYPF, a unique framework that is versatile and extensible is proposed in this research

(*Continued*)

Table 14.1 (Continued)

Author	*Year*	*Model*	*Features used*	*Evaluation parameter*	*Inference*
[7]	2019	Clustering, Bayesian, ANN (artificial neural network)	Soil types, pressure, suitable weather, crop type	Algorithm result (accuracy of the datasets), matrix such as true positive, true negative, etc.	The chapter's major emphasis was on agricultural yield and crop cost prediction algorithms
[8]	2022	Linear regression, decision tree, decision tree regression	Historical dataset, or information on the costs of groundnuts, black gram, and spices like ginger and pepper during the last five years	Accuracy score	The model provides an estimated value and depicts the outcomes using graphs and percentages to make it simpler for average farmers to understand
[9]	2019	*k*-Nearest neighbour (*k*-NN), naive Bayes	Rainfall, MSP, crop price, maximum trade, yield, cultivation cost, seed cost	–	A web application with a generally user-friendly interface has been developed that uses effective machine learning methods and technology
[10]	2020	Decision trees, XGBoost, linear regression, data analytics, machine learning, prediction, and neural networks	Area harvested, area planted, etc.	RMSE (root mean square error)	Compared to all the other algorithms, XGBoost forecasts the target more accurately
[11]	2021	Different machine learning algorithm (SVM, naive Bayes, decision tree, random forest)	Various soil parameters, rainfall, temperature, humidity, previous crop data	Accuracy of all the algorithms is calculated	Naive Bayes algorithm achieved marginal accuracy as compared to all other algorithms for crop recommendation
[12]	2021	Deep learning, deep CNN, LSTM, Ant colony optimisation (ACO-IDCNN-LSTM)	Different temperatures, humidity, dew, and wind pressure	Accuracy, precision, recall, F-measure, error rate	It is discovered that the suggested ACO-IDCNN-LSTM recommendation model is successful in selecting an appropriate crop

(*Continued*)

Table 14.1 (Continued)

Author	*Year*	*Model*	*Features used*	*Evaluation parameter*	*Inference*
[18]	2021	Random forest, naive bayes, machine learning, support vector machine, logistic regression, decision tree, gradient boosting, *k*-neighbours	Temperature, humidity, pH, rainfall, amount of potassium, nitrogen, phosphorous in soil	Precision, recall, F1-score, support	Random forest and naive Bayes algorithms achieved more accuracy (about 99.4%) as compared to all other algorithms
[19]	2022	XGBoost regressor, ridge regression, LGBM classifier, hyperparameter tuning	District, crop, average temperature, rainfall and area, production (yield)	RMSE (root mean squared error)	LGBM classifier with hyperparameter tuning achieve best results for predicting the yield of crop
[20]	2021	SVM (support vector machine)	Location, pH level, amount of nutrients like NPK weather and temperature, kind of soil, nutrient value of that region's soil, frequency of that region's rainfall, and soil composition	Mean accuracy, mean AUC, F1-score	Based on the pre-processed crop data, the support vector machine technique aids in the more accurate prediction of the crop (around 90.01%)
[21]	2019	*K*-means clustering, ontology, fertiliser recommendation, random forest algorithm	Soil nutrients like nitrogen, phosphorus, potassium (NPK) and previous year's research data	Precision	On their mobile phones, the proposed system offers crop recommendations based on geography, crop type, and fertiliser's recommendations according to different soil nutrient quantity
[22]	2020	Analysis of the soil, use of the Java API, data analytics, and recommendation	Nitrogen, phosphorus, potassium (NPK) contents of soil	Accuracy, precision, recall, F-score	The suggested approach determines the NPK ratio of a soil sample taken from a farm and suggests the right amount of fertiliser for the specified crops

(*Continued*)

Table 14.1 (Continued)

Author	*Year*	*Model*	*Features used*	*Evaluation parameter*	*Inference*
[23]	2020	MATLAB, WSN, ANN, SVM, Arduino, Zigbee	Soil profile and PH samples, nitrogen, potassium, phosphate, depth, temperature, precipitation	–	The suggested approach uses soil and PH sample inputs to anticipate the crops that can be suggested as acceptable for the soil and the fertilisers that may be utilised as the website's answer

pre-processing, and soil analysis. The NPK ratio of soil nutrients is measured using a two-electrode NPK sensor on an IoT device. Sensor data is then formatted for pre-processing, and so is machine learning algorithm. Find an efficient fertiliser. This approach lets farmers choose the best fertiliser to boost productivity early on. The suggested method determines the NPK ratio in a soil sample from an agricultural field and recommends the best fertiliser for the crop. The suggested method lets farmers boost crop cycle yield without changing soil composition or other properties. This reduces crop over-fertilisation, ensuring their health. This robotics fertiliser recommendation system allows autonomous mobile robots to spray the right fertiliser based on the NPK value. Smart and precision farming can be improved by computing NPK values more precisely. This recommendation method helps the government analyse soil conditions and farmer needs to maximise land output. In Ref. [23], the authors discussed how to measure soil nutrients and environmental components tracked by crop and habitat fertilisation indicators. A major challenge in agriculture is choosing the right crop for the soil and seeding it for maximum yield. The recommended approach takes soil samples and PH as inputs and predicts suitable crops and fertiliser as a website solution. Sensors capture soil data, which the Arduino sends to MATLAB via Zigbee and WSN for Artificial Neural Network analysis and crop recommendations.

14.2.1 Findings from literature survey

- A significant challenge faced throughout the advancement of crop recommendation systems is the restricted availability of data. To ensure the successful implementation of crop recommendation systems, it is crucial to have access to significant amounts of data related to diverse parameters such soil composition, meteorological conditions, water availability, crop yield, and market values. Obtaining this data can pose difficulties in terms of both intricacy and expenditure.
- Another challenge arises in the creation of crop recommendation systems that demonstrate resilience in response to variations in climatic patterns and market dynamics. The agricultural sector has been confronted with more difficulties in precisely predicting crop yields due to the influence of climate change. This has also coincided with a rise in market price volatility. To ensure the delivery of accurate suggestions to farmers, it is crucial for crop recommendation systems to include the ability to adapt to the ever-changing nature of agricultural practices.
- Given the significant challenges at hand, it is important to acknowledge that crop recommendation systems have the potential to profoundly revolutionise the field of agriculture. By providing support to agricultural practitioners in the strategic choice of crops for cultivation, these innovative solutions possess the capacity to boost agricultural production, reduce costs, and eventually bolster the economic sustainability of farmers.

- It is expected that there will be breakthroughs in crop recommendation systems in the near future, resulting in improved accuracy, robustness, and user-friendliness. These agricultural systems will have the capacity to integrate a wider range of factors, including climate change, market dynamics, and farmers' preferences. Moreover, these novel solutions will demonstrate improved ease of use and increased availability, accommodating the varied requirements of farmers operating at different levels.
- The soil testing process in India is carried out by the central government in collaboration with state governments. Subsequently, a soil health card is issued every three to five years, which provides an assessment of various soil nutrients and offers recommendations solely pertaining to fertilisers.

14.3 PROPOSED METHODOLOGY

The block diagram for the methodology used has been shown in Figure 14.1 and is described in this section.

Step 1: The inputs to the crop recommendation interface will be given by the user and are classified into two categories: first category is based on their geographical location, and the second is soil analysis. The parameters encompass a multitude of variables, specifically N, P, K, and pH are provided according to the soil analysis; and the parameters linked with the geographical location are humidity, temperature, and rainfall. In order to determine the necessary inputs for optimal

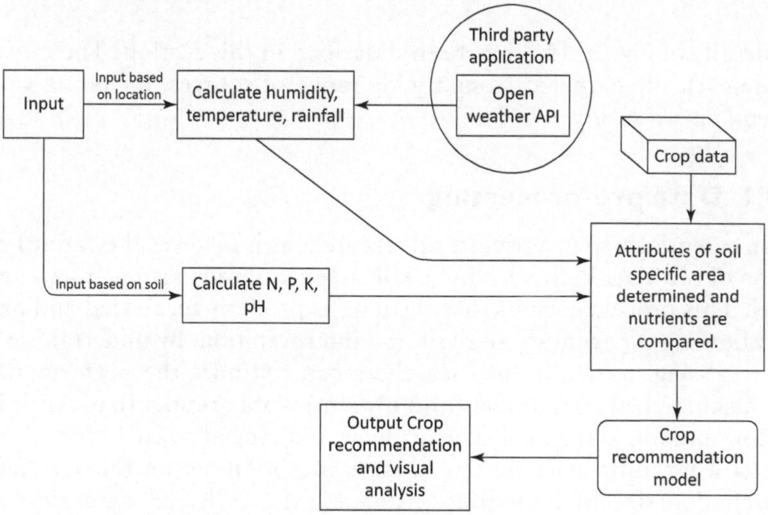

Figure 14.1 Crop recommendation interface.

soil conditions, it is imperative to employ a specialised instrument to gather crucial data regarding the levels of nitrogen, potassium, phosphorus, and pH in the soil. The soil characteristics within a particular region are assessed, and a comprehensive analysis of nutrient composition is carried out. The soil test data is utilised as an input parameter for the model, in conjunction with location-specific input obtained through a weather API. The input provided encompasses location-specific variables, including but not limited to humidity levels, temperature variations, and precipitation patterns. Following this, the final input is compared and contrasted.

Step 2: The computation of inputs, contingent upon the geographical location, necessitates the assessment of pertinent data pertaining to humidity, temperature, and precipitation. The aforementioned parameters are acquired by leveraging a third-party application, namely the Open Weather API. The utilisation of the Open Weather API facilitates the acquisition of meteorological data via an API invocation. Through the specification of the targeted season and geographical location, valuable insights can be acquired pertaining to the projected patterns of precipitation, atmospheric moisture levels, and ambient temperature.

With the available crop data, the proposed system aims to generate a predictive output that effectively identifies suitable crops for cultivation based on various factors and parameters.

Step 3: The crop recommendation model is an innovative machine learning model that has been meticulously crafted to fulfil the noble objective of accurately forecasting the most appropriate crops that can be cultivated in a given agricultural setting; the detailed flow is shown in Figure 14.2.

The details of Figure 14.2 has been described in this section. The *raw data* is systematically gathered from a wide range of sources, including various academic institutions.

14.3.1 Data pre-processing

It is an essential step in agricultural research, as it involves the careful preparation of raw data before it can be effectively utilised in a machine learning model. This process ensures that the data is properly formatted and organised, allowing for accurate analysis and interpretation. By undertaking data pre-processing, agricultural researchers can optimise the performance of their machine learning models and obtain reliable results that can inform decision-making in the field. The primary and crucial phase in the advancement of a machine learning model is its inception within the agricultural research domain. In numerous instances, the collected data showcases varying levels of noise, incorporates instances with incomplete values, and

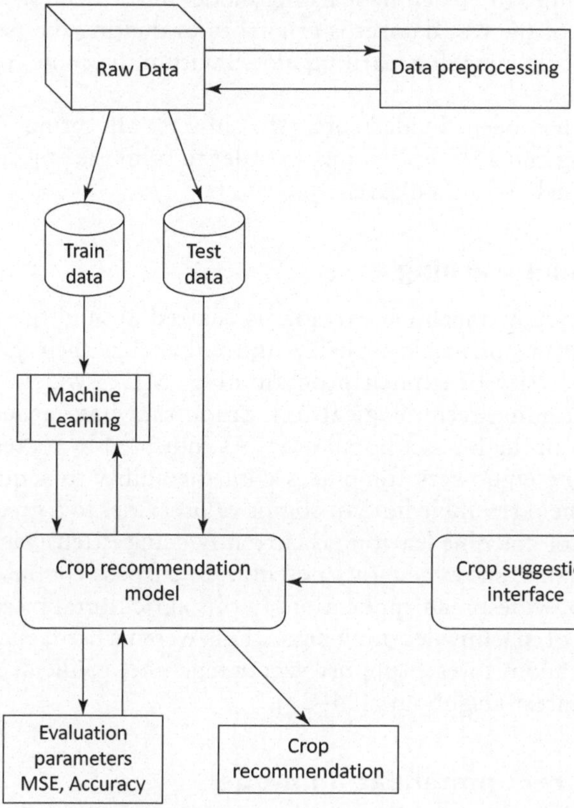

Figure 14.2 Machine learning model.

may be presented in a format that lacks compatibility with direct utilisation by machine learning models. Data pre-processing is a crucial step in the research process, specifically in the field of agriculture, as it involves the purification and enhancement of data to make it appropriate for utilisation in machine learning models. This process additionally improves the precision and effectiveness of the model.

14.3.2 Data partitioning

In order to carry out this procedure, it is imperative to partition the dataset into two separate subsets, specifically referred to as the training data and the testing data. The training dataset is the initial subset used for fitting the model. The training process of the model does not include the second subset. In this context, an input element is received from the dataset and predictions are generated. These predictions are then compared to the expected values for analysis. The nomenclature "test data file" pertains to the subsequent data file in question. The train dataset is employed for the purpose of training the machine learning model in the field of agriculture. The utilisation of the test dataset is crucial in evaluating the performance and precision of the machine learning model within the context of agricultural research.

The dataset has been divided into two subsets, allocating 75% of the data for training and 25% for testing, in order to fulfil the objectives of this research in the field of agriculture.

14.3.3 Machine learning

A discipline of study, machine learning, is centred around the facilitation of computers in acquiring knowledge and enhancing their performance without the necessity of explicit programming. ML is widely recognised as a highly intriguing technological innovation that has garnered significant interest from the human population. As indicated by its terminology, machine learning empowers computers with the ability to acquire knowledge, making them resemble human cognitive processes to a greater extent. The utilisation of machine learning is currently being extensively employed in various domains, showcasing its potential to surpass common expectations in terms of widespread application. In this agricultural research study, a diverse range of machine learning algorithms were utilised, encompassing naive Bayes, random forest, support vector machine, artificial neural network, and *k*-nearest neighbours [24–27].

14.3.4 Crop recommendation model

The crop recommendation model was formulated by employing advanced machine learning techniques to propose suitable crops by considering a diverse set of input variables. The *crop suggestion interface*, as depicted in Figure 14.1, offers a comprehensive representation of the model's capabilities, including specific inputs and subsequent analysis.

14.3.5 Evaluation parameters

These parameters encompass the utilisation of various evaluation measures to assess the effectiveness of a machine learning model and determine its strengths and weaknesses. In the preliminary phases of the research endeavour, it is of utmost importance to evaluate the efficacy of the model. The assessment of models also serves a crucial function in the surveillance of models within the agricultural research domain. The utilisation of a confusion matrix is a valuable technique employed within the realm of agricultural research and statistical analysis to assess the efficacy of a classification model. The assessment of classification accuracy is a crucial aspect within the realm of agricultural research. The logarithmic loss metric is a significant measure utilised in agricultural research. The measurement of the total area enclosed by a curve is of great interest in agricultural research. The F1-score is a widely employed metric within the realm of academic research and evaluation, specifically in the field of agriculture, to evaluate the efficacy and proficiency of a classification model. The mean absolute error (MAE) is a widely employed metric within the field of agricultural research. It serves as a valuable tool for quantifying the average magnitude of errors present in a given set of predictions or forecasts. The mean squared error (MSE) is a metric commonly used in agricultural research to evaluate the accuracy of predictive models.

Crop recommendation is the result of a predictive model that assesses various soil and weather factors to determine the optimal crop for cultivation.

14.4 RESULTS AND DISCUSSION

The implementation of machine learning algorithms has been optimised to offer recommendations for crop selection that aims to maximise yield under specific environmental conditions. The MATLAB Classification Learner programme was employed to facilitate the development of models for data categorisation. Different computational models, such as decision tree, naive Bayes, support vector machine, *k*-NN, neural network, and other relevant models, were trained utilising the crop recommendation sample dataset.

After conducting a thorough analysis of the models, it has been observed that the naive Bayes algorithm and the discriminant analyser have demonstrated the most promising results in terms of accuracy. The recorded level of precision stands at approximately 99.1%, surpassing that of all other models, as depicted in Table 14.1. The model's predictive capabilities can be attributed to its training utilising a diverse range of machine learning techniques. This enables it to effectively analyse and assess multiple factors, including temperature, humidity, rainfall, pH, potassium, phosphorus, and nitrogen levels in the soil. As a result, the model can accurately determine the most appropriate crop for a given agricultural scenario. Table 14.2 provides the comparison of machine learning algorithms for crop recommendation.

The confusion matrix depicting the performance of the naive Bayes algorithm in relation to the predicted values has been presented in Figures 14.3 and 14.4.

Figure 14.3 shows that the confusion matrix was built using a trained naive Bayes algorithm. It is evident from the analysis that the naive Bayes Classifier, upon computation of true positive (TP), true negative (TN), false positive (FP), and false negative (FN) values, attained an accuracy score of 99.45%, indicating a near-perfect performance. After doing a comprehensive analysis of all the trained models, it has been determined that the naive Bayes classifier exhibits the highest level of accuracy in terms of the test score.

Table 14.2 Comparison of machine learning algorithms for crop recommendation

Model number	*Model type*	*Accuracy (validation)*	*Total cost (validation)*	*Accuracy (test)*	*Total cost (test)*
1	Tree	98.79%	20	98.00%	11
2	**Discriminant**	**99.58%**	**7**	**99.09%**	**5**
3	**Naive Bayes**	**96.61%**	**7**	**99.09%**	**5**
4	SVM	83.94%	23	98.73%	7
5	*k*-NN	97.88%	265	85.45%	80
6	Kernel	98.30%	35	97.82%	12
7	Ensemble	94.91%	28	98.18%	10
8	Neural network	98.24%	84	95.27%	26
9	*k*-NN	96.79%	53	97.82%	12

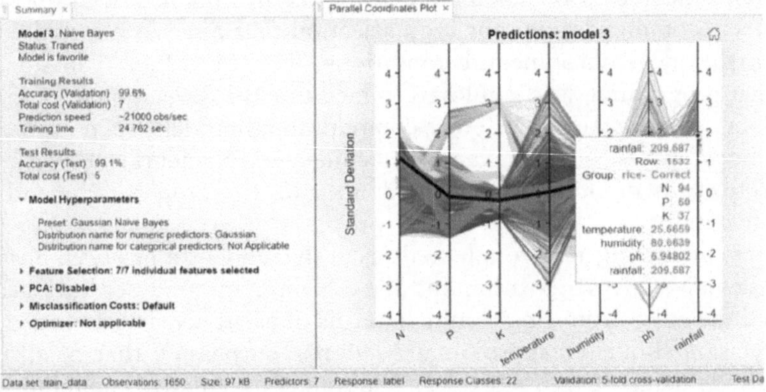

Figure 14.3 Summary and parallel coordinates plot of trained naive Bayes model.

14.5 CONCLUSION AND FUTURE WORK

This study presents a detailed assessment of various machine learning methods within the domain of crop recommendation, which is a critical component of intelligent agriculture. The principal aim of this study is to ascertain the optimal machine learning model for the purpose of crop recommendation. Following this, a software interface referred to as the "crop suggestion interface" is created. This interface makes use of the most effective model that has been established, in order to offer ideal recommendations for crops based on the current environmental conditions.

The utilisation of a crop recommendation system functions as a beneficial resource for agricultural practitioners, assisting them in the enhancement of crop selection and management strategies. This will assist the improvement of agricultural production, promote profitability, and decrease the environmental impact.

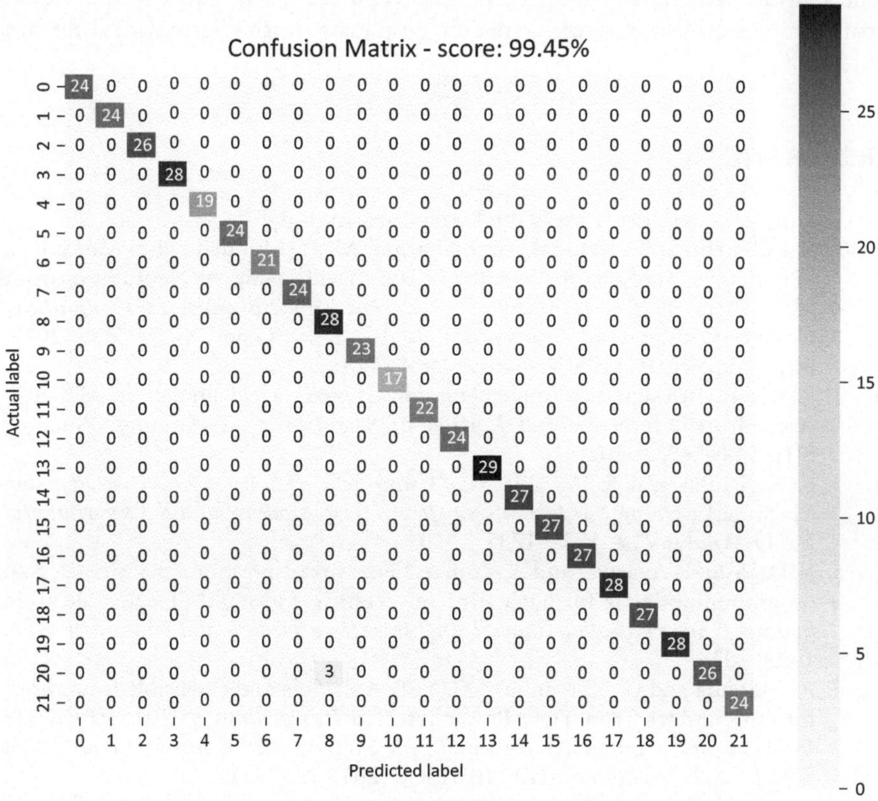

Figure 14.4 Naive Bayes confusion matrix.

Furthermore, a crop recommendation interface is developed using the naive Bayes algorithm. Currently, our team is conducting preliminary research on the creation of a crop recommendation model using a collection of samples.

The suggested technique involves the utilisation of real-time agricultural datasets for model training, which will then be deployed on a cloud-based platform.

The aim of our study is to construct resilient models with the intention of providing recommendations for fertilisers, estimating agricultural yields, and forecasting crop prices. The construction of these models will rely on datasets that are both dependable and accurate.

In addition, our ongoing efforts encompass engaging in active talks with a leading company specialising in soil testing devices. The objective is to provide a comprehensive range of services that encompass thorough soil testing techniques. In addition, expert recommendations will be provided regarding appropriate crop selections, optimal fertilisers, projected yields, and crop pricing. The advice provided will be customised to account for the distinct attributes of different geographical areas and the individual soil samples collected, with a specific emphasis on the agricultural terrain of India.

REFERENCES

1. "https://en.wikipedia.org/wiki/Agriculture_in_India."
2. S. Kulkarni, S. N. Mandal, G. S. Sharma, M. R. Mundada, and Meeradevi, "Predictive Analysis to Improve Crop Yield using a Neural Network Model," in *2018 International Conference on Advances in Computing, Communications and Informatics, ICACCI 2018*, pp. 74–79, Nov. 2018. https://doi.org/10.1109/ICACCI.2018.8554851.
3. S. Reddy, "Design and Implementation of Crop Yield Prediction Model in Agriculture," *International Journal of Scientific & Technology Research* 8(1), 544–549, 2020.
4. S. A. Mulla and S. A. Quadri, "Crop-yield and Price Forecasting using Machine Learning." *International Journal of Analytical and Experimental Modal Analysis* 12, 1731–1737, 2020.
5. V. D. Joshi, P. Agarwal and A. Kumar, "Fuzzy transportation planning: a goal programming tactic for navigating uncertainty and multi-objective decision making," *International Journal on Interactive Design and Manufacturing*, 1–29, 2023.
6. A. Manjula and G. Narsimha, "XCYPF: A flexible and extensible framework for agricultural Crop Yield Prediction," in *Proceedings of 2015 IEEE 9th International Conference on Intelligent Systems and Control, ISCO 2015*, Sep. 2015. https://doi.org/10.1109/ISCO.2015.7282311.
7. "Smart Farming Prediction Using Machine Learning." Available at: https://medium.com/sponsor-a-farmer-farmguide/7-facts-you-need-to-k
8. S. Vighnesh Dhanraj and M. M. Wakwadi, "Prediction the best price for crops and spices using Machine Learning," *International Journal of Scientific Research and Engineering Development.* https://ijsred.com/volume4/issue4/IJSRED-V4I4P23.pdf
9. P. S. Rachana and S. R. Kousar, "Crop price forecasting system using supervised machine learning algorithms," *International Research Journal of Engineering and Technology* 6, 4805, 2008.
10. P. Samuel, B. Sahithi, T. Saheli, D. Ramanika, and N. A. Kumar, "Crop Price Prediction System using Machine learning Algorithms," *Quest Journals Journal of Software Engineering and Simulation* 2, 9288, 2020.
11. A. U. Aryamol, P. G. Raji, and S. G. Benjamin, "A Survey Paper on Crop Prediction using Machine Learning," *International Journal of Creative Research Thoughts* 9, 451–456, 2021.
12. K. Mythili and R. Rangaraj, "Crop Recommendation for Better Crop Yield for Precision Agriculture Using Ant Colony Optimization with Deep Learning Method," *Annals of the Romanian Society for Cell Biology* 25(4), 4783–4794, 2021.
13. A. Kumar, D. K. Rajak, P. Kumar, and A. Kumar (Eds.), *Advances in Sustainable Biomaterials: Bioprocessing 4.0, Characterizations, and Applications* (1st ed.). CRC Press, 2024. https://doi.org/10.1201/9781003434313.

14. A. Kumar, Y. Liu, and R. Kumar (Eds.), *Handbook of Intelligent and Sustainable Manufacturing: Tools, Principles, and Strategies.* CRC Press, 2024. https://doi.org/10.1201/9781003405870.
15. A. Kumar, P. Kumar, and Y. Liu, *Industry 4.0 Driven Manufacturing Technologies.* Springer Nature, 2024. https://doi.org/10.1007/978-3-031-68271-1.
16. D. V. Burande, K. Kalita, R. Gupta, A. Kumar, J. S. Chohan and D. Kumar, "Machine learning metamodels for thermo-mechanical analysis of friction stir welding," *International Journal on Interactive Design and Manufacturing* 19(1), 597–615, 2024.
17. A. Kumar, S. Rani, S. Rathee and S. Bhatia (Eds.), *Security and Risk Analysis for Intelligent Cloud Computing: Methods, Applications, and Preventions.* CRC Press, 2023.
18. S. Anusha, M. Bindu, and G. Navya, "Crop Recommendation System with Comparative Analysis of Different Machine Learning Algorithms," *International Research Journal of Engineering and Technology*, 2021.
19. D. Mani and R. Edinburgh, "Crop-Yield Prediction and Crop Recommendation System." Available at: https://ssrn.com/abstract=4111856.
20. A. Palaniraj, A. S. Balamurugan, R. Durga Prasad, and P. Pradeep, "Crop and Fertilizer Recommendation System using Machine Learning," *International Research Journal of Engineering and Technology* 8, 2021.
21. A. Chougule, V. K. Jha, and D. Mukhopadhyay, "Crop suitability and fertilizers recommendation using data mining techniques," *Advances in Intelligent Systems and Computing* 714, 205–213, 2019. https://doi.org/10.1007/978-981-13-0224-4_19.
22. K. Suba Raja Subramanian and D. Suba Raja, "Design and Implementation of Fertilizer Recommendation System for Farmers Conversion of non-audible murmur to normal speech through Wi-Fi transceiver for speech recognition based on GMM model View project Kanaga Suba Raja Publication View project Design and Implementation of Fertilizer Recommendation System for Farmers", [Online]. Available: https://www.researchgate.net/publication/345904482.
23. G. Preethi, V. Rathi Priya, S. M. Sanjula, S. D. Lalitha, and B. Vijaya Bindhu, "Agro-based crop and fertilizer recommendation system using machine learning," *European Journal of Molecular & Clinical Medicine* 7(4), 2043–2051, 2020.
24. J. Giri, T. Sathish, T. Sheikh, N. Sunehriya, P. Giri, R. Chadge, C. Mahatme, and A. Parthiban, "Automatic liver segmentation using U-Net deep learning architecture for additive manufacturing," *Deleted Journal* 245(1), 90, 2024. https://doi.org/10.1007/s10751-024-01927-9.
25. P. D. Kamble, J. Giri, E. Makki, N. Sunheriya, S. B. Sahare, R. Chadge, C. Mahatme, P. Giri and H. Panchal, "An application of hybrid Taguchi-ANN to predict tool wear for turning EN24 material," *AIP Advances* 14(1), 015206, 2024. https://doi.org/10.1063/5.0186432.
26. C. Mahatme, J. Giri, H. A. Al-Lohedan, F. Mohammad, N. Sunheriya, R. Chadge, P. Giri, S. Mallik, and M. Dhayalan, "Performance evaluation of a mechanical ventilation simulation model for diverse respiratory complications," *AIP Advances* 14(3), 035333, 2024. https://doi.org/10.1063/5.0189452
27. C. Mahatme, J. Giri, R. Chadge, and A. Bhagat, "Comparative analysis of different lattice topologies for cellular structure optimization in additive manufacturing," *Materials Today: Proceedings* 62, 1591–1595, 2022. https://doi.org/10.1016/j.matpr.2022.03.280.

Chapter 15

Analysis of creditworthiness and risk assessment management for enterprises using machine learning

Harsh Agarwal, Subhash C. Patel, Prithviraj Mutalik, and Deepti Dave

15.1 INTRODUCTION

The major reasons for credit risk include a sovereign's incapacity or unwillingness to meet debts, concentration of exposure to a person or industry, and credit failure. Credit risk evaluation is a continual activity throughout the loan process, beginning with the underwriting of a prospective loan and ending with payment collection after the loan is onboarded. With global economic uncertainty increasing [1–4], it is critical to ensure that the correct execution of credit risk evaluation is done. Default prediction and legal and political will to deal with defaulters are all required [5]. In general, working capital, the volume to repay, loans' conditions and contracts, history of potential assets, and associated collateral are the parameters to measure credit risk. For a fee, credit risk departments are relied upon by banks and financial services. Various agencies like Morningstar Inc., Moody's, "Fitch Group," and "Standard and Poor's (S&P)" are engaged in micro-financial services. The history of credit on financial contracts of a customer is focused on by various credit scoring companies such as Experian, Equifax, and CIBIL. Generally, a person reviews credit risk evaluation based on the credit score of a new loan, using numerous and unique parameters as given in Table 15.1. The input type rating implies that input will be taken in the format of "0"

or "1" or "2" or "3" or "4" or "5," where a large number indicates a positive response (Table 15.2). The input-type numerical implies a large number will get selected as input for the user. Input "type" categorical implies that input will be taken as a string for the user and will be validated.

If a person or an organization conducts a credit assessment, then two key questions arise: the first one is "Do they (institution or the syndicate) have the necessary adroitness and strategies for gauging credit risk properly," and second question states, "will the valuation be independent and impartial?" For example, if the test is entrusted to a paid retailer according to the amount and volume of the loan granted, then there is an obvious hazard of underestimating the risks [6], that is, to be optimistic when considering the financial situation of the applicants to increase the business objective. The

 DOI: 10.1201/9781003512585-17

Table 15.1 Parameter and their types for credit risk evaluation

Parameter	*Type*
Total assets	Rating (numerical)
Review portals	Rating (numerical)
Daily sales	Rating (numerical)
Competition	Rating (numerical)
Rent and wages given to employees	Rating (numerical)
Size of shop	Rating (numerical)
Location	Rating (numerical)
Supply chain, stock availability, and warehouse space	Rating (numerical)
Number of customers visiting the shop (approximately)	Rating (numerical)
Frequency of supplier payments	Rating (numerical)
Nature of retailer (based on customer point of view)	Rating (numerical)
Number of legal cases	Numerical
Applicant income	Numerical
Family income	Numerical
Loan amount	Numerical
Loan amount term (in months)	Numerical
Total workers	Numerical
Education	Categorical
Business type	Categorical (yes/no)
Credits given to customers	Categorical (yes/no)

situation becomes nasty when there is some correspondence, which may be drastic, between the seller and the applicant. Such a situation, unfortunately, is not possible.

Credit assessment is a challenging task for supervised classification. The supervised classification approach retrieves historical data from the data set, such as input and a preset output value (target value), and then builds a model that can use training data. Such structures point to suitable classifications for future candidates. Micro-financial institutions use several popular "classification methods," such as "data mining and machine learning algorithms," to discover fraudulent patterns that transpose swiftly or to target specific risk zones. These organizations construct their credit assessment model's logic and knowledge store using their acquired application data. The authors in Refs. [7, 8] indicated the future of artificial intelligence, IoT, and other technologies, and companies are focusing on artificial intelligence in great extent. Predictive models are created using various data mining algorithms such as support vector machine (support vector machine), decision tree, XGBoost, and logistic regression. Artificial neural network is also used in other credit scoring approaches. Since these types require a lot of flexibility and gigantic training and produce output

Table 15.2 Advanced search result for keywords

Sr. No.	*All the words*	*With exact phrase*	*With at least one word*	*Total results*
1	“Credit risk detection (CRD)”	“Machine learning”	“Data science”	533
2	“Loan prediction analysis,” or “machine learning”	“Credit risk detection”	“Loan default prediction” or “loan risk detection” or “credit history” or “loan default analysis” or “loan prediction analysis” or “machine learning”	6,531
3	“Credit Survey” (CS), or “Review,” or “Literature, Review”	“Credit Risk” (CR) or “Credit Score” or “Default” or “Fraud detection”	“Risk” or “Worthiness Analysis” or “Loan” or “Default”	1,822
4	“Credit risk,” or “worthiness analysis,” or “machine learning,” or “deep learning,” “artificial intelligence,” “artificial intelligence”	“Performance”	“Neural network” or “Supervised” or “Unsupervised” or “Classification” or “Regression” or “Random Forest” or “Discriminant Analysis” or “support vector machine”	53
5	“Credit risk,” or “Loan Prediction Analysis,” or “peer to peer,” or “Performance”	“Banking” or “Loan Provider” or “Credit Scoring” or “Fraud Detection”	“Nonperforming Asset” or “Non-Performing Loan” or “Expert System”	26
6	“Credit Risk,” or “Loan Prediction Analysis”	“Loan”	“Radial Basis Function Network” or “Fuzzy Logic”	10
7	“Credit Survey,” Review” or “Literature Review”	“Credit Risk” or “Credit Score” or “Default” or “Fraud Detection”	“Risk” or “Worthiness Analysis” or “Loan” or “Default”	20

that creates interpretative difficulties, according to different studies, such as [RS1] [9], [RS2] [10], and [RS3] [11], applications for dealing with credit scoring (debt) problems are limited.

[B1] [12] advocates that the purpose of credit review is to approve acceptable loan applications, rejecting clients who are likely to fail in repayment in the future. This can be achieved by using a rating system in a variety of ways.

According to [RS1] [9], a credit scoring system that works on logistic regression and discriminant analysis targets the problems of Tunisian banks to provide loans to customers [RS1] [9]. [RS2] [10] used a new method of DDTCA (decision tree credit assessment approach) to answer the problem of credit analysis. The dataset used here is of a car-selling loan company in Taiwan. The algorithm used was decision tree model building.

One paramount algorithm observed was XGBoost [P1] [13], which is a library of optimized distributed gradient boosting meant to be exceedingly efficient, portable, and adaptable, and one of the most notable algorithms this chapter discovered. It applies machine learning methods as part of the gradient-boosting architecture. XGBoost provides parallel tree boosting (also known as GBDT, GBM) to quickly dissolve several data science issues quickly and clearly. The same code can solve problems with over a billion examples in a large distributed system (Hadoop, SGE, MPI).

XGBoost is used for supervised learning algorithms. In this, training data is used along with numerous features x_i to predict target variable y_i. With reference to the mathematical structure from which the prediction y_i is made from x_i, the supervised learning model is expounded. The prediction is given as [14]

$$\hat{y}_i = \sum_j \theta_j x_{ij}$$

a common example of a linear model, a linear combination of weighted input features. Based on charge, that is, classification or regression, there can be different elucidation for prediction values. The undetermined part that needs to be grasped from the data is the parameters. The parameters are the coefficients in linear regression problems (customarily used to designate the parameters). One major share algorithm found was the "tree ensemble model," which consists of a set of CART. The classification is terminated into different leaves, and scores are assigned to each corresponding leaf. The leaf contains barely decision values; this is where CART is a bit different from decision tree. With each of the leaves, a real score is associated with CART, which gives richer exegesis that goes beyond classification. Usually, to be used in practice, a single tree is not adequate. The sum of multiple trees together is used in an ensemble model. The prediction scores of each solitary tree are summed up to get the final score. It is written as [15]

$$\hat{y} = \sum_{k=1}^{K} f_k(x_i),\ f_k \in F$$

where K belongs to the number of trees, f is the function in function space (F), and F is the set of every possible CARTs. The objective function (optimized) is given by [16]

$$obj(\theta) = \sum_{i}^{n} l(y_i, \hat{y}_i) + \sum_{k=1}^{k} \copyright(f_k)$$

Now, the question arises what is the difference between random forest and boosted trees? Here, boosted trees and random forest are the same models; the difference comes into existence by the way the model is trained. This means that only one needs to be written for both random forest and gradient-boosting trees if a predictive service for tree ensembles is depicted.

The proposed framework comprises data analysis and preprocessing, label encoding, and comparison of different models like logistic regression [P2] [17], decision tree classifier [P3] [18, 19], random forest classifier and XGB regressor, prediction and scoring. Real data is obtained from the Kaggle website to predict creditworthiness, which was slightly manipulated to increase some parameters in the data.

Figure 15.1 presents systematic adapted steps taken to complete the study. The study addresses what are the various machine learning methods adopted for the assessment of risk, which datasets are being used by most of the papers, what are the issues faced/challenges in credit risk evaluation, which specific machine learning algorithms are better for processing, and what are the current research trends in credit risk evaluation? An evaluation that comprises papers that do not extensively cover the approach, challenges, and datasets has not been found in any consolidated paper.

15.2 METHODOLOGY

15.2.1 Research

There is extensive literature on credit scoring and credit risk evaluation. The growing concerns of payment default provided us the motivation to conduct this survey.

These research questions will be answered, as well as the reason (Ra) for including them in our research work will be provided,

RQ1. What are the various machine learning methods adopted for the assessment of risk?

Ra. It is a rapidly evolving discipline that encompasses a wide range of problem-solving methods carried out by machines. The insights into various

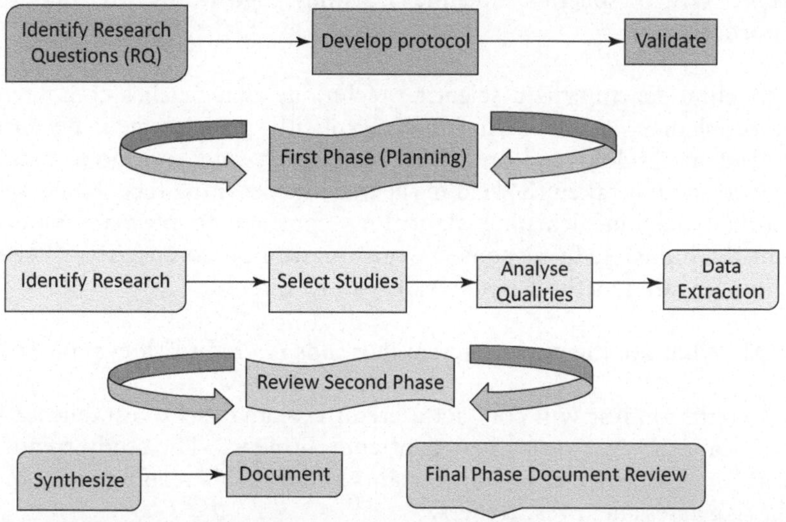

Figure 15.1 Flowchart of research methodology.

machine learning approaches, including ensemble-based, optimization, neural networks, and probabilistic, can be provided by this research question for credit risk evaluation. If there are any common parts like datasets, feature extraction, and models within credit scoring, they are expected to be understood. In response to this question, the scoring techniques (binary or a scale) can be provided for credit risk evaluation.

RQ2. Which datasets are being used by most of the papers?

Ra. A dataset is the summation of cases that have common features. To learn different actions, machine learning models have to be given by training datasets in the machine learning algorithm, after which validation data (or test data) ensures that the given model understands the data correctly. Once these training and validation sets are fed into the system, additional datasets can be used to refine the machine learning models. The more data you give the machine learning system, the faster it improves and learns. Several dataset sources can be provided in answer to this query.

RQ3. What are the issues/challenges faced in credit risk evaluation?

Ra. If the training set is small, or if the data is abnormal, noisy, or polluted from unnecessary features, the system will not function well. Such obstacles would be encountered while building machine learning algorithms in this survey.

RQ4. Which specific machine learning algorithms are better for processing?

Ra. Machine learning is a scientific technique that teaches computers to solve problems without having to be explicitly programmed. Because of improved methodologies, processing capability, and enormous datasets, deep learning has taken the lead in the machine learning race. Nonetheless, traditional machine learning algorithms continue to play an important role in the industry. In response to this question, a comparison of several machine learning methods and results can be offered.

RQ5. What are the current research trends in credit risk evaluation?

Ra. A credit analyst will conduct a credit risk analysis to establish a borrower's capacity to satisfy their debt commitments. The study trends are crucial because they can provide insights and possibly recommendations to credit risk assessment researchers.

15.2.2 Keywords for searching relevant articles

To commence our evaluation, this work necessitates downloading and accessing a colossal number of research papers, survey articles, and varied reports. In order to do the needful, the work adopted two varied methods. First, the paper was searched in the repository of journals like Elsevier, Springer, IEEE (Institute of Electrical and Electronics Engineers), etc. Second, an advanced search using search engines to manually extract the papers was performed. This step is utterly important to come in credence that all the relevant information, sources, and resources had been done in our paper. All the papers and reports have been unsheathed from reputed and trustworthy journals. For searching the article, the implementation of profuse keywords that are associated with the topic of our fascination was done. The most important keywords which were used in every type of search were "Credit Survey," "Review," "Literature Review," "Worthiness Analysis," "Peer to Peer," "Performance," "Loan," "Default," "Peer to Peer," "Fraud Detection," "credit score," "Loan default prediction," "loan risk detection," "credit history," "Loan Default analysis," "Loan prediction analysis," "machine learning in loan prediction," "Loan Prediction analysis," "loan defaulter analysis using machine learning," "loan prediction analysis using ML," "nonperforming asset," "NPA," "nonperforming loan," "NPL," "expert system," "Banking," "Loan Provider," "Credit Scoring," and "Fraud Detection." The paper also used methods of machine learning as keywords along with these elementary keywords stated above for more varied papers. The machine learning keywords are "Deep Learning," "artificial intelligence," "Artificial Intelligence," "neural network," "supervised," "Unsupervised," "Classification," "regression," "Random Forest," "discriminant Analysis," "support vector machine," and "Support Vector Machine," and "Fuzzy logic."

Within all the datasets and journals, searches had been applied using these keywords. For each of the different datasets, innumerable keywords were used. However, to get as many resources as possible, the paper repeated the corresponding keywords in different journals along with novice keywords. Also, query searching format is not common in all the databases, so bigger queries were split into smaller ones. The keywords and searches associated with various journals are given below.

Springer. The query implemented for Springer are "Credit Survey" OR "Literature Review," OR "Performance," OR "Peer to Peer," OR "Fraud Detection," OR "Loan Default analysis," OR "non-performing asset" OR "Deep Learning" OR "Credit Scoring," OR "Fraud Detection," OR "non-performing loan," OR "NPL," OR "expert system," OR "Banking," OR "Loan Provider" OR "Nonperforming asset."

Elsevier. The query accomplished for Elsevier are "Credit Survey" OR "Literature Review," OR "Performance," OR "Peer to Peer," OR "Fraud Detection," OR "Loan Default analysis," OR "NPA" OR "Deep Learning" OR "Credit Scoring," OR "Fraud Detection," OR "nonperforming loan," OR "NPL," OR "expert system," OR "Banking," OR "Loan Provider" OR "Non-performing asset," OR "Unsupervised," OR "Classification," OR "regression," OR "Random Forest," OR "discriminant Analysis," OR "support vector machine," OR "Support Vector Machine," OR "Fuzzy logic."

IEEE. The keywords associated with IEEE Xplore are "Credit Survey," OR "Review," OR "Literature Review," OR "Worthiness Analysis," OR "Peer to Peer," OR "Performance," OR "Loan," OR "Default," OR "Peer to Peer," OR "Fraud Detection," OR "credit score," OR "loan default prediction," OR "loan risk detection," OR "credit history," OR "Loan Default analysis," OR "Loan prediction analysis," OR "ML in loan prediction," OR "Loan Prediction analysis," "Deep Learning," OR "artificial intelligence," OR "Artificial Intelligence," OR "neural network," OR "supervised," OR "Unsupervised," OR "Classification," OR "regression," OR "Random Forest."

15.2.3 Inclusion and exclusion criteria

It is altogether essential to measure the standard of the papers and to identify the relevant papers. This step is quite important for quality appraisal and assessment. The criteria for the literature to be surveyed are as follows:

- The paper discusses novice techniques or improvement of the previous techniques for the credit-worthiness analysis which may include credit scoring, XGBoost, nonperforming assets, and ratings.
- The articles are under the year of publication from 2011 to 2021. This ensures that modern techniques are reviewed rather than old and outdated techniques. Modern artificial intelligence is hence reviewed in our paper.
- The keywords, searches, heading, and abstract contain the information that is required to conduct the survey. If the information seems to fulfil the requirements, it is chosen for our survey.

The paper is rejected if the following are found in the paper:

- If the language used is different from English. It is quite strenuous to review papers written in other languages. Even extensively, almost all the major contributions are recorded in the textual form of English.
- If by any reason the full text and content of the paper is not obtainable or accessible to review.
- Papers which are not published in reputed journals and conferences of computer science.
- Paper that does not extensively cover the approach of credit scoring by using machine learning, deep learning, or any other related field along with proper proofs and validation.
- The review excludes posters, doctorate symposium papers, theses or dissertations, grey literature, and short papers.
- Papers that contain the total amount of information in less than or equal to five pages.
- To maintain the scope of this paper the article addresses only the full research papers.

15.3 ANALYSIS AND DISCUSSION

The initial step in the analysis process is to apply all queries in reputed journals and search engines. After that, we integrated all the inclusion and exclusion criteria and removed unnecessary and irrelevant papers. In these steps, scrutinized papers based on title, abstracts, keywords, results, and conclusions happen. Then we performed a quality and quantitative assessment by thoroughly reading all the selected papers to ensure that they meet and contribute to the risk evaluation. We then analyzed all the papers based on the research questions selected and found the answers to these questions. We extracted papers strictly on the basis of the inclusive and exclusive property. The extraction of papers is scrutinized only and published between the years 2011–2023. This is done to ensure that only relevant and updated

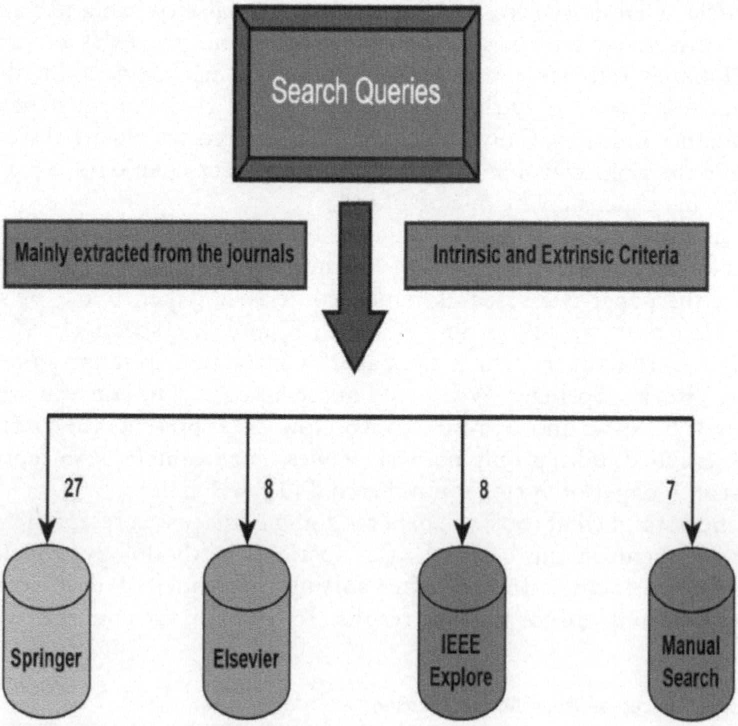

Figure 15.2 Criteria of literature review.

technology is included in the analysis. This process is continued several times to ensure that only the best and relevant papers are included in the study.

A total of 51 papers had been selected for the examination after the tough process. Figure 15.2 shows the publishers from which the papers were extracted.

We mainly got journals from Springer and Elsevier as it has a wide repository of good papers rather than the journals of ACM, Procedia, Wiley, Taylor Francis, Nature Journals, PubMed, and a lot more due to various reasons. The issues faced are as follows.

We found duplicate book chapters, conference papers, and research papers. There were many papers which seemed to be on the same topic but after a thorough reading found them to be of no use. Also, many potential papers are in the paid version, which is quite difficult to access.

One thing worth mentioning is that there are very few survey articles associated with this topic. Quite a few authors showed their interest in writing survey articles on this topic. However, research journals are extensive on the topic. The number of surveys can be counted on fingers. Many survey articles were not accessible due to their availability being in paid versions. Most surveys focus on financial aspects and processes rather than the techniques and incorporation of computer science (artificial intelligence and machine learning) with the process of credit-risk management in the microfinance industry. Only one paper incorporated systematic review and results in the right way, but their topic was different from ours. So, we can say that this paper may be one of a kind.

The article analyzed the papers on the basis of various factors. To name each journal uniquely we used the formula of CXi; here, C represents whether the paper is a research paper, conference paper, book, or survey paper. Hence, it can take values of R, C, B, and S, respectively. X represents the journals from which the paper is extracted, so it can take IEEE Xplore, Elsevier, Springer, Wiley, and miscellaneous. The consecutive symbols are I, E, S, W, and A, respectively. Now "*i*" represents the number of papers assigned, taking only numeric values. For example, RS8 represents the research paper of Springer numbered 8 (Table 15.3).

To understand what topic the paper is going to discuss, a researcher needs to extract them on the basis of title, abstract, methodology, conclusion, and literature focus, although after sorting the reader needs to read the whole paper in detail for the best results. To focus on the concepts written,

Table 15.3 Concepts discussed and focused

Focused concepts	*Papers*
Neural networks	CI1 [76] CI5 [82], CI6 [96] [107]
Linear regression	RS1 [9], CS3 [39], CS5 [22], CS1 [71], RS6 [56], RS7 [55], RS8 [80], CI2 Sheikh et al. (2020), CI3 [84] [31], CI5 [82], CI6 [96], CI7 [72]
Discriminant analysis	RS1 [9], CI6 [96]
Regression	RS2 [11], CS10 [49]
Peer-to-peer lending	CS1 [37], CS6 [71], RS7 [55], CS7 [43], RE1 [103], RE2 [42], CI4 [76], CI5 [82], CI7 [72]
K-mean clustering	CS1 [37], CS4 [41], CS5 [22], RS4 [40] [47], CI2 Sheikh et al. (2020), CI3 [84], CI6 [96]
Decision tree	CS2 [88], CS3 [39], CS5 [22] RS4 [40], RS9 [56], CS7 [43], CS9 [42], CI6 [96]
Support vector machine	CS3 [39], CS4 [41], CS5 [22], RS4 [40], RS6 [56], RS7 [55], RS9 [56], CS9 [42], CI2 Sheikh et al. (2020)
Naive Bayes	CS4 [41], RS7 [55], RS9 [56], CS9 [42], CI6 [96]
Random forest	CS4 [41], CS5 [22], RS4 [40], RS5 [19], RS9 [56], CS9 [42], RE2 [42], [84], RS7 [55]
AUC	[56], C12 [89] CS14 [85], CI7 [72]
Social bias	RS5 [19], RS12 [83], CS11 [92]
LDA	RS7 [55], CI3 [84]

discussing literature is a good idea. Table 15.1 gives a broad perspective on how different papers discuss and focus methodology and research concepts for creditworthiness.

One can predict that credit scoring and peer-to-peer lending is being covered extensively in majority of the papers. Hence, it becomes quite necessary to include them in our survey.

15.3.1 Credit scoring

In financial institutions, "credit scoring" is a common word. These institutions use it as a statistical and analytical tool for evaluating the creditworthiness of a potential borrower. Many papers like [RS1] [9] admit that models are much better than human decisions for credit evaluation. The credit scoring is calculated differently by using different sets of decision models. Over the past few decades, the field has passed its wings to the sky but a lot of scope in the field is required. Generally, the "borrowers" are small owners, retailers, and other professionals. The literature review by Adegoke et al. evaluates the fairness of credit scoring models in mortgage accessibility for underserved populations. It highlights research on the impact of these models on marginalized communities and underscores the need for policy interventions to promote inclusive homeownership. Kamimurare et al. view recent optimization methods in credit scoring models (CSMs), analyzing 46 studies to identify trends like logistic regression and machine learning. It suggests future models integrate big data and deep learning for improved adaptability and performance in credit scoring [20]. The study conducted by Wang et al. employs horizontal federated learning to safeguard data privacy while training a shared credit scoring model. It proposes the FedKT method to address heterogeneous data distributions, effectively improving model performance across multiple datasets [21].

The trouble arises when institutions need to evaluate the credit worthiness of a potential customer who is not only new to credit but who deals only in cash. Another factor that accentuates this is the difficulty in gauging the actual financial status of a potential customer as the bank statements don't truly reflect the cash flow of the retailer's business. Especially in India, small retailers don't file for GST, which would otherwise help to assess the turnover of the business. This brings the picture of credit scoring. The models of computer science, that is, artificial intelligence, machine learning, and deep learning, are used extensively. In general, they make the decision in the form of "Yes or No" or "accepted or rejected" by asking about certain characteristics such as age, income, assets, marital status, shop size, shop location, total dependencies, and many others [Abid]. The whole focus is only to find out if the potential customer would repay the loan in future or not. "The higher the score, the lower the risk, and the lower the score, the higher the risk."

15.3.2 Answer to RQ1: various machine learning algorithms and methods

Mostly loan prediction analysis is done by a restricted number of models and methods. In this question, one can glance at these limited models, approaches, and algorithms of machine learning, deep learning, financial models, and others. Most of the articles (that are published in reputed journals) use these algorithms, theorems, and models for their papers and research. The article looked at how the algorithm functions in microfinance, why papers incorporate the algorithm, and how the particular model is different from other models. One can also glances at comparative review, architecture, and the role of machine learning.

Machine learning and deep learning models: According to [CS5] [22], there are many machine learning models which are used in various fields apart from the financial sector. Talaat et al. addressed the prediction of credit card defaults by combining deep learning with explainable artificial intelligence techniques to enhance model interpretability. Using a real-world dataset, the authors demonstrate that their method not only achieves high prediction accuracy but also provides clear insights into key factors such as payment delays and outstanding bill amounts. This approach offers a balanced solution for effective and understandable credit risk assessment [23]. Similarly, Jovanovic et al. explore integrating blockchain and explainable artificial intelligence (Xartificial intelligence) within federated learning to enhance credit scoring in the financial sector. The proposed framework improves data utilization, model verification, and explainability, enabling reliable, transparent, and automated credit decisions [24]. Jammalamadaka et al. addressed fairness and interpretability in artificial intelligence using explainable artificial intelligence techniques on the German credit card dataset. Their hyperparameter-tuned XGBoost model balanced F1-score, accuracy, and fairness, mitigating gender and age biases with the disparate impact remover. This study emphasizes responsible artificial intelligence in credit score classification [25]. The study by Chen at al. explores the class imbalance problem in credit scoring, focusing on its impact on the interpretability of machine learning models. Using UK residential mortgage data, the study finds that the stability of popular interpretation methods, LIME and SHAP, decreases with greater class imbalance. The results, confirmed by additional datasets, show that class imbalance adversely affects machine learning interpretability [26]. Similarly, Jiang et al. also address the class imbalance problem in credit scoring, which affects the identification of default records. It compares traditional methods, such as resampling and cost-sensitive strategies, with generative adversarial networks (GANs) for handling imbalanced data. Using several datasets, the study systematically scores and compares these approaches, providing suggestions for future research [27]. The integration of finance and machine learning has led to precision credit scoring, a new method that uses advanced artificial intelligence to improve credit assessments beyond traditional models. By utilizing a broader array of data sources, including social media and transaction histories, the approach by Bhilare et al. enhances accuracy and financial inclusion, while allowing financial institutions to manage risks more effectively and develop customized financial products [28]. Now the question arises of what is meant by machine learning and how it is associated with the financial industry, especially the banks and other important firms. Typically, machine learning can be elucidated as the computer algorithm that updates and improves a system with the help of various datasets and data provided to it. The more data is provided to the machine for learning, the more the machine will work systematically. Machine learning algorithms are utilized in a wide range of areas, including the pharmaceutical business, email filtering, and many more. In the financial sector, machine learning is used in activities such as fraud detection, automating trading activities, and providing financial advisory services to investors. There are mainly three types of categorizations in machine learning:

- Supervised learning
- Unsupervised learning
- Reinforcement learning.

Each category is further classified into various categories and different algorithms become part of it. Figure 15.3 classified machine learning in a wide perspective. However, there are a lot of subclassifications embedded in this classification. As depicted in Figure 15.3, unsupervised learning is divided into regression and classification. The models that can be incubated in regression are support vector machine, naive Bayes, discriminant analysis, and others. In classification, researchers can incorporate linear regression, decision tree, neural networks, and many others. In unsupervised learning, they can incorporate K-mean clustering, *K*-nearest neighbors, anomaly detection, and so on. In reinforcement, they can also add Markov decision process, QLearning, and others.

A detailed analysis of different algorithms is done here.

Supervised learning: Supervised learning learns the model by the use of the systematic data provided for evaluation. Based on the labeled data provided, it predicts the output. It finds functions for the input (say X) to the output (say Y) variables. Due to these qualities, these algorithms are used in many applications, including credit risk evaluation as it provides the required output from previous examples. But it is highly important to check the input. As it splits into testing and training data, providing biased data will lead to discrimination. If data is not labeled and clear, then shift to other algorithms rather than supervised. From Table 15.3, it is predictable that many papers use them frequently.

Unsupervised learning: The reverse of unsupervised is supervised. However, not everything is reversed. It uses unlabeled data without any

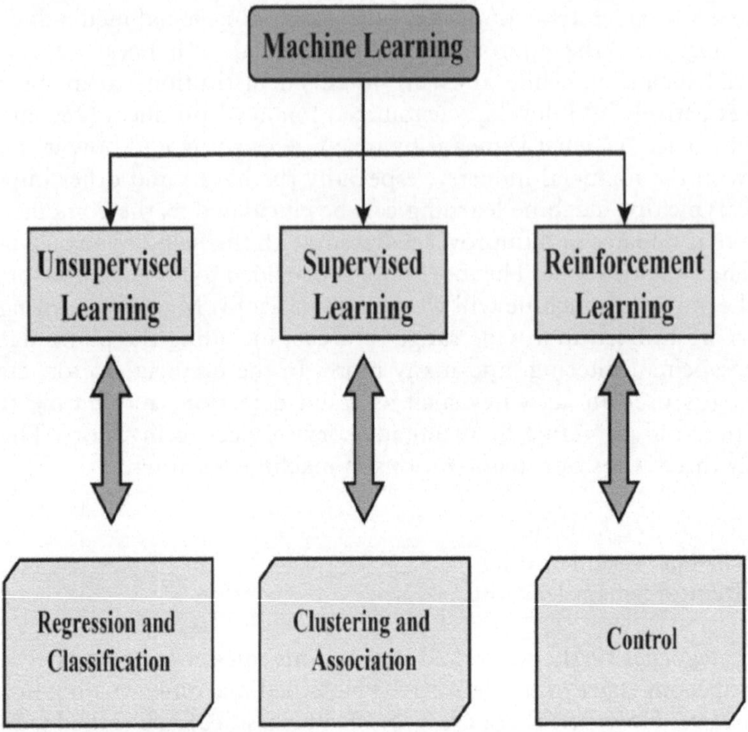

Figure 15.3 Classification of machine learning models.

supervision to predict the output. It completes the full combo of searching the structure of the dataset, grouping them according to analogy, and representing the dataset in a compressed way. These are the best algorithms to find hidden patterns and so are used in the credit risk algorithm. In some cases, the result's accuracy may be low due to unlabeled data.

Reinforcement learning. Reinforcement learning acts "smartly" and these algorithms maximize the benefit. Rather than supervised learning with a proper output, it decides what to do next to maximize the output. The most basic example of reinforcement learning is a chess game. Basically, reinforcement learning is subdivided into positive and negative reinforcement learning depending upon the conditions provided. Due to these fascinating properties (creating new training systems), these algorithms are used in gaming, IT, statistics, and risk management. Creditworthy papers highly prefer these kinds of algorithms for their ease.

Linear regression: This is one of the everlasting and most loved machine learning models used by different journals for prediction in the finance industry. Table 15.1 puts a hallmark on the same. It uses independent variables for predicting the categorical dependent variable. The outcome in this

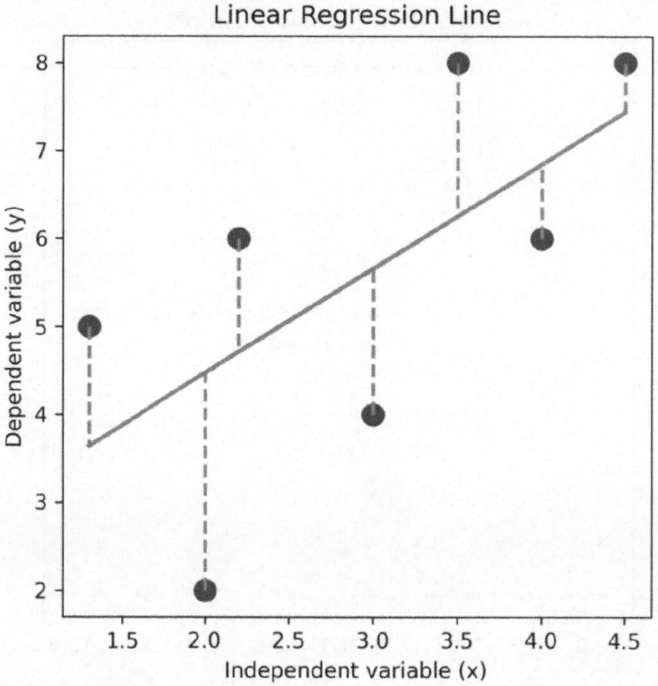

Figure 15.4 Linear regression line.

algorithm needs to be discrete and lie between 0 and 1. The most common algorithm which is quite similar to linear regression is logistic regression. Logistic in general is used to find classification data but linear regression solves regression problems. Some major advantages are that it provides the probability of any event, can differentiate data using discrete and continuous data, and is able to predict the most efficient variable that will be used for the classification. Figure 15.4 presents the linear regression line. The formula for linear regression is given by the formula

$$\ln\left(\frac{y}{1-y}\right) = \beta_0 + \beta_1 x_1 + \beta_2 x_2 + \cdots + \ln\left(\frac{y}{1-y}\right) = \beta_0 + \beta_1 x_1 + \beta_2 x_2 + \cdots \tag{15.1}$$

Figure 15.5 shows logistic regression. Here, *Y* is the probability of bad and *i* the probability of good for the customer. *Xi* is the predictive variable. $\beta i(0 \cdots 4)$ represents the coefficient related to the predictive variable. Based on the categories, linear regression is subdivided into binomial (0 and 1 or pass or fail), multinomial (three or more unordered dependent variable), and ordinary linear regression (three or more ordered pairs). Different journals

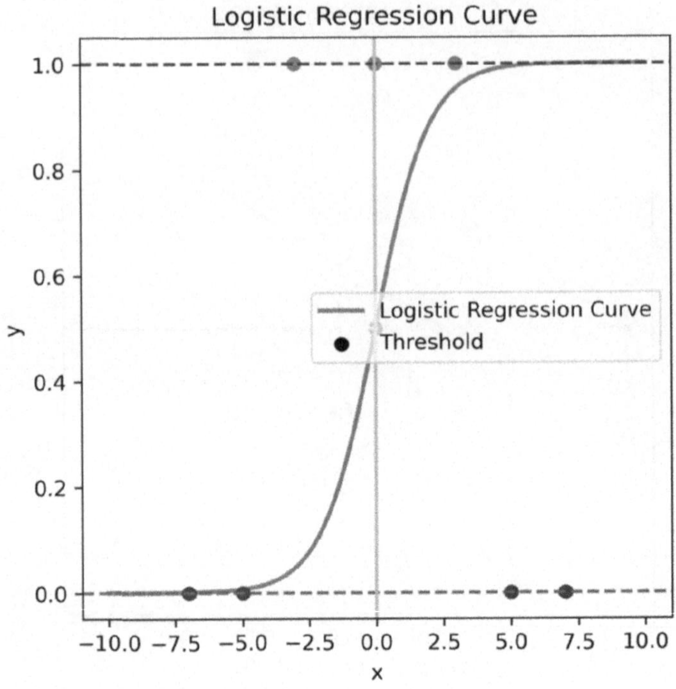

Figure 15.5 Logistic regression.

use the models differently, but the basic data preprocessing steps are fitting linear regression to the training set, predicting the results of the test, confusion matrix (test accuracy), and visualizing the result. He et al. propose a decentralized multiparty method for credit scoring using logistic regression within a vertical federated learning framework. It employs additively homomorphic encryption to preserve privacy and the HE-DPGD algorithm to reduce redundancy. Empirical evaluation shows the method effectively leverages information from multiple sources securely [29].

K-nearest neighbor [30–32]: *K*-nearest neighbor is super easy supervised learning algorithm used for classification and regression. But in real life, it is closer to the classification problems of industries. The same also applies to the financial market. Patil et al. evaluate the performance of the *K*-nearest neighbor (*K*-nearest neighbor) algorithm in stock market forecasting. Using both daily and up-to-the-minute price data, the research applies *K*-nearest neighbor to predict stock prices for large and small capitalizations across three markets. The approach is implemented in Python to enhance the accuracy of stock price predictions [33]. Al-Sayed et al. proposed a credit rating model using 34 accounting variables to assess the credit ratings of Egyptian companies listed on the Egyptian stock exchange from 2016 to 2021. It employs modern artificial intelligence techniques, with results showing that the artificial neural networks model achieved the highest accuracy rate at 89.2%, followed by the support vector machine model at 87.8%, and the *K*-nearest neighbor model at 68.1% [34].

According to the algorithm, the similarity needs to be close together. Hence it has three major advantages over other algorithms. First, it is quite easy to interpret output, even easier from linear regression, CART, and random forest. Second, it has less responding and calculation time; however, it may be equivalent to linear regression in this scenario. Third, prediction power is quite more in this algorithm but not better than random forest. Generally, the K-mean clustering algorithm resembles *K*-nearest neighbor, but they are quite different from each other. K-mean clustering belongs to the unsupervised learning clustering algorithm as compared to *K*-nearest neighbor which belongs to supervised classification and regression. The majority vote of its neighbor classifies as a case. The basic Euclidean distance formula for *K*-nearest neighbor is stated as follows:

$$\sqrt{\sum_{i=1}^{k}(x_i - y_i)2}\sqrt{\sum_{i=1}^{k}(x_i - y_i)2}$$

[14, 35]

The basic Manhattan equation is stated as [15, 36]:

$$\sum_{i=1}^{k} x - i'' - y - i'' \sum_{i=1}^{k} x - i'' - y - i''$$

And the Makowski distance is calculated by [35]

$$\left(\sum_{i=1}^{k}(x - i'' - y - i'')q\right)1/q\left(\sum_{i=1}^{k}(x - i'' - y - i'')q\right)1/q$$

[CS1] [37] where x_i and y_i are two points in space. Depending upon the use and flexibility of use, *K*-nearest neighbor can be subdivided further, such as weighted *K*-nearest neighbor. Due to all these capabilities, *K*-nearest neighbor is used widely in journals for the financial market.

Support vector machine [38]: Support vector machine splits the data into two different classification groups. It comes under supervised learning and uses classification and regression for functioning. Table 15.3 shows the number of papers interested in this major algorithm.

[CS3] [39]. Support vector machine is one of the most robust prediction models which is used in statistical learning frameworks. Generally, to identify each group distinctly, vectors are being used. In this algorithm, each data item is plotted in an *n*-dimensional area [CS5] [22]. The basic idea lies in finding the hyperloop that differentiates them into two best discrete

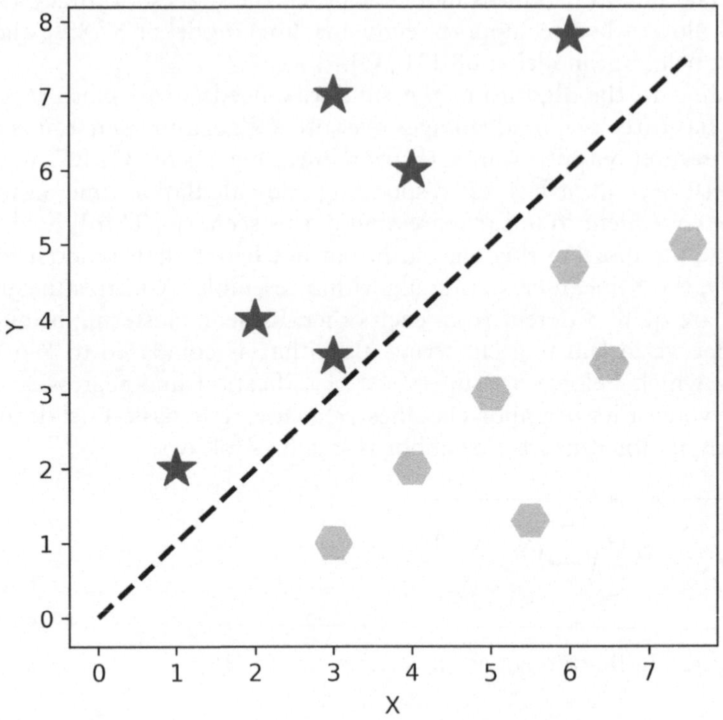

Figure 15.6 Support vector machine.

classes. Maximum elements need to be separated with the help of a hyperloop in support vector machine. Figure 15.6 shows support vector machine.

According to [RS4] [40], one can create a hyperplane for separation in the feature space using the function of Kernel, that is [40],

$$K\left(x_i, x_j\right)$$

Even for the generation of the support vector machine model, the mapping of input vectors is done into new upper-dimensional feature space, which will be denoted by [40]

$$\varnothing : R_d \rightarrow H^f \varnothing : R_d \rightarrow H^f$$

where $d < f$. Here, the Kernel function would be associated with the "linear functions," "Radial Basis Functions (radial basis function)," "Polynomial Functions," or "Sigmoid Functions." Due to this property of separating the data, the financial industry uses this algorithm in the best way possible to maximize the yield.

Decision tree: According to [CS3] [19, 39], a decision tree makes the analysis with the help of tree-like structures to generate the model by using the attribute which is most suitable for the analyses. [CS4] [41] told us that it uses "stream diagrams," which are tree structures, and it uses spreading structure to provide the required result. According to [CS9] [42], the algorithm is used mainly in the discrete-valued function. In this algorithm, all the internal nodes denote some test on the attribute, and every branch of the tree shows the result of the test and each terminal mode provides the class label [CS3] [39]. All the high-ranking attributes provided in the database are assembled in the root node, and the rest are assembled in the leaf node. In every node, a prediction is calculated, and the leaf node provides the answer. All the three buildings would be counted until researchers abstract an outcome in internal modes. [CS9] [42] say that the Gini Index is used for checking the purity of which makes for the homogeneity of the label located at the node. The formula is as follows [42]:

$$Gini(d) = 1 - \sum_{n=1}^{m} P_n^2$$

The varied class is represented by "m," also p_n is the functional probability from the dataset represented as "d," which should belong to the class C_n and must be calculated via "C_n,$dd\$$ over: m" classes.

According to [CS5] [22], decision tree creates the problem of overfitting, but the easier implementation of this algorithm makes a bulging factor for using the algorithm in the industry, especially for creditworthiness. [CS7] [43] shows the advantages of decision tree, which makes it eligible for loan prediction. Since the micro-financial market has a ton of data for analyses, decision tree can handle large datasets with a variety of attributes and can handle data efficiently. The financial market likes straightforwardness at work, and this provides the same as it doesn't require data normalization and scaling.

As the person checking its creditworthiness may/may not have technical background, this algorithm makes the output easy to read and analyze, so that he/she may understand it without any problem. Because of all these advantages, the journals are adopting this model in a very studious way (Table 15.3).

Random forest: According to [CS5] [22], it is unsupervised learning and basic ideology for this is to use ensemble methods for making the multiple decision tree and derive the best model to fit in the frame. One drawback of this algorithm is that it takes a lot of time for execution as it produces a lot of decision tree that automatically takes time. [CS4] [41] agree that RF adopts learning in group procedure by taking numerous calculations to get a diverse outcome of a variety of arrangements, undertakings, and relapsing. He even agrees that each classifier on its own is powerless and when combined can show us tremendous results. [CS9] [42] hence concluded the use of random forest as the one who resolves the overfitting in the decision tree. To get a point in the random forest which makes the use of classification of multiple trees of a special class for which the vote from tree is done, the maximum voted class would be selected by the RF. According to [RE2] [44], the basic formula for RF will be [44]:

$$F_{RF}(x_k) = \text{argmax} \varnothing \sum_{t=1}^{T} w_t h_t(x_k)$$

Here the value of x_k is for the kth instance of the feature vector. Also, $f_{RF}(xf)$ and $h_t(x_k)$ show the results of $x_k$by the help of RF.

It's quite important to get accurate results of any analysis. RF proves to be helpful for it as it provides quite a high accuracy for any problem. Since problems of the financial market can be of classification or regression, RF is flexible enough for both types of algorithms. Even both the categorical and continuous values of any analyses in finance can be analyzed well by RF. Filling missing values and not requiring normalization of data is also a key feature that is loved by the financial market. Loan evaluation in peer-to-peer lending significantly impacts lender profits. Traditional methods are now complemented by machine learning, especially random forest. Ye et al. introduce RFoGAPS, optimizing random forest with a genetic algorithm and a new profit score, to enhance lender profits. Evaluations using Lending Club data show RFoGAPS achieves higher profits than traditional methods [45]. Zhu et al. use the random forest algorithm to build a loan default prediction model with Lending Club data. Employing SMOTE for class imbalance, the results show random forest outperforms logistic regression, decision tree, and other algorithms in predicting defaults [46]. Tables 15.3 and 15.4 show the popularity of random forest among various journals.

Artificial neural networks: According to [RS4] [40, 47], this algorithm was one of the first machine learning techniques to be used for credit risk evaluation in 2020. As agreed by [RE4] [48, 49], layers are used to construct artificial neural network. Each layer can have one or more nodes, and weights are used to connect nodes in different layers. There are numerous algorithms that can be used with artificial neural network. The finest one should be chosen from among the widely utilized backpropagation networks. [RE6] [50] agree that artificial neural network has been claimed to be a reliable method for evaluating corporate credit risk in the credit business.

Artificial neural network is used to underwrite a loan and determine whether it should be approved or rejected [51]. Banks want to reduce loan application failure rates, while increasing returns on loans supplied. Loans granted using artificial neural network have been found to have a lower failure rate than some of the best traditional approaches. Mahbobi et al. enhance default payment predictions in imbalanced data using DNN, support vector machine, *K*-nearest neighbor, and artificial neural network

Table 15.4 Dataset sources

Paper	*Dataset source*
RS1[9]	Created by a Tunisian Bank
RS2 [11]	NA
RS3 [102]	Chinese online loan industry
RS4[40] [47]	UCI repository
RS5 [19]	Kaggle
RS6 [56]	Blockchain technology
RS7 [55]	Peer-to-peer lending platform (Eloan) of China
RS8 [80]	Real-life bank data
RS9 [56]	UCI machine learning repository
RS10 [86]	NA
RS13 [104]	Shenzhen Stock Exchange, Shanghai Stock Exchange
RS14 [111]	UCI machine learning repository
CS1 [37]	Peer-to-peer online lending platform
CS2 [88]	A bank in Taiwan
CS3 [39]	UCI machine repository
CS4 [41]	Kaggle
CS5 [22]	Generated by the banking loan operations by the user
CS6 [71]	Lending club's website
CS7 [43]	PPDai.com
CS8 [93]	Dataset from German, Australian, and Japanese
CS9 [42]	Not included intentionally (project on airline dataset)
CS10 [49]	NA
CS11 [92]	NA
CS12 [89]	UCI machine learning database repository
CS13 [81]	A bank in Taiwan that sells different cards to consumers
CS14 [85]	Artificial data
RE1 [103]	Lending club and Renrendai data on loans in the United States as well as China
RE2 [44]	Lending club platform
RE3 [90]	German credit data
RE4 [48]	UCI machine learning repository
RE5 [75]	Datasets from two peer-to-peer lending institutions
RE6 [50]	Corporate credit datasets
RE7 [78]	US commercial banks
SA1 [4, 68]	Selected banks in Lagos State, Nigeria
SA2 [68]	US bond data
SA3 [100]	NA
SA4 [91]	NA
SA5 [112]	NA
SE1 [74]	Italian bank real-time data
SE2 [94]	NA

(Continued)

Table 15.4 (Continued)

Paper	*Dataset source*
SE3 [15]	Australian, German, and Kaggle credit datasets
SE4 [66]	NA
SS1 [44]	NA
RI1 [110]	Solvent US firms
RI2 [70]	European peer-to-peer lending platform Bondora
RW1 [98][9]	European bank data
RW2 [77]	NA
RW3 [79]	Credit register of Spain
RA1 [65]	European bank
RA2 [98]	Bangladeshi banks: Prime Bank, Agrani Bank, Dhaka Bank, National Credit, Janata Bank & Commercial Bank, Basic bank & Sonali Bank
RA3 [113]	NA
RA4 [109]	Commercial bank in Kenya

classifiers with techniques like SMOTE and ALL-*K*-nearest neighbor on a 30,000-record dataset. Support vector machine with ALL-*K*-nearest neighbor achieves the highest accuracy at 98.6% and the lowest cross-entropy loss of 0.028, improving credit risk assessments [52]. Khalili et al. propose an optimal credit rating model using a new hybrid performance metric called the probability of credit scoring correctness. It evaluates five classifiers – *K*-nearest neighbor, support vector machine, decision tree, AdaBoost, and PNN – along with eight performance metrics, employing genetic algorithms for feature optimization. Tested on Iranian, UCI German, and Australian datasets, the model shows strong performance, helping banks better assess customer risk and make informed loan decisions [53]. Asencios et al. design six profit-scoring models to predict the internal rate of return (IRR) of credit applications using multilayer perceptron, XGBoost, and TabNet algorithms. Key features from a Peruvian credit union dataset include the effective monthly interest rate and delinquency days. XGBoost, particularly with all features considered, shows the best performance, aiding credit analysts in lending decisions [54].

Naive Bayes: According to [RS7] [55] and [RS9] [56], based on Bayesian theory, the naive Bayes classification model argues that characteristics are conditionally independent. The prior probabilities of each feature are calculated using training samples. The Bayesian formula is then used to calculate the likelihood of the sample belonging to each class, and the sample category is ultimately chosen based on the probabilities. If the sample X has a total of v categories, C_v, composed of an attribute vector $A = a_1, a_{2,\ldots,} a_m$, the probability that sample x_i belongs to class c_i is [55]

$$P\left(\frac{C_i}{X}\right) > p\left(\frac{C_j}{X}\right), 1 \le j \le v, i \ne j$$

$$P\left(\frac{C_i}{X}\right) = P(X / C_i)P(C_i) / P(X)$$

Moreover, studies such as one done by Bhargav et al. compare the effectiveness of the random forest and naive Bayes algorithms for loan prediction using Kaggle datasets, with each algorithm applied to ten samples. The novel random forest classifier shows slightly higher accuracy (80.892%) and lower loss (19.108%) compared to naive Bayes, though the statistical difference between their performances is not significant (p = 0.914) [57]. Suryawan et al. provide decision support system using the naive Bayes method for more objective credit evaluations. Analyzing 340 applications, the system found 321 applicants eligible, with an overall accuracy of 94.31%, enhancing transparency and reducing subjectivity in credit decisions [58].

K-means clustering: According to [CS1] [37] and [CS4] [41], supervised methods like K-means clustering can be used to investigate the hazards associated with the online lending platform. The researchers must first set the *K* value, which is the number of clusters in the final dataset before they can calculate K-means. Second, the first data cluster's center point is generated at random. The clustering result is obtained after the first repeat. A new centroid point must then be defined for the next clustering. This new centroid point is the center of the categories in the previous clustering result. The clustering technique can be performed using the new centroid point, and additional clusters can be computed and iterated upon using the new centroid.

By grouping enormous voluminous datasets in banks, clustering enables effective decision-making. An improved risk evaluation of multidimensional risk prediction clustering method is used to evaluate which loan applicants are working efficiently and inefficiently. Reference [59] introduces the clustered support vector machine (Csupport vector machine) for credit scorecard development, addressing computational inefficiencies of traditional nonlinear support vector machine methods. The Csupport vector machine achieves comparable classification performance with reduced computational cost, making it suitable for large historical datasets [59], and improves credit scoring by using class-wise classification as a preprocessing step for ensemble classifiers, addressing the issue of unrepresentative samples. Techniques like neural networks, support vector machine, and Bayesian networks enhance accuracy. The approach also uses the Markov blanket for effective feature selection [60] and [62]. Xiao et al. introduce an ensemble classification approach for credit scoring using supervised clustering to ensure diverse training subsets, improving overall accuracy. Base classifiers are constructed from these subsets and combined by weighted voting. Experimental results show higher diversity and accuracy compared to other methods [61]. With advancements in statistical methods and machine learning, credit scoring now leverages ensemble and hybrid models for better accuracy. The study by Haoting et al. uses logistic regression, support vector machine, neural network, gradient boosting, and random forest. It introduces the CF-GA-Ens model, combining genetic algorithm for classifier selection and unsupervised clustering with fuzzy assignment, tested on multiple datasets, showing improved performance [63].

15.3.3 Answer to RQ2: datasets used by most of the papers

Mainly the work analyzed that there is a shortage of varied datasets. Table 15.4 shows different data sources for different papers. Generally, the papers are from some financial institutions such as "Lending Club," "UCI Repository," "Tunisian and Taiwan Banks," "Kaggle," and others. Many papers also did not show any dataset which is depicted as "NA," making it difficult to analyze the algorithms (Table 15.4).

Some of the datasets are explained here. [RS1] [9] collected data from a Tunisian Bank during 2010–2012. Here, peer-to-peer lending is used. The data was from 603 customers and contained 4 variables. Two point crossover, Tournament selection, and ten-fold cross-validation are adopted in this paper. Dataset contains 36,000 borrowed records from Lending Club, 17 primitive features, of which 3 are nominal and 14 are numerical.

[RS5] [64] used datasets from UCI machine learning repository and Kaggle dataset contained 597 data points and was divided in a ratio of 75:25 for training and testing. [RS7] [55] obtained a dataset from a peer-to-peer institution in China named Eloan as a knowledge provider. The dataset was big data with 137,000 records and 27 attributes. [RS11] [10] used data from a car loan-providing company in Taiwan. They separated the data into datasets for testing and training, with training accounting for roughly 66% of the total.

15.3.4 Answer to RQ3: issues faced/ challenges in credit risk evaluation

One of the most important backbones of any economy is its credit market. Especially for developing countries, this market is very important. Even the market serves as oxygen for some economies that are not on track. Any major issue faced by the micro-financials can harm the respective country in a significant way. Since it is such a large market, the adaptation of any algorithm may lead to a lot of issues. However, the human interference in it is quite less due to automation of different algorithms. Like other research work, there are also challenges and issues in credit risk and worthiness assessment. This work discusses these issues in detail.

The challenge is not only restricted to the algorithm's poor performance, algorithm's lack of sophistication, or reliability but also to the factors beyond the wall of algorithms.

One issue faced is biased data by the credit scoring and peer-to-peer platforms. Generally, all the algorithms assume that the nature and pattern of the data will remain the same in the future as it was in the past. The model predicts the output based on human characteristics rather than loan repayment in some cases. This is also known as "bias against groups or classes of people." To cope with this problem, banks and micro-financial institutions need to use a diversified and large volume of data. Sometimes large data is not available as per the requirement, which can create hindrance for the same. Banks also apply techniques such as "demographic blindness" (selecting an uncorrelated collection of features) and "equal odds" (selecting the same amount of false-positives and true-positives for every class).

Second, computers make decisions based on logic rather than emotions or reasoning to understand the cause of any action. Due to this reason, these algorithms are sometimes called "Black Boxes." Many times, due to this drawback companies need to bear a huge number of losses.

Take the example of a credit scoring process that is taking place on a potential customer named John. John is a great businessman, and he has a handsome experience of 20 years in the business. He had survived such a long time due to his skills and ability. But recently his wife was in isolation due to severe disease, and for the same reason, he lent some assets, namely, a car, gold, a shop, and others, to get some money for treatment. Now his wife has recovered from the disease all due to the money he got. But he needs to reestablish his business. Unfortunately, he had to wait to get the assets back from the lender. To sanction a loan, he needs to check his credit score. Due to the lack of any assets and other scenarios, his credit score is quite low. However, if he is given the loan, he will be able to pay back the loan on time due to his skills. This leads the company to lose its potential customer.

Third, the challenge which is faced by the creditworthiness analysis is the difference in the training and testing databases. It is possible that training data is from a different region, micro-financial market, or bank, and testing data can belong to a different place. Also, within the same organization, there are chances of different rules and regulations (according to country), leading to changes in the dataset. Also, due to different policies and adaptability of varied banks, attributes and relations may differ and all these causes combined cause inaccuracy in the sample.

Also, unstructured data is also a big problem that consumes a lot of time for processing data and even leads to loss of accuracy. The data can be unstructured text which needed processing or may be in the form of images. Some data may also need to be extracted and cleaned, which might be useful, such as GDP of the nation and inflation ratio statistics. For a small firm or a start-up and even researchers, getting a big amount of data for processing is also very difficult as many times such large data is privately controlled. To deal with data shortage, oversampling and other approaches are used. The data would no longer be realistic in such cases. Due to noises in the data researchers use data filtering and feature-selection steps. Some new approaches are adopted for good and to become up to date, but every model has its pros and cons, such as peer-to-peer modeling. According to [CS1] [58], improper supervision, low barriers to entry, imperfect market exit mechanisms, and illegal operations are some of the drawbacks of peer-to-peer modeling.

These problems create a bunch of troubles for researchers and firms. But over the passage of time problems are sorted out slowly and gradually. No model is 100% accurate, and the work can also not predict which algorithm is better in all the scenarios. Depending upon the suitability, an individual firm or researcher copes with the problem as per their best solution.

15.3.5 Answer to RQ4: better algorithm

Each machine learning model for credit risk detection has different performance characteristics. The estimation for choosing the top two or one of the best models from various models that were developed can be accomplished by using resampling techniques such as cross-validation.

When researchers have a new dataset, they employ several different approaches to assess accuracy to select one or two of the available machine learning algorithms, such as decision tree classifier, logistic regression, support vector machine, XGB regressor, random forest classifier, and so on.

The key to comparing machine learning algorithms is to ensure that each algorithm is evaluated on the same data using the same manner to better analyze the likelihood of loan default. The work compares naive Bayes, logistic regression, *K*-nearest neighbors, decision trees, support vector machine, and boosting algorithms (Table 15.5).

To analyze which algorithm is best, the work analyzed all the papers and got the following result for loan prediction analysis. Table 15.6 distinguishes various machine learning algorithms implemented by various papers.

15.3.6 Answer to RQ5: research trends

Over a few decades, the research trend has been changing according to the changes in the financial market. We can see that structural changes were observed in the global market and, subsequently, an increase in the global risk assessment was observed. Many markets of the world reported

Table 15.5 Advantages and disadvantages of various models

	Advantages and disadvantages				
Model name	*Advantage No. 1*	*Advantage No. 2*	*Advantage No. 3*	*Advantage No. 4*	*Disadvantage*
Artificial neural network (ANN)	Artificial Neural Network is capable of identifying intricate nonlinear correlations between dependent and independent variables in a dataset	Less formal statistical analysis training is needed	The Artificial neural network approach works well for issues involving both regression and classification	Preprocessing relies on independent variables that the financial institutions may supply	Methodologically costly to train the network to handle a complicated categorization issue
Decision tree (DT)	Data preparation is simple in decision tree, and it is capable of handling a wide range of data kinds, such as the typical categories, nominal, and numeric data found in finance	Classes are considered to be mutually exclusive	The resulting categorization tree is simpler to comprehend and interpret	Algorithms are based on the order of attributes or variables that may be present in the trade	Algorithms cannot branch if a nonleafy node's attribute or variable value is missing
K-nearest neighbor	*K*-nearest neighbor is easy to understand algorithm for trading that can quickly classify instances	It is suitable for usage as a multiclass classifier	Because data may be both classed and regressed, it can be used for both tasks. Since all attributes are given the same weight, classification performance may suffer as needed in practical applications	It can handle missing attribute values or noisy instances, which are prevalent in economic data	As the number of attributes grows, it becomes computationally expensive. There is no information on which attributes work well for what kind of classification

(*Continued*)

Table 15.5 (Continued)

	Advantages and disadvantages				
Model name	*Advantage No. 1*	*Advantage No. 2*	*Advantage No. 3*	*Advantage No. 4*	*Disadvantage*
Linear regression (logistic regression)	Linear regression is simple to set up and maintain, and it can be easily updated	The independent variables and observations must be mutually exclusive and exhaustive	It gives a pleasing real-time interpretation of model parameters, which is beneficial for many applications	The distribution of the independent variables are not based on any prior assumptions	When the input variables have complex relationships, the accuracy is poor. It also ignores the linear relationship between variables
Naive Bayes (NB)	Naive Bayes is straightforward and quite useful for massive datasets, as required by businesses. It can make probabilistic predictions and of handling with both discrete and continuous data	Classes must be mutually exclusive	It runs for the issues of both multiclass and binary classification	It is predicated for the normal distribution of numerical attributes	Due to dependency between attributes, it has a terrible effect on the classification performance
Random (RF) forest	For huge datasets, RF performs well. This is an ensemble-based classifier that outperforms its individual base classifiers. Lower chance of overfitting and variance of training data	The number of base classifiers must be specified	In estimating variable importance, it favors variables or qualities that may take big wide varieties of the alternative values	It may predict which qualities and variables are significant in classification	More complex and computationally costly

(Continued)

Table 15.5 (Continued)

Model name	*Advantages and disadvantages*				
	Advantage No. 1	*Advantage No. 2*	*Advantage No. 3*	*Advantage No. 4*	*Disadvantage*
Support vector machines	Support vector machine excels at classifying semi-structured or unstructured data such as words, photos, and so on	Can work with many feature scopes	Less chances of overfitting which is crucial for credit worthiness	It is far greater strong in comparison to linear regression and hence may be useful to the firms	The resulting model, weighting, and impact of factors are frequently difficult to comprehend, difficult to understand. Large and complicated datasets are computationally expensive. When the data is noisy, it does not perform well

Table 15.6 Various machine learning algorithms

Algorithms	*Journals*
Logistic regression	RS1 [9], RS2 [11], RS6 [56], RS7 [55] RS8 [80], RS9 [56], CS3 [39][32], CS4 [41], CS5 [22], CS6[71],CS11 [38] [74], CS12 [89], RE4 [48], RE5 [75], RE6 [50], RE7 [78], RA2 [98]
Linear regression	RS2 [11], RA2 [98], RS2 [11]
Neural network	RS3 [57] [103], RS14[111], RA3[113], RI1[69], CS14 [85]
Artificial neural network	RS4[40] [47], RE6 [50], RE4 [48] [114]
Random forest	RS5 [76], RS7 [55], RS8 [80], RS9 [56], RA1[65], RE2 [44], CS12 [89], CS9[42], CS5 [22] CS4 [41] [115]
Support vector machine	RS6 [56], RS7 [55] RS14[111], RE4 [48] CS11 [38], CS12 [89], CS9 [35], CS8 [93] [75], CS5 [22], CS3 [39], CS4 [41]
Naive Bayes	RS7 [55], RS9 [10], CS12 [89] [116]
Decision tree	RS9 [56], RS10 [86][105] 9 RS11 [10], RS13 [104], CS2 [88],[106], RA3[90], [98], RE4 [48], RE6 [50], CS9[42], CS7 [43], CS3 [39], CS4 [41], CS5 [22]
Boosting	CS1[37], RA1[65], RE6 [50],CS12 [83]
Deep learning	RA1[65], RE1 [103] [117]
Classification and regression trees (CART)	RE3 [90]
K-nearest neighbor	RS9 [56], RS10 [86], CS1[37], CS13 [81], CS8 [93]
Hybrid sampling, dynamic ensemble	RS12 [83]
Bayesian network	RS10 [86]
NA	SA1[4] [68], SA5[112], SE2 [94], SE3 [15], SE4[66], RW1 [98], RW2 [77], RA4[109]
Regression	RW3 [23] RA1[65]

the number of defaulters growing subsequently over the past few decades leading to a heavy loss to the firms (Table 15.7).

This is the major reason to study credit and credit score analysis for the firms in the world, leading to the increasing number of research in the domain area. The development is quite imperial in the credit risk evaluation. Figure 15.7 represents publications year-wise.

Figure 15.8 shows the number of papers published in different repositories at a specific year that are being chosen for the analysis. The increase in the number of papers clearly shows the interest in the domain of the researchers. The line is specifically drawn to represent the straightness in the curve which shows the increasing awareness of the required topic of financial risks and subsequent analysis to reduce the loss.

The nationality or the country from which the paper is evaluated is a good idea to show which country is focusing on the topic to a greater

Table 15.7 The algorithmic trend in papers

Algorithm	*Percentage (%)*
Logistic	19.10
Decision tree	14.60
Support vector machine	12.40
Random forest	11.20
NA	9.00
K-nearest neighbor	5.60
Neural	5.60
Boosting	4.50
Artificial neural network	3.40
Linear regression	3.40
Naive Bayes	3.40
Deep learning	2.20
Regression	2.20
Bayesian network	1.10
Hybrid sampling	1.10

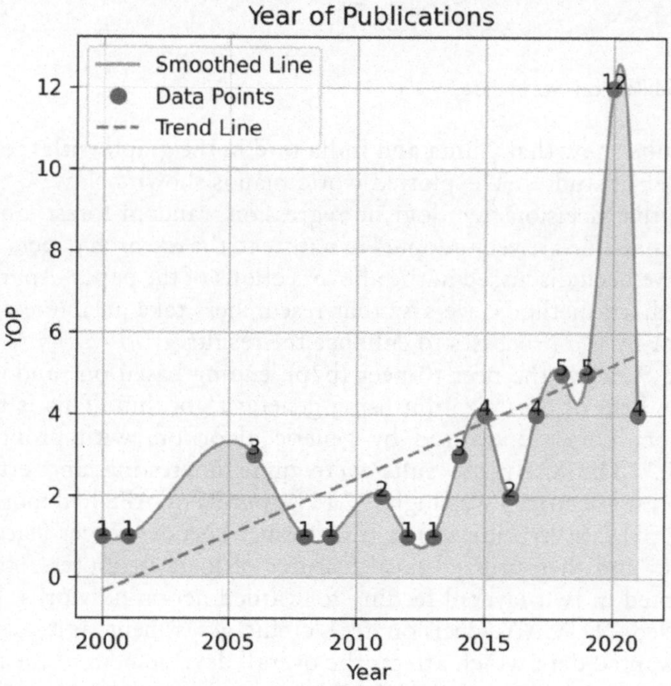

Figure 15.7 Year of publications.

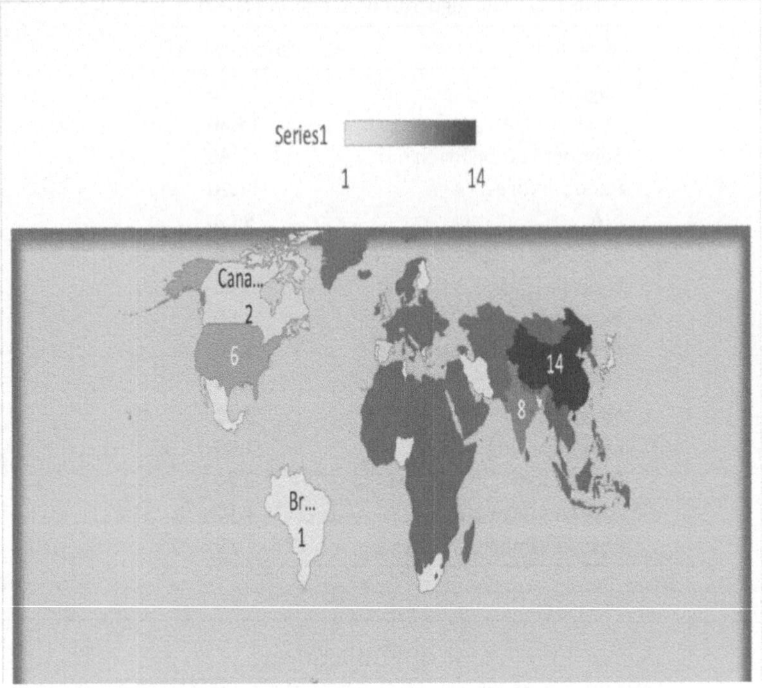

Figure 15.8 Publication worldwide.

extent. Results show that China and India lead in the graph with the maximum number of studies. The plotted world map is shown below.

It is seen that decision tree, logistic regression, random forest, and support vector machine are very adopted in nature to the research papers. These methods have been discussed in the above sections of the paper. Apart from the standardized methods, we saw that researchers take an interest in the optimized or hybrid processes to enhance the results.

Figure 15.9 shows the peer-to-peer (p2p) lending based on random forest with the help of optimization using genetic algorithm. This is termed the "Random Forest optimized by genetic algorithm with profit score (RFoGAPS)." The analysis results were quite interesting and effective. Random forest's accuracy was high (77.14%) but RFoGAPS had more profitability. [RE2] [59] hybridized the transformer encoder to get features in textual form and then utilized hard features of loan attributes. This was then encrypted in two-layered feeding forwarded neural networks. [RS11] [14] proposed "DTCAA (decision tree credit assessment approach)" to extract unwanted data which affects the overall development of the model, specifically the decision tree. [CS1] [58] has optimized the algorithm by collectively using "K-means Cluster Analysis" and "AdaBoost Algorithm"

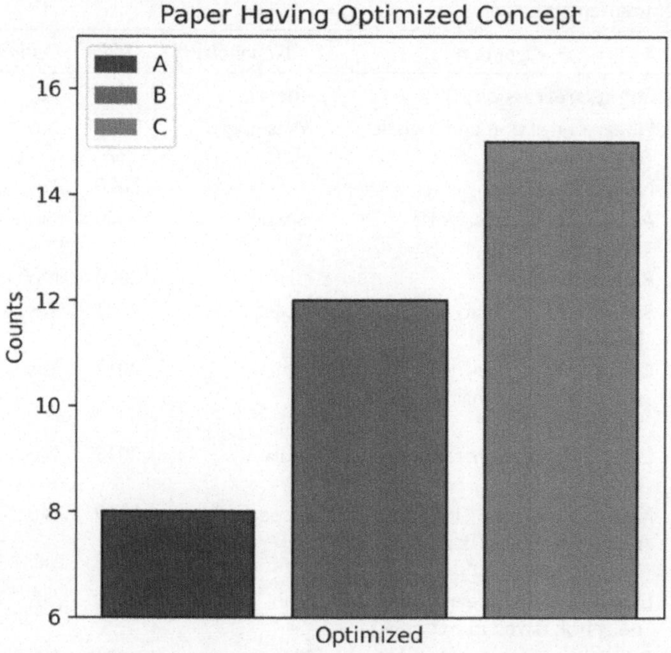

Figure 15.9 Optimization statistics.

and prepared an online loan risk rating model. [CS7] [55] optimized "Classification and Regression Trees" to analyze efficiency in classifying data about peer-to-peer lending (Table 5.8).

15.4 OPEN ISSUES AND FUTURE SCOPE

Some of the issues encountered in different research works are stated as follows.

(1) *Preprocessing of datasets (outliers):* Underestimating outliers can lead to unnecessary low accuracy and computation. Thus, the dataset's pre-processing is crucial to enhancing the performance of any model.
(2) *Optimization of algorithms:* It provides a way to minimize the loss function for deep learning. Applying optimization algorithms wherever needed can drastically boost the performance of the model. The goodness of the model is directly related to how good the predictions can be, that is, for a classification problem, the model should be correctly able to identify things, whereas, for the regression model predictions, the predictions should be close enough to the real values.

Table 15.8 Summarizing studies

Paper	*Algorithm*	*Nationality*	*YOP*	*Optimization*
RS1 [9]	Logistic regression	Tunisia	2016	No
RS2 [11]	Linear regression and logistic regression	Washington, DC	2020	No
RS3 [57]	Neural networks		2020	Yes
RS4 [40] [47]	Artificial neural network	Brazil	2020	No
RS5 [76]	Random forest		2020	No
RS6 [56]	Support vector machine, logistic regression	China	2021	Yes
RS7 [55]	Logistic regression, naive Bayes, support vector machine, random forest	China	2017	Yes
RS8 [80]	Logistic regression, random forest	India	2019	No
RS9 [56]	*K*-nearest neighbor, logistic regression, naive Bayes, decision trees, random forest	Greece	2019	No
RS10 [86]	Decision tree, *K*-nearest neighbor, Bayesian networks	Iran	2019	Yes
RS11 [10]	Decision tree	Taiwan	2021	No
RS12 [83]	Hybrid sampling, dynamic ensemble	China	2020	Yes
RS13 [104]	Decision tree	China	2016	No
RS14 [111]	Neural networks, support vector machine	Mexico, Spain	2014	
CS1 [37]	*K*-nearest neighbor, AdaBoost	China	2020	Yes
CS2 [88]	Decision tree	China	2018	Yes
CS3 [39]	Logistic regression, decision tree, support vector machine	India	2020	No
CS4 [41]	Support vector machine, random forest, logistic regression, decision tree	India	2021	No
CS5 [22]	Logistic regression, decision trees, random forest, support vector machine	India	2021	No
CS6 [71]	Logistic regression	Japan, Taiwan	2019	Yes
CS7 [43]	Decision tree	China	2020	Yes
CS8 [93]	Support vector machine, decision tree, *K*-nearest neighbor	India	2017	Yes
CS9 [42]	Decision tree, naive Bayes, random forest, support vector machine	India	2018	No

(Continued)

Table 15.8 (Continued)

Paper	*Algorithm*	*Nationality*	*YOP*	*Optimization*
CS10 [49]	Regression, GBDT	China	2020	No
CS11 [92]	Logistic regression, support vector machine	Switzerland, Portugal	2017	Yes
CS12 [89]	Logistic regression, naive Bayes, support vector machine, random forest, XGBoost	India	2017	No
CS13 [81]	*K*-nearest neighbor	India	2018	No
CS14 [85]	Neural network	London (UK), Boston	2020	No
RE1 [103]	Deep learning	China	2020	Yes
RE2 [42]	Random forest	China	2018	Yes
RE3 [90]	Classification and regression trees (CART)	Gainesville, USA	2006	No
RE4 [48]	Logistic regression, decision tree, artificial neural network, support vector machine	Hong Kong, China	2011	No
RE5 [75]	Logistic regression	China, USA	2015	Yes
RE6 [50]	Logistic regression, decision tree, artificial neural network, boosting	China, Hong Kong	2011	Yes
RE7 [78]	Logistic regression		2013	
SA1 [4, 68]	NA	Nigeria	2014	Yes
SA5 [112]	NA	Canada, USA	2006	No
SE2 [94]	NA	UK	2000	No
SE3 [15]	NA	South Africa	2020	Yes
SE4 [66]	NA		2015	
RI1 [110]	Neural network	USA	2001	
RI2 [31]	Artificial neural network	China, Finland	2015	
RW1 [98]	NA	Switzerland	2006	No
RW2 [77]	NA	Chicago, USA	2012	No
RW3 [79]	Regression	Spain, Switzerland, London	2014	No
RA1 [65]	Regression, random forest, gradient boosting, deep learning	France	2018	Yes
RA2 [27]	Linear regression, logistic regression	Canada, Bangladesh	2015	No
RA3 [113]	Decision tree, neural network, support vector machine	Taipei	2009	
RA4 [109]	NA		2008	

(3) *Combining machine learning models:* These models are rule-based and statistical into a single tool that would help in credit risk evaluation as per requirements of the financial institutions.
(4) *Deep learning:* It is found that deep learning can be useful in fraud detection and credit scoring (more exploration is required). A better choice for classification problems is support vector machines, as it overcomes the hurdles of local optimum and overfitting in neural network-based models.
(5) *Kernel function and deeper data processing:* More suitable kernel function and deeper data processing will help in increasing the accuracy. As historical data increases, there is a dire need to find computationally cost-effective models.

In the future, the research will expand the literature work to a larger number of models and a larger number of articles; the literature number is currently 51. The article asked five questions and tried to answer them in this work based on our findings. One can ask other questions based on different parameters such as indexing, accuracy, and acceptance of the model in financial institutions. More findings are required in the form of literature so that more research and data can be analyzed properly.

15.5 CONCLUSION

Based on the resources available to us, the work analyzed 7,000+ papers and extracted 51 papers for the reviewing process. Though many survey papers focused completely either on artificial intelligence, on credit risk, or on a certain class of algorithm (if they analyze both credit risk and artificial intelligence), still there has been a gap in studying all three in one intersection. This chapter has tried to solve this problem so that significant knowledge can be added for future researchers. One can believe this literature survey will add more knowledge to the easy understanding of artificial intelligence in the credit risk evaluation.

As one dives deep into the papers, the paper found that there are a lot of problems in the credit risk evaluation field. Every model has its risks and challenges which cannot be dependent totally on the evaluation. As discussed in RQ3, different banks have different rules and regulations even for different regions in the same bank and hence datasets will vary accordingly. Therefore, if researchers train and evaluate the model on the basis which belongs to a domain and then test data on other domain, then there is a high possibility that they may lose accuracy.

This chapter first found the different attributes of credit scoring used by financial institutions. It then asked five crucial questions and tried to answer them, and sorted 51 papers based on the inclusion and exclusion criteria. The chapter discussed different concepts and focused on different surveys. The research work showed the analysis of different papers based on algorithm, nationality, YOP, and optimization. The chapter also showed various database origin sources so that researchers can get insight about these sources. The work summarized advantages and disadvantages of various models and showed various machine learning algorithm being used in different sections. Research trends based on publication worldwide, year of publications graph, and optimization were also analyzed.

This chapter has discussed and evaluated the creditworthiness in detail. One can say that there is a dire need for developing more efficient and reliable tools that can face the issue of continuously changing different domains of datasets (due to changing world) and give relaxation for adding any type of model for evaluating creditworthiness and risk.

REFERENCES

1. Feridun, M., & Güngör, H. (2020). Climate-related prudential risks in the banking sector: A review of the emerging regulatory and supervisory practices. *Sustainability*, 12(13), 5325.
2. Muñoz-Cancino, R., Bravo, C., Ríos, S. A., & Graña, M. (2023). On the dynamics of credit history and social interaction features, and their impact on creditworthiness assessment performance. *Expert Systems with Applications*, 218, 119599.
3. Rani, S., Tripathi, K., Arora, Y., & Kumar, A. (2022, December). A Machine Learning Approach to Analyze Cloud Computing Attacks. In *2022 5th International Conference on Contemporary Computing and Informatics (IC3I)* (pp. 22–26). IEEE.
4. Kumar, A., Singh, H., Kumar, P., & Al Mangour, B. (Eds.). (2023). *Handbook of Smart Manufacturing: Forecasting the Future of Industry 4.0*. CRC Press.
5. Nwanna, I. O., & Oguezue, F. C. (2017). Effect of credit management on profitability of deposit money banks in Nigeria. *IIARD International Journal of Banking and Finance Research*, 3(2), 137–160.
6. Chaudhry, S. M., Ahmed, R., Huynh, T. L. D., & Benjasak, C. (2022). Tail risk and systemic risk of finance and technology (FinTech) firms. *Technological Forecasting and Social Change*, 174, 121191.
7. Rajendra, P., Kumari, M., Rani, S., Dogra, N., Boadh, R., Kumar, A., & Dahiya, M. (2022). Impact of artificial intelligence on civilization: Future perspectives. *Materials Today: Proceedings*, 56, 252–256.
8. Rani, S., Kumar, A., Bagchi, A., Yadav, S., & Kumar, S. (2021, August). RPL based routing protocols for load balancing in IoT network. In Journal of Physics: Conference Series (Vol. 1950, No. 1, p. 012073). IOP Publishing.
9. Abid, L., Masmoudi, A., & Zouari-Ghorbel, S. (2018). The consumer loan's payment default predictive model: an application of the logistic regression and the discriminant analysis in a Tunisian commercial bank. *Journal of the Knowledge Economy*, 9, 948–962.
10. Wang, Z., Xiao, J., Wang, L., & Yao, J. (2024). A novel federated learning approach with knowledge transfer for credit scoring. *Decision Support Systems*, 177, 114084. https://doi.org/10.1016/j.dss.2023.114084

11. Ross, E. J., & Shibut, L. (2021). Loss given default, loan seasoning and financial fragility: evidence from commercial real estate loans at failed banks. *The Journal of Real Estate Finance and Economics*, 63, 630–661.
12. Witzany, J., & Witzany, J. (2017). Rating and Scoring Systems. *Credit Risk Management: Pricing, Measurement, and Modeling*, 19–116.
13. Chen, T., & Guestrin, C. (2016, August). Xgboost: A scalable tree boosting system. In *Proceedings of the 22nd acm sigkdd international conference on knowledge discovery and data mining* (pp. 785–794).
14. Abu Alfeilat, H. A., Hassanat, A. B., Lasassmeh, O., Tarawneh, A. S., Alhasanat, M. B., Eyal Salman, H. S., & Prasath, V. S. (2019). Effects of distance measure choice on k-nearest neighbor classifier performance: a review. *Big data*, 7(4), 221–248.
15. Nayak, S., Bhat, M., Reddy, N. S., & Rao, B. A. (2022). Study of distance metrics on k-nearest neighbor algorithm for star categorization. In *Journal of Physics: Conference Series* (Vol. 2161, No. 1, p. 012004). IOP Publishing.
16. Dastile, X., Celik, T., & Potsane, M. (2020). Statistical and machine learning models in credit scoring: A systematic literature survey. *Applied Soft Computing*, 91, 106263.
17. Wright, R. E. (1995). Logistic regression. In L. G. Grimm & P. R. Yarnold (Eds.), *Reading and understanding multivariate statistics* (pp. 217–244). American Psychological Association.
18. Swain, P. H., & Hauska, H. (1977). The decision tree classifier: Design and potential. IEEE Transactions on Geoscience Electronics, 15(3), 142–147.
19. Kotsiopoulos, T., Sarigiannidis, P., Ioannidis, D., & Tzovaras, D. (2021). Machine learning and deep learning in smart manufacturing: The smart grid paradigm. Computer Science Review, 40, 100341.
20. Adegoke, N. T. I., Ofodile, N. O. C., Ochuba, N. N. A., & Akinrinola, N. O. (2024). Evaluating the fairness of credit scoring models: A literature review on mortgage accessibility for under-reserved populations. *GSC Advanced Research and Reviews*, 18(3), 189–199. https://doi.org/10.30574/gscarr.2024.18.3.0104
21. Wang, Z., Xiao, J., Wang, L., & Yao, J. (2024). A novel federated learning approach with knowledge transfer for credit scoring. *Decision Support Systems*, 177, 114084. https://doi.org/10.1016/j.dss.2023.114084
22. Sharma, M., Madan, A., Shakarwal, A., Singh, A. P., & Kumar, N. (2021). Determining the Predictive Accuracy of Loan Defaulters Using R. In *Proceedings of 3rd International Conference on Computing Informatics and Networks: ICCIN 2020* (pp. 323–331). Springer Singapore.
23. Talaat, F. M., Aljadani, A., Badawy, M. *et al.* (2024). Toward interpretable credit scoring: integrating explainable artificial intelligence with deep learning for credit card default prediction. *Neural Computing and Applications* 36, 4847–4865. https://doi.org/10.1007/s00521-023-09232-2
24. Jovanovic, Z., Hou, Z., Biswas, K., & Muthukkumarasamy, V. (2024). Robust integration of blockchain and explainable federated learning for automated credit scoring. *Computer Networks*, 243, 110303. https://doi.org/10.1016/j.comnet.2024.110303
25. Jammalamadaka, K. R., & Itapu, S. (2023). Responsible AI in automated credit scoring systems. *AI Ethics*, 3, 485–495. https://doi.org/10.1007/s43681-022-00175-3

26. Chen, Y., Calabrese, R., Martin-Barragan, B. (2024). Interpretable machine learning for imbalanced credit scoring datasets. *European Journal of Operational Research*, 312(1), 357–372. https://doi.org/10.1016/j.ejor.2023.06.036
27. Jiang, C., Lu, W., Wang, Z., Ding, Y. (2023). Benchmarking state-of-the-art imbalanced data learning approaches for credit scoring. *Expert Systems With Applications*, 213, 118878. https://doi.org/10.1016/j.eswa.2022.118878
28. Bhilare, M., Wadajkar, V., & Kale, K. (2024). Empowering Financial Decisions: Precision Credit Scoring through Advanced Machine Learning. *International Journal of Intelligent Systems and Applications in Engineering*, 12(13s), 155–167.
29. He, H., Wang, Z., Jain, H., Jiang, C., & Yang, S. (2023). A privacy-preserving decentralized credit scoring method based on multi-party information. *Decision Support Systems*, 166, 113910. https://doi.org/10.1016/j.dss.2022.113910
30. Rani, S., Tripathi, K., Arora, Y., & Kumar, A. (2022, February). Analysis of Anomaly detection of Malware using KNN. In *2022 2nd International Conference on Innovative Practices in Technology and Management (ICIPTM)* (Vol. 2, pp. 774–779). IEEE.
31. Orlova, E. V. (2021). Methodology and models for individuals' creditworthiness management using digital footprint data and machine learning methods. *Mathematics*, 9(15), 1820.
32. Dosalwar, S., Kinkar, K., Sannat, R., & Pise, N. (2021). Analysis of loan availability using machine learning techniques. *International Journal of Advanced Research in Science, Communication and Technology*, 9(1), 15–20.
33. Patil, A., Padole, G., & Sontakke, A. (2023).. Stock Trend Prediction Using KNN Algorithm. *International Journal of Scientific Research in Science and Technology*, 10(3), 13–18. https://doi.org/10.32628/ijsrst5231035
34. Al-Sayed, D. A. A., Awad, W. A. Q., & Salem, M. T. M. (2024).. A comparative study of forecasting corporate credit ratings using artificial neural networks, support vector machine, random forest, the naive bayes, decision tree and -nearest neighbor. *Advances and Applications in Statistics*, 91(2), 125–139. https://doi.org/10.17654/0972361724010
35. Chomboon, K., Chujai, P., Teerarassamee, P., Kerdprasop, K., & Kerdprasop, N. (2015, March). An empirical study of distance metrics for k-nearest neighbor algorithm. In *Proceedings of the 3rd international conference on industrial application engineering* (Vol. 2, p. 4).
36. Abu Alfeilat, H. A., Hassanat, A. B., Lasassmeh, O., Tarawneh, A. S., Alhasanat, M. B., Eyal Salman, H. S., & Prasath, V. S. (2019). Effects of distance measure choice on k-nearest neighbor classifier performance: a review. *Big data*, 7(4), 221–248.
37. Xie, W. (2021). The analysis on the application of machine learning algorithms in risk rating of P2P online loan platforms. In *Big Data Analytics for Cyber-Physical System in Smart City: BDCPS 2020*, 28–29 December 2020, Shanghai, China (pp. 1426–1433). Springer Singapore.
38. Tang, L., Cai, F., & Ouyang, Y. (2019). Applying a nonparametric random forest algorithm to assess the credit risk of the energy industry in China. *Technological Forecasting and Social Change*, 144, 563–572.

39. Rath, G. B., Das, D., & Acharya, B. (2021). Modern approach for loan sanctioning in banks using machine learning. In *Advances in Machine Learning and Computational Intelligence: Proceedings of ICMLCI 2019* (pp. 179–188). Springer Singapore.
40. Aniceto, M. C., Barboza, F., & Kimura, H. (2020). Machine learning predictivity applied to consumer creditworthiness. *Future Business Journal*, 6(1), 1–14.
41. Srinivasa Rao, M., Sekhar, C., & Bhattacharyya, D. (2021). Comparative analysis of machine learning models on loan risk analysis. In *Machine Intelligence and Soft Computing: Proceedings of ICMISC 2020* (pp. 81–90). Springer Singapore.
42. Arora, G., Dubey, A. K., & Jaffery, Z. A. (2018). Performance measure based segmentation techniques for skin cancer detection. In *Data Science and Analytics: 4th International Conference on Recent Developments in Science, Engineering and Technology, REDSET 2017, Gurgaon, India, October 13–14, 2017, Revised Selected Papers 4* (pp. 226–233). Springer Singapore.
43. Wu, C. (2021). Risk Analysis in Online Peer-to-Peer Loaning Based on Machine Learning: A Decision Tree Implementation on PPDai.com. In *The 2020 International Conference on Machine Learning and Big Data Analytics for IoT Security and Privacy: SPIoT-2020* (Vol. 1, pp. 177–186). Springer International Publishing.
44. Ye, X., Dong, L. A., & Ma, D. (2018). Loan evaluation in P2P lending based on random forest optimized by genetic algorithm with profit score. *Electronic Commerce Research and Applications*, 32, 23–36.
45. Ye, X., Dong, L. A., & Ma, D. (2018). Loan evaluation in P2P lending based on Random Forest optimized by genetic algorithm with profit score. *Electronic Commerce Research and Applications*, 32, 23–36. https://doi.org/10.1016/j.elerap.2018.10.004
46. Zhu, L., Qiu, D., Ergu, D., Ying, C., & Liu, K. (2019). A study on predicting loan default based on the random forest algorithm. *Procedia Computer Science*, 162, 503–513. https://doi.org/10.1016/j.procs.2019.12.017
47. Faisal, M. F., Saqlain, M. N. U., Bhuiyan, M. A. S., Miraz, M. H., & Patwary, M. J. (2021, December). Credit approval system using machine learning: Challenges and future directions. In *2021 International Conference on Computing, Networking, Telecommunications & Engineering Sciences Applications (CoNTESA)* (pp. 76–82). IEEE.
48. Wang, G., Hao, J., Ma, J., & Jiang, H. (2011). A comparative assessment of ensemble learning for credit scoring. *Expert systems with applications*, 38(1), 223–230.
49. Goodell, J. W., Kumar, S., Lim, W. M., & Pattnaik, D. (2021). Artificial intelligence and machine learning in finance: Identifying foundations, themes, and research clusters from bibliometric analysis. *Journal of Behavioral and Experimental Finance*, 32, 100577.
50. Wang, G., & Ma, J. (2011). Study of corporate credit risk prediction based on integrating boosting and random subspace. *Expert Systems with Applications*, 38(11), 13871-13878.
51. Phogat, M., Kumar, A., Nandal, D., & Shokhanda, J. (2021, August). A novel automating irrigation techniques based on artificial neural network and fuzzy logic. In *Journal of Physics: Conference Series* (Vol. 1950, No. 1, p. 012088). IOP Publishing.

52. Mahbobi, M., Kimiagari, S., & Vasudevan, M. (2021). Credit risk classification: an integrated predictive accuracy algorithm using artificial and deep neural networks. *Annals of Operations Research*, 330(1–2), 609–637. https://doi.org/10.1007/s10479-021-04114-z
53. Khalili, N., & Rastegar, M. A. (2023). Optimal cost-sensitive credit scoring using a new hybrid performance metric. *Expert Systems With Applications*, 213, 119232. https://doi.org/10.1016/j.eswa.2022.119232
54. Asencios, R., Asencios, C., & Ramos, E. (2023). Profit scoring for credit unions using the multilayer perceptron, XGBoost and TabNet algorithms: Evidence from Peru. *Expert Systems With Applications*, 213, 119201. https://doi.org/10.1016/j.eswa.2022.119201
55. Jiang, C., Wang, Z., Wang, R., & Ding, Y. (2018). Loan default prediction by combining soft information extracted from descriptive text in online peer-to-peer lending. *Annals of Operations Research*, 266(1–2), 511–529.
56. Sariannidis, N., Papadakis, S., Garefalakis, A., Lemonakis, C., & Kyriaki-Argyro, T. (2020). Default avoidance on credit card portfolios using accounting, demographical and exploratory factors: decision making based on machine learning (ML) techniques. *Annals of Operations Research*, 294, 715–739.
57. Kamimura, E. S., Pinto, A. R. F., Nagano, M. S. (2023). A recent review on optimisation methods applied to credit scoring models. *Journal of Economics, Finance and Administrative Science*, 28(56), 352–371. https://doi.org/10.1108/jefas-09-2021-0193
58. Suryawan, I. G. T., Arimbawa, I. K. S., & Sudipa, I. G. I. (2023). Implementation of naive bayes method for granting fisherman business credit. *Jurnal Info Sains : Informatika Dan Sains*, 13(01), 24–32.
59. Harris. T. (2015). Credit scoring using the clustered support vector machine. *Expert Systems With Applications*, 42(2), 741–750. https://doi.org/10.1016/j.eswa.2014.08.029
60. Hsieh, N. C., & Hung, L. P. (2010). A data driven ensemble classifier for credit scoring analysis. *Expert Systems With Applications*, 37(1), 534–545. https://doi.org/10.1016/j.eswa.2009.05.059
61. Xiao, H., Xiao, Z., & Wang, Y. (2016). Ensemble classification based on supervised clustering for credit scoring. *Applied Soft Computing*, 43, 73–86. https://doi.org/10.1016/j.asoc.2016.02.022
62. Hsieh, N. (2005). Hybrid mining approach in the design of credit scoring models. *Expert Systems With Applications*, 28(4), 655–665. https://doi.org/10.1016/j.eswa.2004.12.022
63. Zhang, H., He, H., & Zhang, W. (2018). Classifier selection and clustering with fuzzy assignment in ensemble model for credit scoring. *Neurocomputing*, 316, 210–221. https://doi.org/10.1016/j.neucom.2018.07.070
64. Hassani, B. K. (2021). Societal bias reinforcement through machine learning: a credit scoring perspective. *AI and Ethics*, 1(3), 239–247.
65. Addo, P. M., Guegan, D., & Hassani, B. (2018). Credit risk analysis using machine and deep learning models. *Risks*, 6(2), 38.
66. Akoglu, L., Tong, H., & Koutra, D. (2015). Graph based anomaly detection and description: a survey. *Data Mining and Knowledge Discovery*, 29, 626–688.

67. Alalade, S. A., Binuyo, O., & Oguntodu, A. (2014). Managing credit risk to optimize banks' profitability: A survey of selected banks in Lagos State, Nigeria. *Research Journal of Finance and Accounting*, *5*(18), 76–84.
68. Allen, L., & Saunders, A. (2002, May). A Survey of Cyclical Effects in Credit Risk Measurement Models. NYU Working Paper No. S-CDM-02-04. Available at SSRN: https://ssrn.com/abstract=1295817
69. Atiya, A. F. (2001). Bankruptcy prediction for credit risk using neural networks: A survey and new results. *IEEE Transactions on neural networks*, *12*(4), 929–935.
70. Byanjankar, A., Heikkilä, M., & Mezei, J. (2015, December). Predicting credit risk in peer-to-peer lending: A neural network approach. In *2015 IEEE symposium series on computational intelligence* (pp. 719–725). IEEE.
71. Chen, S., Wang, Q., & Liu, S. (2019, June). Credit risk prediction in peer-to-peer lending with ensemble learning framework. In *2019 Chinese Control and Decision Conference (CCDC)* (pp. 4373–4377). IEEE.
72. Chen, S. F., Chakraborty, G., & Li, L. H. (2019). Feature selection on credit risk prediction for peer-to-peer lending. In *New Frontiers in Artificial Intelligence: JSAI-isai 2018 Workshops, JURISIN, AI-Biz, SKL, LENLS, IDAA*, Yokohama, Japan, November 12–14, 2018, Revised Selected Papers (pp. 5–18). Springer International Publishing.
73. Chen, T., & Guestrin, C. (2016, August). Xgboost: A scalable tree boosting system. In *Proceedings of the 22nd ACM SIGKDD International Conference on Knowledge Discovery and Data Mining* (pp. 785–794).
74. Del Giovane, P., Eramo, G., & Nobili, A. (2011). Disentangling demand and supply in credit developments: a survey-based analysis for Italy. *Journal of Banking & Finance*, 35(10), 2719–2732.
75. Guo, Y., Zhou, W., Luo, C., Liu, C., & Xiong, H. (2016). Instance-based credit risk assessment for investment decisions in P2P lending. *European Journal of Operational Research*, 249(2), 417–426.
76. Hassan, A. K. I., & Abraham, A. (2013, August). Modeling consumer loan default prediction using neural netware. In *2013 International Conference On Computing, Electrical And Electronic Engineering (ICCEEE)* (pp. 239–243). IEEE.
77. He, Z., & Xiong, W. (2012). Rollover risk and credit risk. *The Journal of Finance*, 67(2), 391–430.
78. Imbierowicz, B., & Rauch, C. (2014). The relationship between liquidity risk and credit risk in banks. *Journal of Banking & Finance*, 40, 242–256.
79. Jiménez, G., Ongena, S., Peydró, J. L., & Saurina, J. (2014). Hazardous times for monetary policy: What do twenty-three million bank loans say about the effects of monetary policy on credit risk-taking? *Econometrica*, 82(2), 463–505.
80. Kovvuri, R. S., & Cheripelli, R. (2020). Credit risk valuation using an efficient machine learning algorithm. In *Advances in Decision Sciences, Image Processing, Security and Computer Vision: International Conference on Emerging Trends in Engineering (ICETE)* (Vol. 2, pp. 648–657). Springer International Publishing.
81. Kumar, A., Bhatnagar, R., & Srivastava, S. (2018, January). Analysis of credit risk prediction using ARSkNN. In *International Conference on Advanced Machine Learning Technologies and Applications* (pp. 644–652). Springer International Publishing.

82. Li, W., Ding, S., Chen, Y., & Yang, S. (2018). Heterogeneous ensemble for default prediction of peer-to-peer lending in China. *IEEEAccess*, 6, 54396–54406.
83. Liu, S., Wei, J., Chen, X., Wang, C., & Wang, X. A. (2021). Credit Rating Based on Hybrid Sampling and Dynamic Ensemble. In *Advances in Intelligent Networking and Collaborative Systems: The 12th International Conference on Intelligent Networking and Collaborative Systems (INCoS-2020)* (Vol. 12, pp. 338–347). Springer International Publishing.
84. Maheswari, P., & Narayana, C. V. (2020, October). Predictions of loan defaulter-A data science perspective. In *2020 5th International Conference on Computing, Communication and Security (ICCCS)* (pp. 1–4). IEEE.
85. Mitic, P., & Cooper, J. (2020). Enhanced Credit Prediction Using Artificial Data. In *Intelligent Data Engineering and Automated Learning–IDEAL 2020: 21st International Conference*, Guimaraes, Portugal, November 4–6, 2020, Proceedings, Part II 21 (pp. 44–53). Springer International Publishing.
86. Moradi, S., & Mokhatab Rafiei, F. (2019). A dynamic credit risk assessment model with data mining techniques: evidence from Iranian banks. *Financial Innovation*, 5(1), 1-27.
87. Onyango, J. O. (2010). *A Survey of the Credit Risk Management Practices of Commercial Banks in Kenya* (Doctoral dissertation, University of Nairobi, Kenya).
88. Pang, S., & Yuan, J. (2018). Wt model & applications in loan platform customer default prediction based on decision tree algorithms. In *Intelligent Computing Theories and Application: 14th International Conference, ICIC 2018*, Wuhan, China, August 15–18, 2018, Proceedings, Part I (Vol. 14, pp. 359–371). Springer International Publishing.
89. Patil, P., Aghav, J. V., & Sareen, V. (2018). Two-level diversified classifier ensemble for classification of credit entries. In *Advances in Computer and Computational Sciences: Proceedings of ICCCCS 2016* (Vol. 2, pp. 629–636). Springer Singapore.
90. Piramuthu, S. (2006). On preprocessing data for financial credit risk evaluation. *Expert Systems with Applications*, 30(3), 489–497.
91. Rosenberg, E., & Gleit, A. (1994). Quantitative methods in credit management: a survey. *Operations Research*, 42(4), 589–613.
92. Ruiz, S., Gomes, P., Rodrigues, L., & Gama, J. (2017). Credit scoring in microfinance using non-traditional data. In *Progress in Artificial Intelligence: 18th EPIA Conference on Artificial Intelligence, EPIA 2017*, Porto, Portugal, September 5–8, 2017, Proceedings (Vol. 18, pp. 447–458). Springer International Publishing.
93. Sivasankar, E., Selvi, C., & Mala, C. (2017). A study of dimensionality reduction techniques with machine learning methods for credit risk prediction. In Computational *Intelligence in Data Mining: Proceedings of the International Conference on CIDM, 10-11 December 2016* (pp. 65–76). Springer Singapore.
94. Thomas, L. C. (2000). A survey of credit and behavioural scoring: forecasting financial risk of lending to consumers. *International Journal of Forecasting*, 16(2), 149–172.
95. Thomas, L. C., Oliver, R. W., & Hand, D. J. (2005). A survey of the issues in consumer credit modelling research. *Journal of the Operational Research Society*, 56, 1006–1015.

96. Turkson, R. E., Baagyere, E. Y., & Wenya, G. E. (2016, September). A machine learning approach for predicting bank credit worthiness. In *2016 Third International Conference on Artificial Intelligence and Pattern Recognition (AIPR)* (pp. 1–7). IEEE.
97. Wang, L., & An, F. (2021). Machine Learning Algorithm Credit Risk Prediction Model. In *2020 International Conference on Applications and Techniques in Cyber Intelligence: Applications and Techniques in Cyber Intelligence (ATCI 2020)* (pp. 93–99). Springer International Publishing.
98. Weber, O., Fenchel, M., & Scholz, R. W. (2008). Empirical analysis of the integration of environmental risks into the credit risk management process of European banks. *Business Strategy and the Environment*, 17(3), 149–159.
99. Weber, O., Hoque, A., & Ayub Islam, M. (2015). Incorporating environmental criteria into credit risk management in Bangladeshi banks. *Journal of Sustainable Finance & Investment*, 5(1–2), 1–15.
100. Wilson, N., & Summers, B. (2002). Trade credit terms offered by small firms: survey evidence and empirical analysis. *Journal of Business Finance & Accounting*, 29(3–4), 317–351.
101. Wu, Z. (2022). Using machine learning approach to evaluate the excessive financialization risks of trading enterprises. *Computational Economics*, 59(4), 1607–1625.
102. Xia, H., Liu, J., & Zhang, Z. J. (2020). Identifying Fintech risk through machine learning: analyzing the Q&A text of an online loan investment platform. *Annals of Operations Research*, 1–21.
103. Zhang, W., Wang, C., Zhang, Y., & Wang, J. (2020). Credit risk evaluation model with textual features from loan descriptions for P2P lending. *Electronic Commerce Research and Applications*, 42, 100989.
104. Zhu, Y., Xie, C., Wang, G. J., & Yan, X. G. (2017). Comparison of individual, ensemble and integrated ensemble machine learning methods to predict China's SME credit risk in supply chain finance. *Neural Computing and Applications*, 28, 41–50.
105. Saraswat, D., Bhattacharya, P., Verma, A., Prasad, V. K., Tanwar, S., Sharma, G., ... & Sharma, R. (2022). Explainable AI for healthcare 5.0: opportunities and challenges. *IEEE Access, 10*, 84486–84517.
106. Bhatore, S., Mohan, L., & Reddy, Y. R. (2020). Machine learning techniques for credit risk evaluation: a systematic literature review. *Journal of Banking and Financial Technology*, 4, 111–138.
107. Faisal, M. F., Saqlain, M. N. U., Bhuiyan, M. A. S., Miraz, M. H., & Patwary, M. J. (2021, December). Credit approval system using machine learning: Challenges and future directions. In *2021 International Conference on Computing, Networking, Telecommunications & Engineering Sciences Applications (CoNTESA)* (pp. 76–82). IEEE.
108. Pant, A., & Rajput, R. S. (2019). Linear Regression Analysis Using R for Research and Development. *Writing Qualitative Research Paper of International Standard*, 180–195.
109. Afande, F. O. (2014). Credit risk management practices of commercial banks in Kenya. *European Journal of Business and Management*, 6(34), 21–62.
110. Coenen, L., Abdullah, A. K., & Guns, T. (2020, October). Probability of default estimation, with a reject option. In *2020 IEEE 7th International Conference on Data Science and Advanced Analytics (DSAA)* (pp. 439–448). IEEE.

111. García, V., Marqués, A. I., & Sánchez, J. S. (2015). An insight into the experimental design for credit risk and corporate bankruptcy prediction systems. *Journal of Intelligent Information Systems*, *44*, 159–189.
112. Fatemi, A., & Fooladi, I. (2006). Credit risk management: a survey of practices. *Managerial finance*, *32*(3), 227–233.
113. Hung, C., & Chen, J. H. (2009). A selective ensemble based on expected probabilities for bankruptcy prediction. *Expert systems with applications*, *36*(3), 5297–5303.
114. Giri, J., Sathish, T., Sheikh, T., Sunehriya, N., Giri, P., Chadge, R., Mahatme, C., & Parthiban, A. (2024). Automatic liver segmentation using U-Net deep learning architecture for additive manufacturing. *Deleted Journal*, 245(1), 90. https://doi.org/10.1007/s10751-024-01927-9
115. Kamble, P. D., Giri, J., Makki, E., Sunheriya, N., Sahare, S. B., Chadge, R., Mahatme, C., Giri, P., T., S., & Panchal, H. (2024). An application of hybrid Taguchi-ANN to predict tool wear for turning EN24 material. *AIP Advances*, 14(1), 015206. https://doi.org/10.1063/5.0186432
116. Mahatme, C., Giri, J., Al-Lohedan, H. A., Mohammad, F., Sunheriya, N., Chadge, R., T., S., Giri, P., Mallik, S., & Dhayalan, M. (2024). Performance evaluation of a mechanical ventilation simulation model for diverse respiratory complications. *AIP Advances*, 14(3), 015206. https://doi.org/10.1063/5.0189452
117. Mahatme, C., Giri, J., Chadge, R., & Bhagat, A. (2022). Comparative analysis of different lattice topologies for cellular structure optimization in additive manufacturing. *Materials Today: Proceedings*, *62*, 1591–1595. https://doi.org/10.1016/j.matpr.2022.03.280

Chapter 16

Intersection of artificial intelligence in agriculture sector management

A critical analysis

Deepali Rani Sahoo

16.1 INTRODUCTION

The primary cause of the same issue is neglecting to select a suitable secondary employment because doing so would require additional time and energy. By reducing time consumption and nearly eliminating labor-intensive tasks, AI in agriculture will contribute to solving these sensible issues [1]. The right application of AI will unquestionably lead to a higher yield, more uniform plantation, and proper crop growth, all of which will improve farmers' quality of life. AI in agriculture will assist farmers in developing a backup income stream that will boost their morale and prevent depression and suicide [2]. The issues are slow economic growth and the failure of the benefits of development to filter down to the poorer strata of society. The Indian government has been forced to focus on loan availability due to the country's rapid population growth, rising unemployment, and increasing levels of poverty. For the purpose of enhancing their socioeconomic circumstances, those who depend on agriculture, related businesses, and other priority sectors will be given credit. In rural areas, informal organizations like private moneylenders, friends, relatives, and others predominated and formal organizations only occasionally offered loans [3]. The priority sector in general and agriculture in particular have long been ignored by the commercial banks that make up significant credit agencies. NABARD was established on July 12, 1982, as a result of a parliamentary act. NABARD assumed full responsibility for the Agricultural Refinancing and Development Cooperative's (ARDC) projects as well as the RBI's refinancing duties with regard to State Cooperative Banks (SCBs) and RRBs. NABARD is required to support agricultural and related industries through production and investment finance [4].

16.2 NEED OF THE STUDY

The Indian economy is dominated by the agricultural sector. Gandhi, the country's founder, once claimed that the villages are where "real India"

 DOI: 10.1201/9781003512585-18

is found. Without the expansion of the rural sector, the objectives of economic planning would only be aspirations rather than realities [5]. As a result, the development of the rural economy depends on financial institutions like public sector banks and other businesses. AI has a big influence on how the rural economy develops, which is essential for the socioeconomic advancement of rurally underprivileged people [6]. Agricultural growth has been given top importance in the post-independence era, especially in several Five-Year Plans. The bank provided credit to farmers in the form of medium- and long-term loans to finance capital expenditures in agriculture and related industries, as well as short-term loans for crop production plans. Now, NABARD also meets the short- and long-term loan needs of these organizations. NABARD sparked the growth of agricultural finance in India [7].

16.3 OUTSTATE OF AI AND NABARD

Here, an effort has been made to review the reports and articles on research that have been published in the domain of NABARD finance for the growth of the agricultural sector. The subsequent paragraph reviews and presents the same. In his work "Commercial Bank and Agriculture Development," Mishra (1999) examined the role performed by commercial banks in fostering the development of agriculture in India's hilly and drought-prone regions [8]. He has made an effort to find answers to questions like whether the current cropping pattern is adequate, whether cropping patterns should change in order to advance agriculture in a previously underdeveloped area, whether banking services are sufficient to meet farmer credit demands, whether farmers use the credit they receive from banks, and whether they make timely loan repayments [9]. The investigation discovered that some farmers are in default because their crops failed to produce due to droughts, floods, and so on. Desai (2000) conducted research on the need for institutional financing of agricultural output. With the increasing amount of refinancing for schematic lending and the equitable distribution of the amount across the many classes of farmers and various regions of the country, his study has discovered that there are regional discrepancies in the refinance dispensed by NABARD [10].

16.4 PURPOSE OF THE RESEARCH

The following are the objectives of the study:

1. To research NABARD's involvement in providing institutional agencies in the agriculture sector with ground-level financing.
2. To analyze the role of AI facility in agriculture sector and allied activities to various sector and institutional agencies.

16.5 APPLICATIONS OF AI IN AGRICULTURE

At the Dartmouth Conference in 1955, the term "Artificial Intelligence" (AI) and suggested conducting research on the basis of the notion that "every aspect of learning or any other feature of intelligence can, in principle, be so precisely described that a machine can be made to simulate it" [11]. Artificial intelligence, one of the foundational areas of computer science, has today permeated a number of industries, including manufacturing, healthcare, education, and finance, due to its capacity to address problems that people struggle to handle effectively (G. Banerjee, U. Sarkar, S. Das, and I. Ghosh 2018) [12]. In addition, it is predicted that there will be 9.1 billion people on the planet by 2050, which will require 70% more food to be produced. More funding will be required in addition to the projected agricultural costs since, in the absence of it, 370 million people would go hungry by 2050 [13]. Due to the projected widening disparity between the increasing demand for water and the existing supply, almost 3 billion people are expected to face water stress by 2025 [14]. Even though artificial intelligence is still in its infancy, both the government and experts agree that it is crucial to use it in addition to more traditional approaches [15]. McKinion and Lemmon made the first attempt to apply AI to agriculture when they developed GOSSYM, an expert system-based cotton crop simulation model that maximizes cotton yield by adjusting for climate, irrigation, fertilization, weed control, cultivation, and other factors [16]. This chapter aims to describe the current status of artificial intelligence in agriculture by emphasizing three key components and accomplishments: the usage of the Internet of Things (IoT), weed control, and soil management. Additionally, it assesses the urgent issues in this field that still need to be resolved, such as the predictable unequal distribution of mechanization across various regions, concerns about privacy and security, and the adaptability of algorithms in practical applications where large datasets and other factors must be processed and plants are physically heterogeneous [17]. This chapteris devoted to the advancement of agricultural robots. It does this by outlining the history of this specific subject, providing accurate examples, and emphasizing crucial points. Although weed-inhibiting spray is a common tool, overuse of it can be detrimental to both the environment and the general public [18].

16.6 DIFFICULTIES FOR AI-BASED SOLUTIONS TO BE USED EFFECTIVELY IN AGRICULTURE

There could be an uneven distribution of mechanization in the future. While most AI systems are internet-dependent, their application may be restricted in remote or isolated areas because of a lack of web services and insufficient understanding (Ogwugwam E C., Eli-Chukwu N. 2019)

[19, 20]. Agriculture has long experimented with robotics in an attempt to improve productivity, accuracy, and dependability [21, 22]. This would replace labor-intensive automated tasks with a large reduction in the quantity of physical labor needed. Automation is the key to solving significant societal concerns like population aging and decline, but it will never be easy to automate the labor-intensive, precise operations that farmers have traditionally completed to guarantee superior products [23, 24]. When scientists began investigating the use of robots in agriculture in the 1980s, Japan was the first to develop a robot that could apply pesticide. It could go on its own or be directed remotely, doing specialist tasks in greenhouses that usually needed a lot of physical labor. In actuality, human operators are especially vulnerable to pesticides, fungicides, and other chemical agents in the heated and poorly ventilated greenhouse climate [25]. This has led to skin problems, chronic ailments, and even fatalities among those who controlled the robots (Mandow A., Gomez-De-Gabriel J. M., Martinez J. L. 2016). A tractor that employs machine vision to follow a crop line after getting instructions from a program that shows the route to go, or from the global positioning system (GPS), is an early example of an agricultural robot. A Cartesian coordinate system was used in 2000 by robots designed especially to harvest apples to try to locate apple fruit for manipulation [26]. The nonlinear least squares method was used to store the distribution of apples in both horizontal and vertical directions. The manipulator of the apple-harvesting robot can be built using this technique (Bechar A, 2016) (Murakami A, Bulanon D M, Kataoka T, 2021). Regarding the weed management issue that was previously discussed, a design from 2003 concentrated on the four-wheel system's mobility and usability. The system's functionality is primarily achieved through the use of embedded controllers and industry-standard communication protocols. The plan was to test a robotic platform for population mapping of weeds (Jakobsen H., Bak T. 2004) [27].

16.7 LITERATURE REVIEW

Historically, agriculture only produced crops and food because artificial intelligence was not employed in the field (Eli-Chukwu, N. C., 2019). In any case, changes have been made to crop production, processing, marketing, and distribution over the past 20 years. Agriculture has recently been the main driver of improvements in GDP, standard of living, and so on (Oyakhilomen, O., & Zibah, R. G. (2014). It is held that the goal of agricultural practices is to maintain and improve human existence [28]. Technology, robots, data analytics, IoT, cameras, low-cost sensors, drones, and widespread internet access are being used to incorporate AI into agriculture. AI systems are able to forecast which products are produced in large quantities each year as well as which agricultural illnesses will strike.

It has been demonstrated that technology can help to decrease the negative effects on natural ecosystems, increase worker safety, lower food prices, and boost food production to meet population needs (Oyakhilomen, O., & Zibah, R. G. (2014) [29]. Applications of manure and compost promote soil porosity and aggregation. Increased aggregation indicates the existence of organic matter, which is important since it keeps soil crusts from forming. There are ways to stop the physical degradation of soil, such as using alternative tillage practices. Applying organic materials is necessary to improve the quality of the soil (Pagliai, M., Vignozzi, N., & Pellegrini, S. 2004). Vegetable and fruit production is often significantly impacted by a number of soil-borne illnesses that need to be managed through soil management (Abawi, G. S., & Widmer, T. L. (2000). Soil erosion is an issue that needs to be considered when evaluating the sustainability of land management systems because of the disparity in soil's ability to endure and recover from change (Syers, J. K. (1997). In 2018, IBM created a small soil testing equipment that can assess five indicators using a colorimetric testing method. Chemical analysis is carried out on the card's microfluidic chip, and artificial neural networks (ANN) can predict the moisture content of soil according to research. Support vector machines (SVMs) are AI-based machine learning algorithms that are supervised. It was used to predict the soil's mean weight diameter (Bhattacharya, P., Maity, P. P., Ray, M., & Mridha, N., 2021) [30]. AI is able to identify the sources and sinks of carbon in different places. In ARIES (2018), various models were used as inputs. ANN proposes the carbon flow model, sink model for possibly stored carbon release, source model for carbon sequestration, and greenhouse gas emission model. Management-oriented modeling (MOM) is an AI-powered method of managing soil. It is a helpful tool to stop nitrogen from leaking (Li, M., & Yost, R. S. (2000). Another essential soil management strategy is the use of decision support systems (SRC-DSS). Through the use of AI's fuzzy logic, it assesses accidental risk in order to detect soil attributes and contaminated soil. The removal of impressions or uncertainty is another advantage of fuzzy logic in soil management [31].

16.8 RESEARCH GAP

A number of literatures on NABARD with regard to agriculture credit and refinance have been published; a few studies conducted so far have not made in-depth investigation into ground-level credit (GLC) and refinance operation of NABARD through institutional agencies, particularly in Odisha. There is a limited number of research studies conducted by researchers on agriculture credit and refinance activities of NABARD. This chapter would fill the following main gaps [32]:

- A few studies conducted so far have not made in depth into ground-level credit and refinance of NABARD through institutional agencies.
- It fills the gap of literature by evaluating the distribution of farm credit and refinance of NABARD
- It highlights the distribution of refinance to various agencies for agricultural sector.

16.8.1 Analysis and interpretation of data

The following sections deal with the analysis and interpretation of collected secondary data.

16.8.2 Overview of ground-level credit (GLC) flow

It has been observed from Table 16.1 that NABARD has provided the credit to both agriculture sector and total priority sector. The GLC under agriculture sector has increased from Rs. 10,454 crores during 2012–2013 to Rs. 19,985 crores in 2016–2017, while GLC under total priority sector has increased from Rs. 21,910 crores to Rs. 39,820 crores in 2016–2017, which has registered an increase of 1.82 times. The average GLC for agriculture sector was Rs. 15,939.4 crores and average GLC for total priority sector was Rs. 30,533.8 crores. On the other hand, the annual growth rate was 17.59% in agriculture sector and 16.11% in case of total priority sector.

Table 16.1 Broad sector-wise credit projection for the state

Sl. No.	*Particulars*	*Amount (in crores)*
A	Farm credit	
I	Crop production, marketing	24,212.00
II	Term loan	9,141.72
	Subtotal	33,353.72
B	Agriculture infrastructure	1,642.27
C	Ancillary activities	1,126.66
I	Credit potential (A + B + C)	36,122.64
II	MSME	16,695.17
III	Export credit	406.98
IV	Education	1,183.78
V	Housing	4,502.33
VI	Renewable energy	00,110.71
VII	Others	7,292.45
VIII	Bank credit	240.45
	Total priority sector (I–VIII)	66,554.51

Source: STP, Odisha: NABARD.

The maximum percentage of growth in agriculture sector was 44.05%, while 44.81% is in total priority sector during 2015–2016. The coefficient of variation is less in case of agriculture sector. Hence, it is shown that to provide more GLC to agriculture sector than to total priority sector. NABARD has provided credit to both agriculture sector and total priority sector. The GLC are Rs. 19,985 crores and Rs. 39,820 crores to agriculture and total priority sector, respectively, in 2016–2017. This has registered an increase of 1.82 times. The average GLC for agriculture sector is Rs. 15,939.4 crores and average GLC for total priority sector was Rs. 30,533.8 crores. On the other hand, the average annual growth rate was 16.75% in agriculture sector and 13.81% in case of total priority sector. The maximum percentage of growth in agriculture sector is 44.05%, while 44.81% is in total priority sector during 2015–2016. The coefficient of variation is less in case of agriculture sector as compared to the total priority sector. It has been concluded that ground-level credit to agriculture sector should be provided more than the total priority sector in order to have agriculture development. The state focus document that NABARD creates for the entire state paints a thorough picture of the possibilities for development in every area. It lists the restrictions that must be overcome in order to realize the credit potential under the overall priority sector.

16.8.3 Agency-wise flow of ground-level credit

The agency-wise credit disbursed by NABARD is presented in Table 16.1. It has been observed from the table that CBs have achieved Rs. 27,323.99 crores out of total targeted amount of Rs.39,863.95 crores. The average targeted amount of CBs was Rs. 7,972.91 crores, while the average achieved amount stood at Rs. 5,464.79 crores. The maximum percentage of growth in achieved amount was 40.82% during 2015–2016. In this way, the highest percentage of growth in targeted amount was 41.23% during 2013–2014. The average achieved amount of the SCBs stood at Rs. 6,845.81 crores, while average targeted amount was Rs. 7,254.58 crores. The maximum percentage of growth in case of achieved amount was 66.24% during 2015–2016 and the highest percentage of targeted amount was 40.08% during 2015–2016. In this way, RRBs have achieved Rs. 14,458.28 crores out of the total targeted amount of Rs. 15,069.13 crores. The average targeted credit of RRBs is Rs. 3,013.83 crores, while the average of achieved credit stood at Rs. 2,691.65 crores. The maximum percentage of growth in targeted credit was 243.40% during 2015–2016, while the maximum amount of achieved credit stood at Rs. 38.29 crores during 2013–2014. The highest overall annual growth rate of three banks is CBs at 34.29% in case of credit achievement. On the other hand, the overall annual growth rate of SCBs is the highest that stood at 24.99%. The coefficient of variation is less in case of a cooperative bank.

16.9 THE DIFFICULTY WITH AGRICULTURAL ROBOTS

There are presently no commercially feasible robots that can operate in complex agricultural situations, despite major advancements in agricultural robotics research. The main contributing factor was the absence of algorithms that could work in the erratic and chaotic real-world agricultural setting. Other elements, such as the agriculture's unformed seasonality, serve to highlight the contrast between the actual world and the experimental environment in labs [33].

However, partial autonomy will continue to support technological production. The underlying tenet of the Pareto principle, which is relevant to many activities, is that 80% of a process can be automated, while the other 20% is extremely difficult. In other words, 80% of the required manual labor may be replaced by automation. Additionally, through experimenting with software and hardware components, additional expertise will be obtained. It is possible to move autonomous farming systems using the 80% automation (Vigneault (2016) A Bechar [34]).

16.10 FINDINGS AND RESULT

Even though AI has revolutionized and advanced agriculture in amazing ways, there are still certain challenges. The first and most important issue is a lack of AI experience and understanding. Using AI in agriculture requires a certain level of skill set. AI technology consists of several tools such as sensors, hardware, software, and other tools. Farmers need to be properly trained in order to utilize and operate these gadgets. But most framers lack the time and the skilled teachers needed to train them. Another problem is the gap between farmers and AI engineers. Farmers usually do not study IoT, ML, and AI. Conversely, engineers do not usually conduct research work in the fields, or use agricultural methods. Sometimes odd things happen in agriculture. It takes time for AL and ML to evaluate, examine, investigate, and develop a solution [35]. Everything could have gotten out of control by then. The price of technology is a significant additional concern. The machinery itself is the initial expense, and maintenance costs are what comes next. Robots and drones for farming are not inexpensive. To purchase them, farmers must make significant financial investments. As a result, a lot of businesses rent out these tools and technology in exchange for a cut of the sales. Either way, there are still problems with using AI in agriculture, which means that improving agricultural practices is essential [36].

16.11 CONCLUSION

In this chapter, an overview of the use of AI technology in agriculture is provided. Scientists from all over the world have been researching and developing artificial intelligence for years in response to the current state of society, which includes a decline in physical employment, a lack of arable land, and a growing gap between the amount of food produced and the world population. The Turing test is heavily emphasized in the early definitions of AI that are presented in this chapter [37]. The management of soil and weeds with the introduction of the Internet of Things (IoT) is emphasized; IoT is a helpful data analysis and storage technology with broad applicability in agriculture. This evaluation also identifies three main practical challenges for AI in agriculture: first, despite significant recent advancements, much more skepticism is needed when transferring AI-based machines and algorithms from control experiments to a real agricultural environment; second, the uneven distribution of modern technology will limit the application of AI in some contexts; and third, AI in agriculture is still in its infancy [38]. The next section of the chapter summarizes the current status of agricultural robot development. First, a few examples of robots made to perform different tasks in the agricultural sector are given. There are autonomous mobile robots that can spray pesticides. Examples of innovations include apple-picking robots that locate objects using a Cartesian coordinate system, tractors with a preprogrammed travel path that utilize GPS and machine vision, an apple-harvesting machine with a special flexible gripper, and two different kinds of robots that manage weed issues and innovate in different ways [25, 38–42].

REFERENCES

1. McCarthy, J., Minsky, M. L., Rochester, N., & Shannon, C. E. A proposal for the Dartmouth summer research project on artificial intelligence. *AI Magazine*, 2006, 27(4): 12.
2. Banerjee, G., Sarkar, U., Das, S., & Ghosh, I. Artificial intelligence in agriculture: A literature survey. *International Journal of Scientific Research in Computer Science Applications and Management Studies*, 2018, 7(3): 1–6.
3. Michael, C., Fuetc. Google DeepMind's AlphaGo. Or/ms Today, 2016.
4. "Hunger and Food Insecurity." Food and Agriculture Organization of the United Nations, Food and Agriculture Organization of the United Nations, 2020, www.fao.org/hunger/en/.
5. "2050: A Third More Mouths to Feed." Food and Agriculture Organization of the United Nations, Food and Agriculture Organization of the United Nations, 2020, www.fao.org/news/story/en/item/35571/icode/
6. Popa, C. Adoption of artificial intelligence in agriculture. *Bulletin of the University of Agricultural Sciences & Veterinary*, 2011, 68 (1).
7. Mckinion, J. M., & Lemmon, H. E. Expert systems for agriculture. *Computers & Electronics in Agriculture*, 1985, 1(1): 31–40.
8. Gertsis, A. C., Galanopoulou-Sendouca, S., Papathanasiou, G., Symeonakis, A. Use of Gossym-A cotton growth simulation model-to manage a low input cotton production system in Greece. In *First European Conference for Information Technology in Agriculture*, 1997.

9. Kok, J. N., Boers, E. J. W., Kosters, W. A., et al. Artificial intelligence: definition, trends, techniques, and cases. *International Symposium on Southeast Asia Water environment*, 270–299, 2009.
10. Warwick, K. *Artificial Intelligence: The Basics*. Routledge, 2011.
11. Dobrev, D. A Definition of Artificial Intelligence. *arXiv preprint arXiv:1210.1568*, 2012.
12. Nilsson, N. J. *Principles of Artificial Intelligence*. Morgan Kaufmann, 2014.
13. Eli-Chukwu, N., Ogwugwam, E. C. Applications of artificial intelligence in agriculture: A review. *Engineering, Technology and Applied Science Research*, 2019, 9(4): 4377–4383.
14. Elijah, O., Member, S., IEEE, et al. An overview of internet of things (IoT) and data analytics in agriculture: Benefits and challenges. *IEEE Internet of Things Journal*, 2018, PP(99): 1.
15. Partel, V., Charan Kakarla, S., Ampatzidis, Y. Development and evaluation of a low-cost and smart technology for precision weed management utilizing artificial intelligence. *Computers and Electronics in Agriculture*, 2019, 157: 339–350.
16. Patrícioa, D. I., & Riederb, R. Computer vision and artificial intelligence in precision agriculture for grain crops: A systematic review. *Computers and Electronics in Agriculture*, 2018, 153: 69–81.
17. Tzounis, A., Katsoulas, N., Bartzanas, T., et al. Internet of things in agriculture, recent advances and future challenges. *Biosystems Engineering*, 2017, 164: 31–48.
18. Wei, X., Jia, K., Lan, J., Li, Y., Zeng, Y., & Wang, C. Automatic method of fruit object extraction under complex agricultural background for vision system of fruit picking robot. *Optik*, 2014, 125(19): 5684–5689.
19. Mandow, A., Gomez-De-Gabriel, J. M., Martinez, J. L., et al. The autonomous mobile robot AURORA for greenhouse operation. *IEEE Robotics & Automation Magazine*, 1996, 3(4): P.18–28.
20. Bechar, A., & Vigneault, C. Agricultural robots for field operations: Concepts and components. *Biosystems Engineering*, 2016, 149: 94–111.
21. Kataoka, T., Murakami, A., Bulanon, D. M., et al. Estimating apple fruit locations for manipulation by apple harvesting robot. *IFAC Proceedings Volumes*, 2000, 33(29): 67–72.
22. Bak, T., & Jakobsen, H. Agricultural robotic platform with four wheel steering for weed detection. *Biosystems Engineering*, 2004, 87(2): 125–136.
23. Aitkenhead, M. J., Dalgetty, I. A., Mullins, C. E., et al. Weed and crop discrimination using image analysis and artificial intelligence methods. *Computers & Electronics in Agriculture*, 2003, 39(3): 157–171.
24. Baeten, J., Donné, K., Boedrij, S., et al. Autonomous fruit picking machine: A robotic apple harvester. *Springer Tracts in Advanced Robotics*, 2007, p. 42.
25. Eli-Chukwu, N. C. Applications of artificial intelligence in agriculture: A review. *Engineering, Technology & Applied Science Research*, 2019, 9(4): 4377–4383.
26. Oyakhilomen, O., & Zibah, R. G. Agricultural production and economic growth in Nigeria: Implication for rural poverty alleviation. *Quarterly Journal of International Agriculture*, 2014, 53(892-2016-65234): 207–223.
27. Fan, M., Shen, J., Yuan, L., Jiang, R., Chen, X., Davies, W. J., & Zhang, F. Improving crop productivity and resource use efficiency to ensure food security and environmental quality in China. *Journal of Experimental Botany*, 2012, 63(1), 13–24.

28. De Kimpe, C. R., & Morel, J. L. Urban soil management: A growing concern. *Soil Science*, 2000, 165(1): 31–40.
29. Pagliai, M., Vignozzi, N., & Pellegrini, S. Soil structure and the effect of management practices. *Soil and Tillage Research*, 2004, 79(2): 131–143.
30. Abawi, G. S., & Widmer, T. L. Impact of soil health management practices on soilborne pathogens, nematodes and root diseases of vegetable crops. *Applied Soil Ecology*, 2000, 15(1): 37–47.
31. Syers, J. K. Managing soils for long-term productivity. *Philosophical Transactions of the Royal Society of London. Series B: Biological Sciences*, 1997, 352(1356): 1011–1021.
32. Sarkar, A., Maity, P. P., & Mukherjee, A.. Application of AI in soil science. *Technology*, 2021, 62: 131–136.
33. Sanuade, O. A., Adetokunbo, P., Oladunjoye, M. A., & Olaojo, A. A. Predicting moisture content of soil from thermal properties using artificial neural network. *Arabian Journal of Geosciences*, 2018, 11: 1–10.
34. Bhattacharya, P., Maity, P. P., Ray, M., & Mridha, N. Prediction of mean weight diameter of soil using machine learning approaches. *Agronomy Journal*, 2021, 113(2): 1303–1316.
35. Li, M., & Yost, R. S. Management-oriented modeling: optimizing nitrogen management with artificial intelligence. *Agricultural Systems*, 2000, 65(1): 1–27.
36. López, E. M., García, M., Schuhmacher, M., & Domingo, J. L. A fuzzy expert system for soil characterization. *Environment International*, 2008, 34(7): 950–958.
37. Ghanbarian-Alavijeh, B., Liaghat, A. M., & Sohrabi, S. Estimating saturated hydraulic conductivity from soil physical properties using neural networks model. *International Journal of Environmental and Ecological Engineering*, 2010, 4(2): 58–63.
38. Tajik, S., Ayoubi, S., & Nourbakhsh, F. Prediction of soil enzymes activity by digital terrain analysis: comparing artificial neural network and multiple linear regression models. *Environmental Engineering Science*, 2012, 29(8): 798–806
39. Jayaprakash, R., Natarajan, K., Daniel, J. A., Chinnappan, C. V., Giri, J., Qin, H., & Mallik, S. Heuristic machine learning approaches for identifying phishing threats across web and email platforms. *Frontiers in Artificial Intelligence*, 2024, 7: 1414122. https://doi.org/10.3389/frai.2024.1414122
40. Mahatme, C., Giri, J., Al-Lohedan, H. A., Mohammad, F., Sunheriya, N., Chadge, R., T., S., Giri, P., Mallik, S., & Dhayalan, M. Performance evaluation of a mechanical ventilation simulation model for diverse respiratory complications. *AIP Advances*, 2024, 14(3): 015206. https://doi.org/10.1063/5.0189452
41. Mahatme, C., Giri, J., Chadge, R., & Bhagat, A. Comparative analysis of different lattice topologies for cellular structure optimization in additive manufacturing. *Materials Today: Proceedings*, 2022, 62: 1591–1595. https://doi.org/10.1016/j.matpr.2022.03.280
42. Natarajan, M., Pasupuleti, T., Giri, J., Al-Lohedan, H. A., Katta, L. N., Mohammad, F., Sunheriya, N., Chadge, R., Mahatme, C., Giri, P., Mallik, S., & Sathish, T. Optimization of wire spark erosion machining of Grade 9 titanium alloy (Grade 9) using a hybrid learning algorithm. *AIP Advances*, 2024, 14(1): 015319. https://doi.org/10.1063/5.0177658

Chapter 17

A study on management of employee perception

Towards the role of artificial intelligence tools in promoting employee engagement: an empirical study of the banking sector in Kumaun region

Sugandha Joshi, Shilpa Wadhwa, and Amanjeet Singh Sethi

17.1 INTRODUCTION

The basic idea behind artificial intelligence (AI) is to use machine learning and large datasets to find a solution to a problem. It also encompasses a number of fields that are often cited as being part of AI, such as artificial learning and deep learning. Embedded within these sections are artificial intelligence algorithms that aim to create trained machines capable of making input-based forecasts or taxonomies. In the last few years, the field of artificial intelligence has undergone many changes and revolutions, but even the transformations in this field have left them spellbounded. Artificial intelligence in the present times is not just limited to the field of computer sciences, it has also entered almost every industry and domain. The use of AI is increasing with each passing day and the world has just begun to reconnoiter the manner in which these can be used [1–5].

Human resource management theorists use the phrase "employee engagement" to describe workers' enthusiasm and dedication to their jobs. Involved employees care about their work and the success of their company, and they think their small contributions add up to a significant impact. An invested employee sees themselves as more than just a cog in the wheel, and they take pride in the fact that their well-being is directly tied to their productivity and, by extension, the bottom line. Since employee engagement is strongly associated with both job happiness and worker confidence, it can play a crucial role in a company's growth. Motivated employees are more likely to go above and beyond in their work, which is why good communication is crucial for building and maintaining engagement. The engaged workers frequently show a superior degree of commitment to the organization mission and vision [6–8].

DOI: 10.1201/9781003512585-19

Artificial intelligence continues to find greater mileage in each and every facet of a common person life and even in all the departments of a business organization. Companies regardless of their size are frequently making efforts to use the expertise of AI in some arrangement or another, and with the swift leap of technological developments, there is no better time than now for companies to completely implement the concept of AI to their benefit, particularly to the concept of employee engagement as this aspect has troubled all of them to a greater extent in the last few years [9–11].

17.2 REVIEW OF LITERATURE

G. George and M.R. Thomas (2019) state that technological revolutions are endless and have affected almost everyone, that is, individuals, communities, governments, business organizations, among others. One such technological revolution that is competent of changing the ecosphere, the most articulated, deliberated, and executed in various domains, is artificial intelligence. Artificial intelligence is a computer application which can think rationally, the same as how a rational human being contemplates. As per few researches, AI is divided into four segments: it thinks like a human being, think judiciously, act like a human being, and act reasonably. AI is now present in most of the computer applications and is cohesive into many business areas. One such business area is the incorporation of AI, which is taking place at a rapid speed in the area of human resource management. This research emphasizes on qualitative research and intends to elucidate how AI has been cohesive into various areas of human resources and its influence on companies, workers, and HR managers. The authors conclude that in the times to come, AI will not take over the role of a HR manager but will assist them to focus on more important tasks and monotonous and low-value responsibilities will be taken care by it [12].

The optimistic opinion that each worker has of the organization and its worth is what R.E. Sari et al. (2020) mean when they say that employees are engaged. If artificial intelligence can help managers discover intangible effects like employee engagement level and provide evidence of what components impact them and how managers can enhance them, then this chapter will be a complete success. To compare the amount of participation from employees before and after the AI program was put into place, the researchers polled both upper-level management and a subset of workers. According to the results, upper-level management can benefit greatly from AI programs since they allow them to track employees' engagement levels and even predict their actions and attitudes using extrapolative gauges. This will then help the organization to proactively retain the workers [13].

R. Chhatwal (2021) is of the view that companies have always made efforts to take care of their most important asset, that is, employees. HR managers need to be better equipped with to bring revolution and how they care for their employees, so that they can better align to the new veracities across the important ranges of people, processes, and technology. The author made an effort to study those skills that were on the focus of every company such as developing sales leadership, young product leader, and budding revenue leaders. The author designed an Industrial Relations (IR) model with elements such as New Labor Codes and Draft Rules, Discipline and Grievance Handling, Union Negotiations, and Contract Labor Management. The author concluded by stating that all kinds of health are important, which will help the HR managers to take care of forthcoming encounters related to artificial intelligence, machine learning, and digital marketing [14].

M. Rozman et al. (2022) aimed to increase employee engagement and company performance by developing a multidimensional talent management model that integrated AI into HR procedures. A total of 317 managers and owners from Slovenian businesses participated in the research. Various aspects of AI in HR-related business events make up the multidimensional talent management model. The results showed that AI helped with things like finding and keeping talented workers, building good teams, fostering a positive company culture, assisting with leadership, and lowering employee workloads, all of which had a positive effect on the company's output and engagement levels. Executives will be able to use the results to create an AI-enhanced productive workplace, which will boost engagement and productivity. The authors end by saying that the outputs will shed light on the many elements that need to be considered to boost employee engagement and the firm's competitive edge, and that this will aid in the proper adoption of AI in a company [15].

The impact of trust on the HRM of the Saudi Arabian banking industry is investigated in the research of E.H. Al-Qahtani and M.A.K. Alsmairat (2023), which focuses on the drivers of AI adoption and the arbitrating findings of trust. The data was collected from 261 bankers and professionals working in different banks in Saudi Arabia using a survey-based questionnaire. Structural equation modeling with partial least squares was used for data analysis. The authors stressed the positive and statistically significant benefits of trust and perceived usefulness on the uptake of artificial intelligence in healthcare. Also, there were notable indirect impacts of trust between perceived usefulness and AI in HRM for the same reason [16]. Table 17.1 represents outcomes of literature analyzed.

17.3 AI TOOLS: EMPLOYEE ENGAGEMENT

In the present times, it is no wonder that numerous organizations have implemented AI as a way to engage and encourage their employees. Figure 17.1 shows the role of AI tools in employee engagement. Some of the main factors identified in the review of literature have been discussed [17–19]:

Table 17.1 Review of literature outcome

Authors	*Year*	*Study's objectives*	*Methodology/techniques used*	*Outcomes*
G. George and M.R. Thomas	2019	To demonstrate how AI has been introduced into many aspects of HR and its impact on companies, employees, and managers	Qualitative techniques	AI will help HR managers by handling repetitive and low-priority activities, allowing them to focus on more important obligations
R.E. Sari, et al.	2020	To determine whether AI can help managers recognize intangible aspects such as employee engagement levels, identify the sources that influence these factors, and suggest ways to improve them	Interviewing senior management and employees to measure the level of employee involvement before and after implementing the AI program	AI can help managers to assess individual employee engagement levels and forecast their perceptions and behaviors, allowing the firm to proactively retain personnel
R. Chhatwal	2021	To evaluate the important skills for employees, that is, sales leadership, young product leader, and budding revenue leaders and investigate how HR managers may effectively support their workforce	Developing an innovative model that incorporates new labor legislation and guidelines, disciplinary and grievance management, union negotiations, and contract labor supervision	Emphasizing the importance of skills for HR managers in solving forthcoming difficulties linked with AI, machine learning, and digital marketing
M. Rozman, et al.	2022	Creating a comprehensive talent management framework that incorporates various components of AI into HR processes to improve both employee engagement and company success	Conducted a study including 317 managers/owners from Slovenian enterprises, investigating the varied applications of artificial intelligence in HR tasks	AI helps talent acquisition and retention, effective training and development, team creation, organizational culture enhancement, and leadership improvement, thereby improving employee engagement and company performance
E.H. Al-Qahtani and M.A.K Alsmairat	2023	To explore the factors influencing the adoption of AI and the function of trust as a mediator in HRM within the Saudi banking industry	A survey-based questionnaire was given to 261 bankers and professionals, and the analysis was performed using partial least squares structural equation modeling	Perceived usefulness and trust had a strong beneficial effect on the adoption of AI in HRM. Trust was discovered to buffer the association between perceived usefulness and AI adoption in HRM

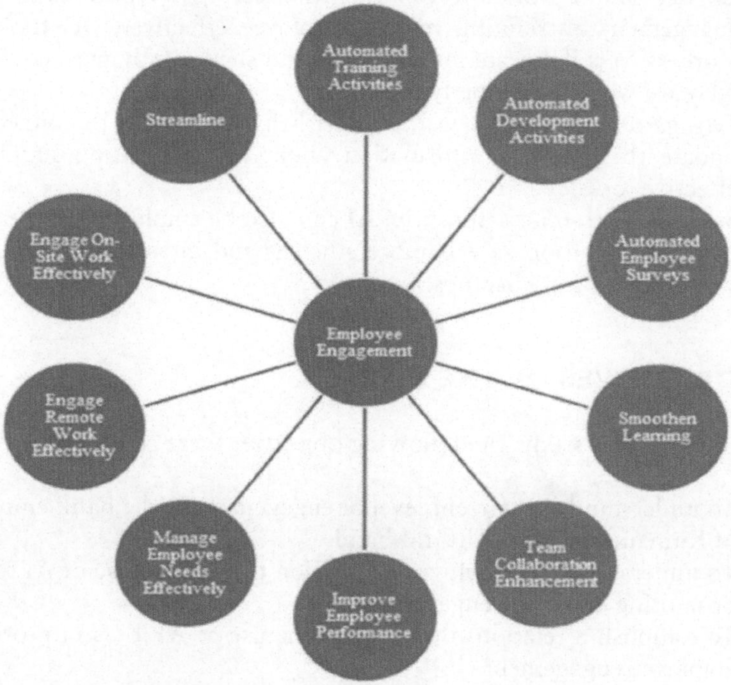

Figure 17.1 AI tools: employee engagement.

- *Automated training activities:* AI helps by automating onboarding and learning activities as it saves time and makes training quicker and well-organized.
- *Automated development activities:* AI can automate employee engagement by designing e-development programs and other interactive ingredients such as orientation, upgradation, and soft skills exercises.
- *Automated employee surveys:* AI can automate employee surveys creation and circulation procedure, thus eradicating the requirement for human intervention.
- *Smoothen learning:* AI can also be used to evolve a learning program for the workers as this will assist them to learn new skills more swiftly and efficiently.
- *Team collaboration enhancement:* AI facilitates the categorization and grouping of related concepts, facilitating the assignment of specialized project tasks to team members.
- *Improve employee performance:* AI aids employees in goal-setting, provides timely feedback on their progress, and finds relevant training courses to enhance services.
- *Manage employee needs effectively:* AI helps employers to better recognize their workers' requirements. AI can recognize trends that would otherwise be problematic to perceive.

- *Engage remote work effectively:* AI is affecting the future of employee engagement by refining remote employee effectively. AI tools help workers to collaborate more rapidly and successfully with coworkers who are working remotely.
- *Engage on-site work effectively:* AI helps employees to connect and update the essential information about tasks by using simple and effective tools.
- *Streamline conflict resolution:* AI can also be employed to streamline conflict resolution as it is more efficient and impartial than humans when supervising disputes.

17.4 OBJECTIVES OF THE STUDY

In conducting this study, the following objectives were considered:

- To understand the current level of engagement in the bank employees of Kumaun region of Uttarakhand.
- To understand the employee perception toward the use of AI tools in promoting employee engagement.
- To establish a relationship between the use of AI tools in promoting employee engagement.
- To identify the factors affecting the use of AI tools and employee engagement.

17.5 HYPOTHESES OF THE STUDY

The following hypotheses were taken for this research study:

- There is no significantly positive relationship between employee perception and the use of AI tools in employee engagement.
- There is no significant relationship between the use of AI tools and employee engagement.
- There is no positive relationship between the use of AI tools and employee engagement.

17.6 RESEARCH METHODOLOGY

- *Population of the study:* People who work in banking in the Kumaun area of Uttarakhand make up the study's population.
- *Sample of the study:* This research made use of a stratified sampling strategy. A total of 109 replies were obtained from the bank employees when the researcher distributed a questionnaire [20].

- *Unit of the sample*: Everyone who works for HDFC, ICICI, or Axis Bank was included in the sample.
- *Study instruments:* Two sections of a survey-based questionnaire were utilized to gather information from the participants. There were five questions about demographics in the first section: bank, gender, branch, age, and designation. The second part contained questions related to AI tools in employee engagement (training, development, surveys, learning, collaboration, performance, needs, remote working, on-site working, and conflict) [21].
- *Study reliability:* The reliability was done by employing Chronbach's Alpha and its result was 0.756, which means the data had a high degree of reliability [22, 23].
- *Data collection:* Both primary data and secondary data were used to collect data [24].
- *Measurement tools:* Likert five scales were used to measure the variables which were organized as strongly agree (1) to strongly disagree (5) [25].
- *Statistical analysis:* Descriptive statistics were analyzed with IBM® SPSS® Software [26].
- *Tools (demographic data):* In order to examine demographic data, one can use tools like frequency, percentage, mean, and standard deviation. The former allows one to assess the responses provided by the respondents, while the latter shows the degree of dispersion of the answers from the mean [27].
- *Tools (Likert scales):* The Mann–Whitney U-test is used to examine the Likert scale questions. These statistical tests were utilized: U test, Wilcoxon W test, Z test, P valve, to determine if two samples are likely to be drawn from the same population, Z test, to compare the sample mean μ to the population mean $\mu 0$, and P valve, to accept or reject the null hypothesis [28–30].

17.7 LIMITATIONS OF THE STUDY

The study contains a number of limitations that should be worth noting. First, it focused on a narrow location and a small number of respondents, which may limit the findings' generalizability. The 109 respondents in the sample may not represent the whole population of bank employees in Uttarakhand's Kumaun. Second, the study used self-reported data, which could create response bias. Participants may have given responses that they thought were socially acceptable rather than reflecting their genuine beliefs or experiences.

Third, the study's cross-sectional design limits the capacity to determine causality. Furthermore, the study did not account for the impact of other factors that may influence employee engagement, such as organizational

Table 17.2 Demographic details

Parameter	*Variables*	*No.*	*%*	$\bar{x}$	σ
Bank	HDFC	47	43	2.21	0.625
	ICICI	39	36		
	Axis	23	21		
Gender	Male	64	59	1.47	0.501
	Female	45	41		
Branch	Urban	68	62	1.50	0.502
	Rural	41	38		
Age	18–30	43	40	2.41	0.760
	31–50	32	29		
	51–60	21	19		
	61+	13	12		
Designation	Executive	33	30	3.38	0.960
	Assistant Manager	41	38		
	Manager	21	19		
	Senior Manager	14	13		

Source: Primary data.

culture, leadership styles, or external factors such as economic situations. Finally, the study concentrated on AI technologies in the banking sector, which may limit the findings' applicability to other sectors or industries [31–34].

17.8 RESULTS AND INTERPRETATION

The demographic and study sample results are provided in Table 17.2.

The results in Table 17.2 demonstrate the demographic details of the employees working in the banking sector of the Kumaun Division of Uttarakhand. These show the details regarding bank's name, gender, bank's branch, age, and designation of employees. It can be seen that out of the sample size, 47 (43%) respondents work in HDFC bank, 39 (36%) respondents work in ICICI bank, and 23 (21%) respondents work in Axis bank with a mean (x) of 2.21 and a standard deviation (σ) of 0.625. The number of male respondents is 64 (59%) and female respondents is 45 (41%) with a mean (x) of 1.47 and a standard deviation (σ) of 0.501. The number of respondents who work in urban branches are 68 (62%) and 41 (38%) respondents work in rural branches with a mean (x) of 1.50 and a standard deviation (σ) of 0.502. The age categorization is 43 (40%) respondents belong to 18–30 age category, 32 (29%) respondents belong to 31–50 age category, 21 (19%) respondents belong to 51–60 age category, and 13 (12%) respondents are above 60 years of age with a mean (x) of 2.41 and a standard deviation (σ) of 0.760. The designation categorization is as follows: 33 (30%) respondents' designation is Executive, 41(38%) respondents' designation is Assistant Manager, 21(19%) respondents' designation is Manager, and 14 (13%) respondents' designation is Senior Manager with a mean (x) of 3.38 and a standard deviation (σ) of 0.960.

The above data demonstrates the AI tools used by the different banks for employee engagement and the employee's perception regarding the same. Figure 17.2 represents the P value obtained with the utilization of AI tools. The details of Table 17.3 are discussed in the following lines:

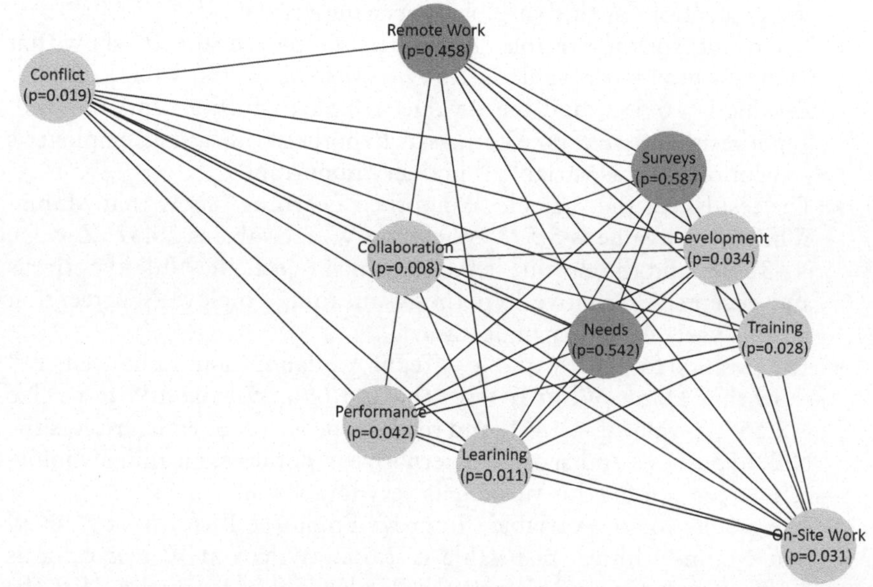

Figure 17.2 AI tools and P-value.

Table 17.3 AI tool: employee engagement[a]

Variable	*Mann–Whitney U*	*Wilcoxon W*	*Z-value*	*P-value*
Training	859.3	2.835	−3.801	0.028
Development	3.256	2.448	−2.563	0.034
Surveys	1.569	2.853	−0.562	0.587
Learning	8.595	2.452	−3.908	0.011
Collaboration	1.096	2.557	−2.841	0.008
Performance	1.062	2.190	−3.253	0.042
Needs	1.753	2.505	−0.856	0.542
Remote work	1.204	2.350	−0.975	0.458
On-site work	1.257	2.653	−1.209	0.031
Conflict	8.558	2.354	−3.789	0.019

[a]Grouping variable: Banks.

- The results for the variable "Automated Training Activities" show that Mann–Whitney test value is 859.3, Wilcoxon W test value is 2.835, Z-value is -3.801, and the P-value is 0.028, which rejects the Null hypothesis and accepts Alternative hypothesis, meaning employee's perception for this variable is very important.
- The results for the variable "Automated Development Activities" show that Mann–Whitney test value is 3.256, Wilcoxon W test value is 2.448, Z-value is -2.563, and the P-value is 0.034, which rejects the Null hypothesis and accepts Alternative hypothesis, meaning employee's perception for this variable is very important.
- The results for the variable "Automated Employee Survey" show that Mann–Whitney test value is 1.569, Wilcoxon W test value is 2.853, Z-value is -0.562, and the P-value is 0.587, which accepts the Null hypothesis and rejects Alternative hypothesis, meaning employee's perception for this variable is not very important.
- The results for the variable "Smoothen Learning" show that Mann–Whitney test value is 8.595, Wilcoxon W test value is 2.452, Z-value is -3.908, and the P-value is 0.011, which rejects the Null hypothesis and accepts Alternative hypothesis, meaning employee's perception for this variable is very important.
- The results for the variable "Team Collaboration Enhancement" show that Mann–Whitney test value is 1.096, Wilcoxon W test value is 2.557, Z-value is -2.841, and the P-value is 0.008, which rejects the Null hypothesis and accepts Alternative hypothesis, meaning employee's perception for this variable is very important.
- The results for the variable "Improve Employee Performance" show that Mann–Whitney test value is 1.062, Wilcoxon W test value is 2.190, Z-value is -3.253, and the P-value is 0.042, which rejects the Null hypothesis and accepts Alternative hypothesis, meaning employee's perception for this variable is very important.
- The results for the variable "Manage Employee Needs Effectively" show that Mann–Whitney test value is 1.753, Wilcoxon W test value is 2.505, Z-value is -0.856, and the P-value is 0.542, which accepts the Null hypothesis and rejects Alternative hypothesis, meaning employee's perception for this variable is very important.
- The results for the variable "Engage Remote Work Effectively" show that Mann–Whitney test value is 1.204, Wilcoxon W test value is 2.350, Z-value is -0.975, and the P-value is 0.458, which accepts the Null hypothesis and rejects Alternative hypothesis, meaning employee's perception for this variable is not very important.
- The results for the variable "Engage On-Site Work Effectively" show that Mann–Whitney test value is 1.257, Wilcoxon W test value is 2.653, Z-value is -1.209, and the P-value is 0.031, which rejects the Null hypothesis and accepts Alternative hypothesis, meaning employee's perception for this variable is very important.

- The results for the variable "Streamline Conflict Resolution" show that Mann–Whitney test value is 8.558, Wilcoxon W test value is 2.2354, Z-value is –3.789, and the P-value is 0.019, which rejects the Null hypothesis and accepts Alternative hypothesis, meaning employee's perception for this variable is very important.

As per the employee's perception, they feel that for factors Training, Development, Learning, Collaboration, Performance, On-Site Work, and Conflict, artificial intelligence tools are useful, whereas for factors Surveys, Needs, and Remote Work, artificial intelligence tools have not been useful to them.

17.9 FUTURE SCOPE OF THE WORK

To strengthen the generalizability of its findings, the study's scope could be expanded in the future to include a broader range of banks and areas. Longitudinal studies might also be done to study the long-term effects of AI tools on employee engagement, providing more insight into their impact. Furthermore, future study could look into the development of more advanced AI technologies that are suited to the banking industry's specific issues and requirements. This might include the development of AI-powered platforms for personalized staff training and development, real-time feedback mechanisms, and enhanced analytics for predicting and reducing employee disengagement.

Given the rapid improvements in AI technology, it would be beneficial for the future studies to investigate future AI trends and their possible consequences for employee engagement in the banking industry. This might include combining AI with other technologies like natural language processing and robotics process automation to improve employee engagement tactics.

Furthermore, future research could look into the impact of AI on other aspects of human resource management, including recruitment, performance management, and employee well-being. This would provide a more complete picture of AI's involvement in changing the future of labor in the banking industry and beyond. In summary, the study's future scope includes investigating novel approaches to use AI to boost employee engagement in the banking sector, ultimately leading to improved organizational performance and employee happiness.

17.10 CONCLUSION

The study of problem-solving through the integration of computer science and large datasets is known as artificial intelligence. Employee engagement is a concept in the field of HR which illustrates the level of eagerness and the feeling of an employee toward their work. Companies regardless of their size are frequently making efforts to use the expertise of AI in the field of HR and particularly to the concept of employee engagement as this aspect has troubled all of them. This research was an endeavor to study the employee's perception in the banking sector of Kumaun Division, Uttarakhand. The factors were identified from the previous works done in this field and thus, ten factors were identified. As per the employee's perception, they feel that for factors training, development, learning, collaboration, performance, on-site work, and conflict, artificial intelligence tools are useful; whereas for factors surveys, needs, and remote work, artificial intelligence tools have not been a great help to them. In the end, it can be stated that AI tools have gained greater access to the field of employee engagement, but still many boxes have to be ticked before this becomes a complementary product for the employees working in various industries across the globe.

REFERENCES

1. https://www.ibm.com/topics/artificial-intelligence#:~:text=At%20its%20simplest%20form%2C%20artificial,in%20conjunction%20with%20artificial%20intelligence. Retrieved on July 29, 2023.
2. Yawalkar, Vivek. "A Study of Artificial Intelligence and its role in Human Resource Management." *International Journal of Research and Analytical Reviews* 6, no. 1 (2019): 20–24.
3. Venkatasubramanian, Venkat. "The Promise of Artificial Intelligence in Chemical Engineering: Is It Here, Finally?" *AIChE Journal* 65, no. 2 (2019): 466–478. https://doi.org/10.1002/aic.16489
4. Kumar, Ajay, Yang Liu, & Rakesh Kumar (Eds.). *Handbook of Intelligent and Sustainable Manufacturing: Tools, Principles, and Strategies.* CRC Press, 2024. https://doi.org/10.1201/9781003405870
5. Kumar, Ajay, Parveen Kumar, & Yang Liu. *Industry 4.0 Driven Manufacturing Technologies.* Springer, 2024. https://doi.org/10.1007/978-3-031-68271-1
6. https://www.investopedia.com/terms/e/employee-engagement.asp. Retrieved on July 29, 2023.
7. Kashyap, Vaneet & Arora, Ridhi. "Decent Work and Work–Family Enrichment: Role of Meaning at Work and Work Engagement." *International Journal of Productivity and Performance Management* 7, no. 1 (2020): 316–336. https://doi.org/10.1108/IJPPM-03-2020-0133
8. Schaufeli, Wilmar. "Work Engagement. What Do We Know and Where Do We Go?" *Romanian Journal of Applied Psychology* 14, no. 1(2012): 3–10.
9. https://blog.shrm.org/sasia/blog/artificial-intelligence-and-employee-engagement-connecting-the-dots. Retrieved on July 30, 2023.
10. Bakker, Arnold & Albrecht, Simon. "Work Engagement: Current Trends." *Career Development International* 23, no. 1 (2018): 4–11. https://doi.org/10.1108/CDI-11-2017-0207

11. Berente, Nicholas, Gu, Bin, Recker, Jan & Santhanam, Radhika. "Managing Artificial Intelligence." *Management Information Systems Quarterly* 45, 3 (2021): 1433–1450. https://doi.org/10.25300/MISQ/2021/16274
12. George, Ginu & Thomas, Mary. "Integration of Artificial Intelligence in Human Resource." *International Journal of Innovative Technology and Exploring Engineering* 9, no. 2 (2019): 5069–5073. https://doi.org/10.35940/ijitee.L3364.129219
13. Sari, Ria, Siu Min, Hiskia Purwoko, Asnan Furinto, & Dewi Tamara. "Artificial Intelligence for a Better Employee Engagement." *International Research Journal of Business Studies* 13, no. 2 (2020): 173–188. https://doi.org/10.21632/irjbs.13.2.173-188
14. Chhatwal, Ruby. "Role of Artificial Intelligence in HR." *Embracing Change & Transformation-Breakthrough Innovation and Creativity* 8, no. 3 (2021): 351–355.
15. Rozman, Maja, Dijana Oreški, & Polona Tominc. "Integrating Artificial Intelligence into a Talent Management Model to Increase the Work Engagement and Performance of Enterprises." *Frontiers in Psychology* 13, no. 1 (2022): 44–62. https://doi.org/10.3389/fpsyg.2022.1014434
16. Al-Qahtani, Eman & Mohammad Alsmairat. "Assisting Artificial Intelligence Adoption Drivers in Human Resources Management: A Mediation Model." *International Scientific Journal about Logistics* 10, no. 1 (2023): 141–150. https://doi.org/10.22306/al.v10i1.371
17. https://enlear.academy/how-artificial-intelligence-is-impacting-the-future-of-employee-engagement-e1665ae4c681. Retrieved on July 31, 2023.
18. Fitria, Tira Nur. "Artificial Intelligence (AI) In Education: Using AI Tools for Teaching and Learning Process." *Senin* 1, no. 1 (2021): 134–147.
19. Lebovitz, Sarah, Natalia Levina, & Hila Lifshitz-Assaf. "Is AI Ground Truth Really True? The Dangers of Training and Evaluating AI Tools Based on Experts' Know-What." *Management Information Systems Quarterly* 45, no. 3 (2021): 1501–1525. https://doi/10.25300/MISQ/2021/16564
20. https://www.investopedia.com/terms/stratified_random_sampling.asp#:~:text=Stratified%20random%20sampling%20is%20a,as%20income%20or%20educational%20attainment. Retrieved on July 31, 2023.
21. Aithal, Architha & P. S. Aithal. "Development and Validation of Survey Questionnaire & Experimental Data – A Systematical Review-based Statistical Approach." *International Journal of Management, Technology and Social Sciences* 5, no. 2 (2020): 233–251. https://doi.org/10.2139/ssrn.3724105
22. Taber, Keith. "The Use of Cronbach's Alpha When Developing and Reporting Research Instruments in Science Education." *Research in Science Education* 48, no. 1 (2018): 1273–1296. https://doi.org/10.1007/s11165-016-9602-2
23. https://www.spsstests.com/2015/02/how-to-test-reliability-method-alpha.html. Retrieved on August 1, 2023.
24. Mazhar, Syeda, Rubi Anjum, Ammar Ibne Anwar, & Abdul Aziz Khan. "Methods of Data Collection: A Fundamental Tool of Research." *Journal of Integrated Community Health* 10, no. 1 (2021): 6–10. https://doi.org/10.24321/2319.9113.202101

25. https://liveinnovation.org/likert-scales-how-to-use-it-to-measure-perceptions-and-behaviors/#:~:text=Here%20are%20a%20few%20common,Slightly%20acceptable%20Acceptable%20%E2%80%93%20Perfectly%20Acceptable. Retrieved on August 1, 2023.
26. https://opentextbc.ca/introductorybusinessstatistics/chapter/descriptive-statistics-and-frequency-distributions-2/. Retrieved on August 1, 2023.
27. https://sphweb.bumc.bu.edu/otlt/mph-modules/bs/bs704_nonparametric/bs704_nonparametric4.html. Retrieved on August 2, 2023.
28. Gupta, Madhulika. "A Study on Employees Perception Towards Employee Engagement." *Globsyn Management Journal* 9, no. 1 (2015): 45–51.
29. https://pressbooks.library.upei.ca/montelpare/chapter/computing-the-wilcoxon-mann-whitney-2-group-comparison-test/. Retrieved on August 2, 2023.
30. Sethi, Amanjeet. "An Investigation into Local Resident Perspectives on the Environmental and Economic Implications of Tourism in Kumaun Division, Uttarakhand." *ACME: Journal of Management & IT* 14, no. 1 (2019): 15–27.
31. Jayaprakash, R., K. Natarajan, J. A. Daniel, C. V. Chinnappan, J. Giri, H. Qin, & S. Mallik. "Heuristic machine learning approaches for identifying phishing threats across web and email platforms." *Frontiers in Artificial Intelligence* 7 (2024): 1414122. https://doi.org/10.3389/frai.2024.1414122
32. Mahatme, C., J. Giri, H. A. Al-Lohedan, F. Mohammad, N. Sunheriya, R. Chadge, P. Giri, S. Mallik, & M. Dhayalan. "Performance evaluation of a mechanical ventilation simulation model for diverse respiratory complications." *AIP Advances* 14 no. 3 (2024): 035333. https://doi.org/10.1063/5.0189452
33. Mahatme, C., J. Giri, R. Chadge, & A. Bhagat. "Comparative analysis of different lattice topologies for cellular structure optimization in additive manufacturing." *Materials Today: Proceedings* 62 (2022): 1591–1595. https://doi.org/10.1016/j.matpr.2022.03.280
34. Natarajan, M., T. Pasupuleti, J. Giri, N. Sunheriya, L. N. Katta, R. Chadge, C. Mahatme, P. Giri, S. Mallik, & K. Ray. "Machinability of Titanium Grade 5 Alloy for Wire Electrical Discharge Machining Using a Hybrid Learning Algorithm." *Information* 14 no. 8 (2023): 439. https://doi.org/10.3390/info14080439

Index

B

U

V

W

X

Z